# Embedded Software Design

## A Practical Approach to Architecture, Processes, and Coding Techniques

Jacob Beningo

**Apress**®

*Embedded Software Design: A Practical Approach to Architecture, Processes, and Coding Techniques*

Jacob Beningo
Linden, MI, USA

ISBN-13 (pbk): 978-1-4842-8278-6                    ISBN-13 (electronic): 978-1-4842-8279-3
https://doi.org/10.1007/978-1-4842-8279-3

Managing Director, Apress Media LLC: Welmoed Spahr
Acquisitions Editor: Steve Anglin
Development Editor: James Markham
Coordinating Editor: Mark Powers

Cover designed by eStudioCalamar

Cover image by Doodlebug on Clean Public Domain (cleanpublicdomain.com/)

Distributed to the book trade worldwide by Apress Media, LLC, 1 New York Plaza, New York, NY 10004, U.S.A. Phone 1-800-SPRINGER, fax (201) 348-4505, e-mail orders-ny@springer-sbm.com, or visit www. springeronline.com. Apress Media, LLC is a California LLC and the sole member (owner) is Springer Science + Business Media Finance Inc (SSBM Finance Inc). SSBM Finance Inc is a **Delaware** corporation.

For information on translations, please e-mail booktranslations@springernature.com; for reprint, paperback, or audio rights, please e-mail bookpermissions@springernature.com.

Apress titles may be purchased in bulk for academic, corporate, or promotional use. eBook versions and licenses are also available for most titles. For more information, reference our Print and eBook Bulk Sales web page at http://www.apress.com/bulk-sales.

Any source code or other supplementary material referenced by the author in this book is available to readers on GitHub (https://github.com/Apress). For more detailed information, please visit http://www. apress.com/source-code.

Printed on acid-free paper

*To my family, friends, and mentors.*

# Table of Contents

# About the Author

**Jacob Beningo** is an embedded software consultant with around 20 years of experience in microcontroller-based, real-time embedded systems. Jacob founded Beningo Embedded Group in 2009 to help companies modernize how they design and build embedded software through software architecture and design, adopting Agile and DevOps processes and development skills. Jacob has worked with clients in over a dozen countries to dramatically transform their businesses by improving product quality, cost, and time to market in the automotive, defense, medical, and space systems industries. Jacob holds bachelor's degrees in Electrical Engineering, Physics, and Mathematics from Central Michigan University and a master's degree in Space Systems Engineering from the University of Michigan.

Jacob has demonstrated his leadership in the embedded systems industry by consulting and training at companies such as General Motors, Intel, Infineon, and Renesas, along with successfully completing over 100 embedded software consulting and development projects. Jacob is also the cofounder of the yearly Embedded Online Conference, which brings together managers and engineers worldwide to share experiences, learn, and network with industry experts.

Jacob enjoys spending time with his family, reading, writing, and playing hockey and golf in his spare time. In clear skies, he can often be found outside with his telescope, sipping a fine scotch while imaging the sky.

# About the Technical Reviewer

**Jack Ganssle** has written over 1000 articles and six books about embedded systems, as well as a book about his sailing fiascos. He started developing embedded systems in the early 1970s using the 8008. He's started and sold three electronics companies, including one of the bigger embedded tool businesses. He's developed or managed over 100 embedded products, from deep-sea navigation gear to the White House security system... and one instrument that analyzed cow poop! He now lectures and consults about the industry and works as an expert witness in embedded litigation cases.

# Acknowledgments

The book you now hold in your hand, either electronically or in paper form, is the product of 12 months of intense writing and development. The knowledge, techniques, and ideas that I share would not have been possible without over 20 years of support from family, friends, mentors, and teachers who helped me get to this point in my career. I humbly thank them for how they have enriched my life and pushed me to be the best version of me.

I would also like to thank all of you who have corresponded with me and even been my clients over the years. Your thoughts, suggestions, issues, challenges, and requests for help have helped me hone my skills and grow into a more mature embedded software consultant. I've learned how to more effectively help others, which has helped me continue to provide more value to my clients and the embedded systems industry.

I owe a big thank you to Jack Ganssle for reviewing this book. I know a lot of the review occurred during the summer while he was sailing the globe. That didn't stop him from asking great questions and giving feedback that provided me with new insights, ideas, and guidance that helped me dramatically improve the book's content.

There are also many, many others who have helped to bring this book into existence. The acknowledgments could go on indefinitely, but instead I'll simply say thank you to Apress, Steve Anglin, James Grenning, Jean Labrosse, John Little, James McClearen, Mark Powers, and Tomas Svitek.

# Preface

## Successful Delivery

Designing, building, and delivering an embedded product to market can be challenging. Products today have become sophisticated, complex software systems that sit upon an ever-evolving hardware landscape. At every turn, there is a challenge, whether it's optimizing a system for energy consumption, architecting for configurability and scalability, or simply getting a supply of the microcontrollers your system is built. Embedded software design requires chutzpah.

Despite the many challenges designers, developers, and teams often encounter, it is possible to successfully deliver embedded products to market. In fact, I'll go a step further and suggest that it's even possible to do so on time, on budget, and at the quality level required by the system stakeholders. It's a bold statement, and, alas, many teams fall short. Even the best teams stumble, but it is possible to reach a level of consistency and success that many teams can't reach today.

*Embedded Software Design*, this book, is about providing you and your team with the design philosophies, modern processes, and coding skills you need to deliver your embedded products successfully. Throughout the book, we cover the foundational concepts that will help you and your team overcome modern challenges and help you thrive. As you can imagine, the book is not all-inclusive in that every team will have different needs based on their industry, experience, and goals. *Embedded Software Design* will help get you started, and if you get stuck or need professional assistance, I'm just an email away.

Before we dive into the intricate details of *Embedded Software Design*, we need to initialize our minds, just like our embedded systems. We need to understand the lens through which we will view the material in this book and how we can apply it successfully to our unique situations. This chapter will explore how this book is organized, essential concepts on how to use the material, and how to maximize the value you receive from it.

**Tip** At the end of each chapter, you will find exercises to help you think through and apply the material from each chapter. "Action Items" will come in the form of questions, design experiments, and more. To maximize your value, schedule time in your calendar to carefully think through and execute the Action Items.

# The Embedded Software Triad

Many teams developing embedded software struggle to deliver on time, on budget, and at a quality level that meets customer expectations. Successful development is often elusive, and software teams have all kinds of excuses for their failure.[1] However, many teams in the industry are repeatedly successful even under the direst conditions. I've found that these teams have mastered what I call the embedded software triad.

The embedded software triad consists of

- Software Architecture and Design

- Agile, DevOps, and Processes

- Development and Coding Skills

Software architecture and design consists of everything necessary to solicit the requirements, create user stories, and design the architecture. The software architecture defines the blueprint for the system. It's important to note that the architecture should be evolvable and does not necessarily need to be developed up front.[2] How the architecture is designed follows the processes used by the team.

Agile, DevOps, and Processes are the procedures and best practices that are used to step-by-step design, build, test, and deliver the software. Processes provide the steps that allow a team to deliver successfully based on the budget, time, and quality parameters provided to them consistently. In today's software development environment, processes

---

[1] Yes, I know, a bit harsh right off the bat, but it's necessary.

[2] There are many arguments about this based on the processes used. We will discuss further later in the book.

are often based on various agile methodologies. Therefore, the processes help to guide the implementation under the parameters and disciplines defined by the team. I've also called out DevOps specifically because it is a type of process often overlooked by embedded software teams that offers many benefits.

Development and coding skills are required to construct and build the software. Development entails everything necessary to take the architecture and construct the system using the processes! At the end of the day, the business and our users only care about the implementation which is completed through development and coding skills. The implementation should result in a working system that meets agreed-upon features, quality, and performance characteristics. Teams will organize their software structure, develop modules, write code, and test it. Testing is a critical piece of implementation that can't be overlooked that we will be discussing throughout the book.

To be successful, embedded software designers, developers, and teams must not just master the embedded software triad but also balance them. Too much focus on one area will disrupt the development cycle and lead to late deliveries, going over budget, and even buggy, low-quality software. None of us want that outcome. This brings us to the question, "How can we balance the triad?"

# Balancing the Triad

Designers, developers, and teams can't ignore any element of the embedded software triad. Each piece plays a role in guiding the team to success. Unfortunately, balancing these three elements in the real world is not always easy. In fact, in nearly every software audit I have performed from small to large companies, there is almost always some imbalance that can be found. In some teams, it's minor and adjustments can result in small efficiency improvements. In others, the imbalance is major and often crippling to the company.

The relationships between the elements in the triad can be best visualized through a Venn diagram, as shown in Figure 1. When a team masters and balances each element, marked by the number 4 in Figure 1, it is more likely that the team will deliver their software on time, on budget, and at the required quality level. Unfortunately, finding a team that is balanced is relatively rare. It is more common to find teams that are hyperfocused on one or two elements. When a team is out of balance, there are three regions of Figure 1 where the teams often fall, denoted by 1, 2, and 3. Each area has its own symptoms and prescription to solve the imbalance.

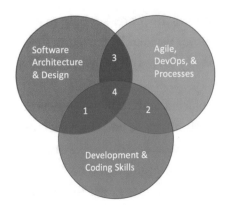

1 - Late, Inconsistent, Quality Issues
2 - Late, Rework, Lost / Meandering
3 - Never completed
4 - Successful Delivery

*Figure 1.* *The balancing act and results caused by the embedded software triad*

---

**Caution**   Being out of balance doesn't ensure failure; however, it can result in the need for herculean efforts to bring a project together successfully. Herculean efforts almost always have adverse effects on teams, costs, and software quality.

---

The first imbalance that a team may fall into is region 1. Teams fall into region 1 when they focus more on software architecture and implementation (denoted by a 1 in Figure 1) than on development processes. In my experience, many small- to medium-sized teams fall within region 1. In smaller teams, the focus is almost always delivery. Delivery equates to implementation and getting the job done, which is good, except that many teams jump right in without giving much thought to what they are building (the software architecture) or how they will implement their software successfully (the processes) repeatedly and consistently.[3]

Teams that operate in region 1 can be successful, but that success will often require extra effort. Deliveries and product quality will be inconsistent and not easily reproducible. The reason, of course, is that they lack the processes that create consistency and reproducibility. Without those processes, they will also likely struggle with quality issues that could cause project delays and cause them to go over budget. Teams don't necessarily need to go all out on their processes to get back into balance, but only need to put in place the right amount of processes to ensure repeatability.

---

[3] We often aren't interested in a one-hit wonder delivery. A team needs to consistently deliver software over the long term. Flukes and accidental success don't count! It's just dumb luck.

**Tip** If you spend 20% or more debugging your software, you will most likely have a process problem. Don't muscle through it; fix your processes!

The second unbalanced region to consider is region 2. Teams that fall into region 2 focus on development processes and implementation while neglecting the software architecture (denoted by a 2 in Figure 1). These teams design their system on the fly without any road map or blueprint for what it is they are building. As a result, while the team's software quality and consistency may be good, they will often still deliver late because they constantly must rework their system with every new feature and requirement. I often refer to these teams as lost or meandering because they don't have the big picture to work from.

The final unbalanced region to consider is region 3. Teams in region 3 focus on their software architecture and processes with little thought given to implementation (denoted by a 3 in Figure 1). These teams will never complete their software. They either lack the implementation skills or bog down so much in the theory of software that they run out of money or customers before the project is completed.

There are several characteristics that a balanced team will exhibit to master the embedded software triad. First, a balanced team will have a software architecture that guides their implementation efforts. The architecture is used as an evolving road map that gets the team from where they are to where their software needs to be. Next, a balanced team will have the correct amount of processes and best practices to ensure quality software and consistency. These teams won't have too much process and not too little. Finally, a balanced team will have development and coding skills to construct the architecture and leverage their processes to test and verify the implementation.

# Successful Embedded Software Design

Successful software delivery requires a team to balance the embedded software triad, but it also requires that teams adopt and deploy industry best practices. As we discuss how to design and build embedded software throughout this book, our focus will discuss general best practices. Best practices "are procedures shown by research and experience to produce optimal results and establish or propose a standard for widespread adoption."[4] The best practices we focus on will be general to embedded software, particularly for microcontroller-based systems. Developers and teams must carefully evaluate them to determine whether they fit well within their industry and company culture.

Applying best practices within a business requires discipline and an agreement that the best practices will be adhered to no matter what happens. Too often, a team starts following best practices, but when management puts on the heat, best practices go out the window, and the software development process decays into a free for all. There are three core areas where discipline must be maintained to successfully deploy best practices throughout the development cycle, as shown in Figure 2.

***Figure 2.*** *Best practice adoption requires discipline, agreement, and buy-in throughout a business hierarchy to be successful*

Developers form the foundation for maintaining best practices. They are the ones on the frontline writing code. If their discipline breaks, no matter what else is going on in the company, the code will descend into chaos. Developers are often the ones that also identify best practices and bring new ideas into the company culture. Therefore, they must adhere to agreed-upon best practices no matter the pressures placed upon them.

Developers may form the foundation for adhering to best practices, but the team they work on is the second layer that needs to maintain discipline. First, the team must identify the best practices they believe must be followed to succeed. Next, they need to reinforce each developer, and the team needs to act as a buffer with upper management when the pressure is on. Often, a single developer pushing back will fail, but an entire team will usually hold some sway. Remember, slow and steady wins the race, as counterintuitive as that is to us intellectually and emotionally.

---

[4] www.merriam-webster.com/dictionary/best%20practice

---

**Definition**   Best practices "are procedures shown by research and experience to produce optimal results and establish or propose a standard for widespread adoption."[5]

---

Finally, the management team needs to understand the best practices that the team has adopted. Management must also buy into the benefits and then agree to the best practices' value. If they know why those best practices are in place, they will realize that it is to preserve product quality, minimize time to market, or some other desired benefit when the team pushes back. The push to cut corners or meet arbitrary dates will occur less but won't completely go away, given its usually market pressure that drives unrealistic deadlines. However, if there is an understanding about what is in the best interest of the company and the customers, a short delay for a superior product can often be negotiated.

# How to Use This Book

As you might have guessed from our discussions, this book focuses on embedded software design best practices in each area of the embedded software triad. There are more best practices within the industry than could probably be written into a book; however, I've tried to focus on the best practices in each area that is "low-hanging fruit" and should provide high value to the reader. You'll need to consult other texts and standards for specific best practices in your industry, such as functional safety.

This book can either be read cover to cover or reviewed on a chapter-by-chapter basis, depending on the readers' needs. The book is broken up into four parts:

- Software Architecture and Design

- Agile, DevOps, and Processes

- Development and Coding Skills

- Next Steps and Appendixes

---

[5]www.merriam-webster.com/dictionary/best%20practice

I've done this to allow the reader to dive into each area of the embedded software triad and provide them with the best practices and tools necessary to balance their development cycles. Balancing can be sometimes accomplished through internal efforts, but often a team is too close to the problem and "drinking the Kool-Aid" which makes resolution difficult. When this happens, feel free to reach out to me through www.beningo.com to get additional resources and ideas. *Embedded Software Design* is meant to help you be more successful.

A quick note on the Agile, DevOps, and Processes part. These chapters can be read independently, but there are aspects to each chapter that build on each other. Each chapter walks you through getting a fundamental piece of a CI/CD pipeline up and running. For this reason, I would recommend reading this part in order.

I hope that whether you are new to embedded software development or a seasoned professional, you'll find new or be reminded of best practices that can help improve how you design and develop embedded software. We work in an exciting industry that powers our world and society can't live without. The technologies we work with evolve rapidly and are constantly changing. Our opportunities are limitless if we can master how to design and build embedded software effectively.

## ACTION ITEMS

To put this chapter's concepts into action, here are a few activities the reader can perform to start balancing their embedded software activities:

- Which area do you feel you struggle the most in?

  - Software Architecture and Design

  - Agile, DevOps, and Processes

  - Development and Coding Skills

- Review Figure 1. Which region best describes you? What about your team? (If you are not in region 4, what steps can you take to get there?)

- Take Jacob's design survey to understand better where you currently are and how you can start to improve (www.surveymonkey.com/r/7GP8ZJ8).

- Review Figure 2. Do all three groups currently agree to support best practices? If not, what can be done to get the conversation started and get everyone on the same page?

# PART I

# Software Architecture and Design

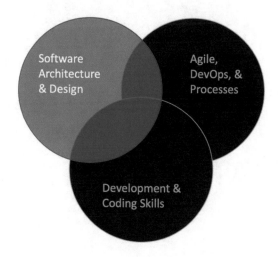

# CHAPTER 1

# Embedded Software Design Philosophy

The first microprocessor, the Intel 4004, was delivered to Busicom Corp in March 1971 for use in their 141-PF calculators.[1] The 4004 flaunted a 4-bit bus running at 0.74 MHz with up to 4 KB of program memory and 640 bytes of RAM.[2] The 4004 kicked off a hardware revolution that brought more computing power to the world at lower costs and led to faster, more accessible software development. In time, fully integrated processors designed for real-time control in embedded systems were developed, including the microcontroller.

The first microcontroller, the Texas Instruments TMS1000,[3] became commercially available in 1974. The TMS1000 boasted a 4-bit bus running at 0.3 MHz with 1 KB x 8 bits program memory and 64 x 4 bits of RAM. The critical difference between a microprocessor and a microcontroller is that a microcontroller integrates ROM, RAM, and combined input/output (I/O) onto a single chip, whereas microprocessors tend to be multichip solutions. Thus, microcontrollers are intended for embedded systems with a dedicated function and are not general computing devices.[4] Examples include automobile controllers, home appliances, intelligent sensors, and satellite subsystems.

---

[1] https://spectrum.ieee.org/tech-history/silicon-revolution/chip-hall-of-fame-intel-4004-microprocessor

[2] https://en.wikichip.org/wiki/intel/mcs-4/4004#:~:text=The%204004%20was%20a%204, and%20640%20bytes%20of%20RAM

[3] https://en.wikipedia.org/wiki/Texas_Instruments_TMS1000

[4] High-end, modern microcontrollers are looking more and more like general computing devices …

© Jacob Beningo 2022
J. Beningo, *Embedded Software Design*, https://doi.org/10.1007/978-1-4842-8279-3_1

Early in embedded system history, the industry was dominated by a hardware-centric mindset. A product's differentiator wasn't the software but whether the system was built on the latest hardware. After all, the first microcontrollers were heavily resource-constrained and had a limited feature set. The more advanced the hardware, the more features one could pack into the product. Early on, there was only 4 KB of ROM for software developers to work with, which doesn't seem like a lot by today's standards, but remember, that was 4 KB of hand-coded assembly language!

The software was viewed as a necessary evil; it's still sometimes considered an essential evil today, even though the hardware is relatively agnostic. Microcontrollers come in a wide variety of flavors with bus architectures including 8-bit, 16-bit, and 32-bit. One might be tempted to believe that 32-bit microcontrollers are the dominant variants in the industry, but a quick look on Digikey or Mouser's website will reveal that 8-bit and 16-bit parts are still just as popular.[5] Microcontroller clock speed varies widely from 8 MHz all the way up to 268 MHz in Arm Cortex-M4 parts. Although there are now offerings exceeding 1 GHz. Finding parts with sporting 512 KB program memory and 284 KB RAM is fairly common. In fact, there are now multicore microcontrollers with 1000 MHz system clocks, over 1024 KB of program memory, and 784 KB of RAM!

A key point to realize in modern embedded system development is that the software is the differentiator and the secret sauce. Microcontrollers can supply nearly as much processor power and memory as we want! The software will make or break a product and, if not designed and implemented correctly, can drive a company out of business. The hardware, while important, has taken the back seat for many, but not all, products.

Successfully designing, building, and deploying production embedded systems is challenging. Developers today need to understand several programming languages and how to implement design patterns and communication interfaces and apply digital signal processing, machine learning, and security, just to name a few. In addition, many aspects need to be considered, ranging from the software and hardware technologies to employ all the way through ramifications for securely producing and performing firmware updates.

As a consultant, I've been blessed with the opportunity to work with several dozen engineering teams worldwide and across various industries for over a decade. I've noticed a common trend between successful groups and those that are not. The

---

[5] www.embedded.com/why-wont-the-8-bit-microcontroller-die/

successful design, implementation, and deployment of an embedded system depend on the design philosophy the developer or team employs for their system (and their ability to execute and adhere to it).

---

**Definition**    *A design philosophy* is a practical set of ideas about how to[6] design a system.

---

Design philosophies will vary based on the industry, company size, company culture, and even the biases and opinions of individual developers. However, a design philosophy can be considered the guiding principles for designing a system. Design philosophies should be practical and improve the chances of successfully developing the product. They should not handcuff the development team or add a burden on the developers. I often see design principles as best practices to overcome the primary challenges facing a team. To develop your design philosophy, you must first examine the embedded system industry's challenges and then the challenges you are facing in your industry and at your company.

---

**Best Practice**    Many discussions in this book are generalizations based on industry best practices and practical industry experience. Don't take it as gospel! Instead, carefully evaluate any recommendations, tips, tricks, and best practices. If they work and provide value, use them! If not, discard them. They will not work in every situation.

---

# Challenges Facing Embedded Developers

Every developer and team faces inherent challenges when designing and building their system. Likewise, every industry will have unique challenges, varying even within the same company between the embedded software developers and the hardware designers. For example, some teams may be faced with designing an intuitive and easy-to-use

---

[6]www.merriam-webster.com/dictionary/philosophy

graphical user interface. Another may face tough security challenges that drive nearly every design aspect. However, no matter the team or developer, there are several tried and true challenges that almost all teams face that contribute to the modern developers' design philosophy. These fundamental challenges include

- Cost (recurring and nonrecurring)

- Quality

- Time to market

- Scalability

- Etc.

The first three challenges are well known and taught to every project manager since at least the 1950s.[7] For example, Figure 1-1 shows the project management "iron triangle,"[8] which shows that teams balance quality, time, and cost for a given scope in every project.

***Figure 1-1.***  *The project management iron triangle demonstrates that teams balance the demands of quality, delivery time, cost, and scope in any project*

You may have heard project managers say that they can pick any two between quality, cost, and delivery time, and the third must suffer. However, what is often overlooked is that scope is just as important in the trade-off. If a project is going over budget or past the deadline, scope reduction can be used to bring them back into

---

[7] Atkinson, Roger (December 1999). "Project management: cost, time and quality, two best guesses and a phenomenon, its time to accept other success criteria." International Journal of Project Management. 17 (6): 337–342. doi:10.1016/S0263-7863(98)00069-6.

[8] www.pmi.org/learning/library/beyond-iron-triangle-year-zero-6381

balance. So these properties are trading off against each other. Unfortunately, this model, in general, is NOT sufficient alone, but it at least visually provides teams with the primary business challenges that every team is grappling with. So, let's quickly look at some challenges and see how they can impact developers.

Every team that I encounter is designing their product under a limited budget. I don't think I have ever met a team with an unlimited budget and was told to spend whatever it takes to get the job done. (If you work for one of those rare companies, my contact information is jacob@beningo.com.) For the most part, businesses are always trying to do more with smaller budgets. This makes sense since a business's core goal isn't to change the world (although they will tell you it is) but to maximize profits! Therefore, the smaller the development budget and the more efficient the engineers are, the higher the potential profit margins.

Many teams face problems with development costs because today's systems are not simple. Instead, they are complex, Internet-connected feature beasts with everything plus the kitchen sink thrown in. As feature sets increase, complexity increases, which requires more money up front to build out the development processes to ensure that quality and time-to-market needs are carefully balanced. The more complex the system becomes, the greater the chances that there will be defects in the system and integration issues.

Budgets can also affect software packages and tools, such as network stacks, compilers, middleware, flash tools, etc. Because it's "free," the big push for open source software has been a major, if not sometimes flawed, theme among teams for the last decade or more (we will discuss this in more detail later in the book). I often see teams unwilling to spend money on proper tools, all in an attempt to "save" money. Usually, it's a trade-off between budget and extra development time.

Product quality, in my mind, is perhaps the most overlooked challenge facing development teams. Product quality is simply whether the product does what it is supposed to when it is supposed to and on time. Unfortunately, today's systems are in a constant bug fix cycle. The real challenge, I believe, is building a product that has the right amount of quality to meet customer expectations and minimize complaints and hassle on their part. (It always drives me crazy when I need to do a hard reset on my smart TV.)

The time to market for a product can make or break a company. There is a careful dance that needs to be performed between delivering the product and marketing the product. Market too soon, and the company will look foolish to its clients. Market too late, and the company may not generate sufficient cash flow to maintain its business. Delivering a product on time can be a tricky business. The project scope is often far more extensive and more ambitious than is reasonable for time and budget and can often result in

- Missed deadlines

- More debugging

- Integration woes

---

**Lesson Learned**    Successful and experienced teams are not immune to delivering late, going over budget, and scope creep! Therefore, one must always be diligent in carefully managing their projects.

---

The first three challenges we have discussed have been known since antiquity. They are project management 101 concepts. Project managers are taught to balance these three challenges because they are often at odds with each other. For example, if I am in a delivery time crunch and have the budget and a project that can be parallelized, I may trade off the budget to get more developers to deliver the project quicker. In *The Mythical Man Month*, Fred Brooks clearly states that "adding people to a late project makes it later."

Teams face many other challenges, such as developing a secure solution, meeting safety-critical development processes, and so forth. The last challenge that I believe is critical for teams, in general, is the ability to scale their solutions. Launching a product with a core feature set that forms the foundation for a product is often wise. The core feature set of the product can be considered a minimum viable product (MVP). The MVP helps get a product to market on a reasonable budget and timeline. The team can then scale the product by adding new features over time. The system needs to be designed to scale in the field quickly.

Scalability is not just a need for a product in the field; it can be a need for an entire product line. Systems today often have similar features or capabilities. The ability to have a core foundation with new features added to create a wide variety of products is necessary. It's not uncommon for a single code base to act as the core for a family of several dozen products, all of which need to be maintained and grown over a decade. Managing product lines over time can be a dramatic challenge for a development team to face, and a few challenges that are encountered include

- Tightly coupled code

- Vendor dependency

- Inflexible architecture

We should also not forget that scalability is essential because developers often work on underscoped projects! In addition, requirements change as the project is developed. Therefore, the scalable and flexible software architecture allows the developer to minimize rework when new stakeholders or requirements are passed down.

The challenges we have discussed are ones we all face, whether working for a Fortune 500 company or just building an experimental project in our garages. In general, developers start their projects a day late and a dollar short, which is a significant reason why so many products are buggy! Teams just don't have the time and money to do the job right,[9] and they feel they need to cut corners to try to meet the challenges they are facing, which only makes things worse!

A clearly defined design philosophy can help guide developers on how they design and build their systems, which can help alleviate some of the pain from these challenges. Let's now discuss the design philosophy I employ when designing software for my clients and projects that you can easily adopt or adapt for your own purposes.

# 7 Modern Design Philosophy Principles

As we discussed earlier, a design philosophy is a set of best practices meant to overcome the primary challenges a team faces. Of course, every team will have slightly different challenges and, therefore, a somewhat different set of design philosophies they follow. What is crucial is that you write down what your design principles are. If you write them down and keep them in front of you, you'll be more likely to follow them and less likely to abandon them when the going gets tough.

---

[9] I believe this is a perception issue, not a reality.

I believe every team should follow seven modern design philosophy principles to maximize their chances for success. These include

- Principle #1 – Data Dictates Design

- Principle #2 – There Is No Hardware (Only Data)

- Principle #3 – KISS the Software

- Principle #4 – Practical, Not Perfect

- Principle #5 – Scalable and Configurable

- Principle #6 – Design Quality in from the Start

- Principle #7 – Security Is King

Let's examine each principle in detail.

---

**Best Practice**   Create a list of your design principles, print them out, and keep them in a visible location. Then, as you design your software, ensure that you adhere to your principles!

---

# Principle #1 – Data Dictates Design

The foundation of every embedded system and every design is data. Think carefully about that statement for a moment. As developers, we are often taught to focus on creating objects, defining attributes, and being feature-focused. As a result, we often get hung up on use cases, processes, and deadlines, quickly causing us to lose sight that we're designing systems that generate, transfer, process, store, and act upon data.

Embedded systems are made up of a wealth of data. Analog and digital sensors are constantly producing a data stream. That data may be transferred, filtered, stored, or used to actuate a system output. Communication interfaces may be used to generate system telemetry or accept commands that change the state or operational behavior of the system. In addition, the communication interfaces may themselves provide incoming and outgoing data that feed a primary processing system.

The core idea behind this principle is that embedded software design is all about the data. Anything else focused on will lead to bloated code, an inefficient design, and the opportunity to go over budget and deliver late. So instead, developers can create an efficient design by carefully identifying data assets in the system, the size of the data, and the frequency of the operations performed with the data.

When we focus on the data, it allows us to generate a data flow diagram that shows

- Where data is produced

- Where data is transferred and moved around the application

- Where data is stored (i.e., RAM, Flash, EEPROM, etc.)

- Who consumes/receives the data

- Where data is processed

Focusing on the data in this way allows us to observe what we have in the system and how it moves through the system from input to output, enabling us to break down our application into system tasks. A tool that can be very useful with this principle is to develop a data flow diagram similar to the one shown in Figure 1-2.

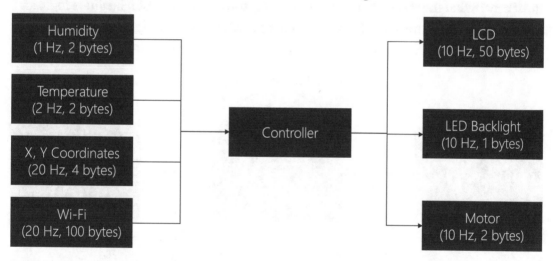

**Figure 1-2.**  *A simple data flow diagram can help developers identify the data assets in their system and how they move through the application*

# Principle #2 – There Is No Hardware (Only Data)

There is no hardware is a challenging principle for embedded developers to swallow. Most embedded software developers want to run out and buy a development board as quickly as possible and start writing low-level drivers ASAP. Our minds are engrained that embedded software is all about the hardware, and without the hardware, we are downriver without a paddle. This couldn't be further from the truth and puts the cart before the horse, leading to disaster.

I often see teams buying development boards before creating their design! I can't tell you how many times I've sat in on a project kickoff meeting where teams on day one try to identify and purchase a development board so they can get started writing code right away. How can you purchase and start working on hardware until the design details have been thought through and clear requirements for the hardware have been established?

Embedded software engineers revel in writing code that interacts with hardware! Our lives are where the application meets the hardware and where the bits and bytes meet the real world. However, as we saw in principle #1, the data dictates the design, not the hardware. Therefore, the microcontroller one uses should not dramatically impact how the software is designed. It may influence how data is generated and transferred or how the design is partitioned, but not how the software system is designed.

Everything about our design and how our system behaves is data driven. Developers must accept that there is no hardware, only data, and build successful business logic for the software! Therefore, your design at its core must be hardware agnostic and only rely on your data assets! A general abstraction for this idea can be seen in Figure 1-3.

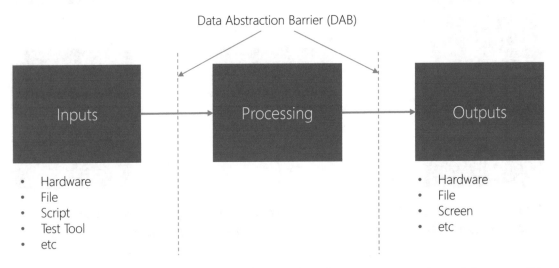

***Figure 1-3.*** *Application design should view hardware as agnostic and focus on the system data's inputs, processes, and outputs*

This idea, at first, can be like Neo from *The Matrix*, having to accept that there is no spoon! For us developers, there is no hardware! Only data! Data acts as an input; it's then processed and used to produce system outputs (remember, it might also be stored). Before you do anything drastic with that pitchfork in your hand, let me briefly explain the benefits of thinking about design in this way.

First, a data-centric and hardware-agnostic software system will naturally abstract itself to be more portable. It is not designed for a specific hardware platform, making the code more "mobile." Second, the software will more readily integrate with test harnesses because the interfaces are designed to not be hardware dependent! This makes it much easier to connect them to a test harness. Third, the abstractions make it much easier to simulate the application! Developers can run the application on a PC, target hardware, or any convenient computing machine even if the hardware designers haven't yet produced prototypes. (And let's be honest, even if you have a prototype, you can't always trust that it will work as intended.) Fourth, suppose we decouple the application from hardware. In that case, we can rapidly develop our application by removing tedious compile, program, and debug cycles by doing all the debugging in a fast and hosted (PC) environment! I'm just warming up on the benefits, but I think you get the point.

Look again at Figure 1-3. You can see how this principle comes into practice. We design tasks, classes, modules, and so forth to decouple from the hardware. The hardware in the system is either producing data, processing that data, or using the data to produce an output. We create data abstraction barriers (DABs), which can be abstractions, APIs, or other constructs that allow us to operate on the data without direct access to the data. A DAB allows us to decouple the hardware and then open up the ability to use generators, simulators, test harnesses, and other tools that can dramatically improve our design and how we build systems.

# Principle #3 – KISS the Software

Keep It Simple and Smart (KISS)! There is a general tendency toward sophisticated and overly complex software systems. The problem is that developers try to be too clever to demonstrate their expertise. As everyone tries to show off, complexity increases and code readability decreases. The result is a design and system that is hard to understand, maintain, and even get working right.

The KISS software principle reminds us that we should be trying to design and write the simplest and least complex software possible. Minimizing complexity has several advantages, such as

- More accessible for new developers to get up to speed

- Decrease in bugs and time spent debugging

- More maintainable system(s)

It's easy to see that the more complex a design is, the harder it will be to translate that design vision into code that performs as designed.

I often have told my clients and attendees at my workshops that they should be designing and writing their software so that an entry-level engineer can understand and implement it. This can mean avoiding design patterns that are difficult to understand, breaking the system up into smaller and more manageable pieces, or even using a smaller subset of a programming language that is well understood.

Keep the software design simple and smart! Avoid being clever and make the design as clear as possible. Brian W. Kernighan stated, "Debugging is twice as hard as writing the code in the first place. Therefore, if you write the code as cleverly as possible, you are, by definition, not smart enough to debug it."

---

**Definition**    A design pattern is a reusable solution to a commonly occurring problem in software design.[10]

---

# Principle #4 – Practical, Not Perfect

There are many reasons why developers are writing embedded software. Still, the primary purpose for many reading this book is professional engineers developing a product, an entrepreneur starting a business, or those perhaps interested in building a do-it-yourself (DIY) project and looking to improve their skills. In general, the purpose is to create a product that creates value for the user and generates profit for a business.

Developers and entrepreneurs often lose sight of the purpose of developing their software. The excitement or the overarching vision starts to cloud their judgment, and they begin to fall into perfectionism. Perfectionism is "a flawless state where everything is exactly right. But, it can also be the act of making something perfect."[11] There are two categories of perfectionism that we need to discuss:

- "One system to rule them all."

- "It still needs tweaking."

---

[10] https://en.wikipedia.org/wiki/Software_design_pattern

[11] www.vocabulary.com/dictionary/perfection

In "one system to rule them all," the team has become solely focused on creating a single product that can offer themselves and their users every possible feature imaginable. The product is a Swiss Army knife on steroids and, once completed, will not need to be modified or rewritten for centuries to come. Of course, it's perfectly acceptable to have a glorious product vision. Still, an excellent way to go out of business is to build the final vision first, rather than slowly working up to that final version. After all, Apple didn't wait until they had version X of their iPhone to launch their product; they found the minimal set of features that would be a viable product and launched that! Over time, they added features and tweaked hardware and software, evolving their product into what it is today.

Early in my career, I worked at a start-up that was creating revolutionary in-dashboard products. Unfortunately, the company never shipped a product in its five years of existence. Every time we had something that was close to shipping, the owner would change the requirements or have some new hardware feature that is a must-have before the product could be shipped. As I write this, it's nearly 15 years later, and the "must-have" features still cannot be found in similar products today!

The second perfectionist category is the "it still needs tweaking" group. This group of engineers is never quite happy with the product. For example, sensor readings are within 1% and meet the specification, but we should be able to get within 0.5%. That code module is working but needs to be rewritten to refactor, add more documentation, etc. (Sometimes, that is a fair statement.) Yes, features A and B are working, but we should include C and D before we even consider launching. (And once C and D are complete, there's E and F, then G and H, then ....)

---

**Best Practice**   The 80/20 rule states, "80% of software features can be developed in 20% of the time."[12] Focus on that 80% and cut out the rest!

---

It's helpful to mention right now that there is no such thing as a perfect system. If you were to start writing the ideal Hello World program right now, I could check in with you when I retire in 30 years, at age 70, and find that you almost have it, just a few more tweaks to go. Developers need to lose the baggage that everything needs to be perfect. Anyone judging or condemning that the code isn't perfect has a self-esteem issue, and I would argue that any developer trying to write perfect code also has one.

---

[12] https://en.wikipedia.org/wiki/Pareto_principle

Systems need to be practical, functional, and capable of recovering from faults and errors. The goal isn't to burn a company's cash reserves to build a perfect system. It's to make a quality product with a minimal feature set that will provide customers with the value they need and generate a profit for the company. The faster this can be done, the better it is for everyone!

# Principle #5 – Scalable and Configurable

Since our goal is not to build "one system to rule them all" or build the perfect system, we want to set a goal to at least make something scalable and configurable. We create a product starting with our minimum feature set, deploy it, and then continue to scale the system with new features and make it adaptable through configuration. Software needs to be scalable and configurable for a product to survive the rapidly evolving business conditions developers often find themselves in.

A scalable design can be easily adapted and grown. For example, as customers use a product, they may request new features and capabilities for the software that will need to be added. Therefore, a good design will be scalable, so developers can quickly and easily add new features to the software without tearing it apart and rewriting large pieces from scratch.

A configurable design allows the developer and/or the customer to change the system's behavior. There are two types of configurations: build time and runtime. Build time configuration allows developers to configure the software to include or exclude software features. This can be used to customize the software for a specific customer, or it can be used when a common code base is used for multiple products. In addition, the configuration can be used to specify the product and the target hardware.

Runtime configurations are usually changeable by the end customer. The customer can change settings to adjust how the features work to best fit their end needs. These changes are often stored in nonvolatile memory like EEPROM. Therefore, changing a setting does not require that the software be recompiled and deployed.

Designing a scalable and configurable software system is critical for companies that want to maximize code reuse and leverage their software assets. Often, this principle requires a little bit of extra up-front work, but the time and money saved can be tremendous on the back end.

# Principle #6 – Design Quality in from the Start

Traditionally, in my experience, embedded software developers don't have the greatest track record for developing quality systems. I've worked with over 50 teams globally so far in my career, and maybe five of them had adequate processes that allowed them to build quality into their product from the start. (My sampling could be biased toward teams that know they needed help to improve their development processes.) I think most teams perform spot checks of their software and simply cross their fingers that everything will work okay. (On occasion, I'm just as guilty as every other embedded developer, so this is not a judgmental statement, I'm just calling it how I see it.) This is no way to design and build modern embedded software.

When we discuss designing quality into software, keep in mind at this point that quality is a "loaded" term. We will define it and discuss what we mean by quality later in the book. For now, consider that building quality into a system requires teams to

- Perform code reviews

- Develop a Design Failure Mode and Effects Analysis

- Have a consistent static analysis, code analytics process defined

- Consistently test their software (preferably in an automated fashion)

It's not uncommon for teams to consider testing to be critical to code quality. It certainly is important, but it's not the only component. Testing is probably not the foundation for quality, but it could be considered the roof. If you don't test, you'll get rained on, and everything will get ruined. A good testing process will ensure

- Software modules meet functional requirements.

- New features don't break old code.

- Developers understand their test coverage.

- Tests, including regression testing, are automated.

The design principles we discussed all support the idea that testing should be tightly integrated into the development process. Therefore, the design will ensure that the resultant architecture supports testing. Building the software will then provide interfaces where test cases can simultaneously be developed, as shown in Figure 1-4.

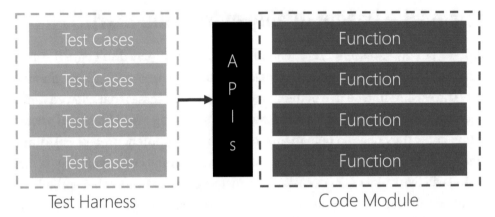

**Figure 1-4.** *Embedded software can be designed to easily interface to a test harness that can automatically execute test cases by interfacing to a module's APIs*

Testing has several important purposes for developers. First, it ensures that we produce a higher-quality product by detecting defects. It doesn't guarantee quality, but can improve it. Second, testing can find bugs and software not behaving as intended. Next, testing can exercise every line and conditional statement in a function and every function in a module. Third, writing tests while developing software makes it more likely to approach 100% code coverage. Finally, testing can confirm that new code added to the system has not compromised software already in the system.

# Principle #7 – Security Is King

The IoT has been taking over nearly every business and industry. Practically everything is in the process of being connected to the Internet: vehicles, home appliances, and even human beings. Even the simple mousetrap is now an Internet-connected device.[13] (Don't believe me, check out the footnote link!) All this connectivity leads to a crucial point and a principle that many developers can no longer ignore: security is king.

Many developers cannot design and build an embedded system without considering security and privacy concerns. For example, a device connected to the Internet needs security! It may not need to be Fort Knox, but it needs to protect company and user data and ward off common attacks so that the device does not become a pawn in nefarious activity.

---

[13]www.zdnet.com/article/victor-aims-to-use-iot-to-build-a-better-mousetrap/

**Best Practice**    Security needs to be designed into the system, **NOT** bolted onto the end. Security must come first, then the rest of the design follows.

This brings me to an important point; you cannot add security to a system after being designed! A secure system requires that the security threats be identified up front to set security requirements that direct the software design. To perform that analysis, designers need to identify their data assets and how they may be attacked. This is an interesting point because, if you recall principle #1, data dictates design! To secure a system, we need to identify our data assets. Once we understand our data, then we can properly design a system by

- Performing a threat model security analysis (TMSA)

- Defining our objectives

- Architecting secure software

- Building the software

- Certifying the software

We will talk more about this process in later chapters.

If you design a completely disconnected system and security is not a concern, then principle #7 easily drops off the list. However, you can think up a new seventh principle that makes more sense for the system you are working on and add it.

Remember, these seven principles are designed to guide designers to overcome the common challenges that are being faced. These may not apply across the board, and I highly encourage you to give each some thought and develop your design principles to guide your development efforts. However, as we will see throughout the book, these principles can go a long way in creating a well-defined, modern embedded software design and development process.

**Resource**    You can download a printable copy of the seven modern design principles at **https://bit.ly/3Oufvf0**.

# Harnessing the Design Yin-Yang

So far, we have discussed how critical it is for developers to understand the challenges they are trying to solve with their design and how those challenges drive the principles we use to design and build software. Success, though, doesn't just depend on a good design. Instead, the design is intertwined in a Yin-Yang relationship with the processes used to build the software.

This relationship between design and processes can be seen easily in Figure 1-5. We've already established that core challenges lead to design solutions. The design and challenges direct the processes, which in turn provide the likelihood of success. This relationship can also be traversed in reverse. We can identify what is needed for success, which directs the processes we use, which can lead our design, telling us what challenges we are solving.

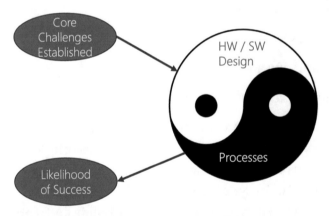

***Figure 1-5.*** *The software design and build processes balance each other to determine the project's likelihood of success and to establish the challenges that are solved by design*

For example, if I am developing a proof-of-concept system, I will most likely constrain my challenge domain to only the immediate challenges I want to solve. My design will focus on that constrained domain, which will likely constrain the processes I use and dramatically improve my chances for success. On the other hand, in a proof of concept, I'm trying to go as fast as possible to prove if something is possible or not, so we end up using the minimal design and processes necessary to get there.

**Tip**    A proof of concept, a.k.a. a prototype, often morphs into a deliverable. That should never happen. Use fast tools that can't be embedded like Excel, Visual Basic, Python, etc., when possible.[14]

If, on the other hand, I want to develop a production system, my challenge domain will likely be much more significant. Therefore, my design will need to solve more challenges, resulting in more software processes. Those processes should then improve our chances of successfully producing a solution that solves our problem domain.

This brings me to an important point: all design and development cycles will not be alike! Some teams can successfully deliver their products using minimal design and processes. This is because they have a smaller problem domain. For example, other teams, maybe ones working in the safety-critical system, have a much larger problem domain, resulting in more development processes to ensure success. The key is to identify your challenges correctly to balance your design and development processes to maximize the chances of success!

**Best Practice**    Periodically review design and development processes! Bloated processes will decrease the chances for success just as much as inadequate processes!

# Traditional Embedded Software Development

Traditional embedded software development is a loaded term if I have ever heard one. It could mean anything depending on a developer's background and industry. However, for our purposes, traditional is the average development processes embedded teams have used over the last decade or two to develop software. This is based on what I saw in the industry when I first started to write embedded software in the late 1990s up until today.

---

[14] A suggestion provided by Jack Ganssle.

Typically, the average software build process looked like Figure 1-6. The process involved hand-coding most software, starting with the low-level drivers and working up to the application code. The operating system and the middleware may have been open source, but teams often developed and maintained this code in-house. Teams generally liked to have complete control over every aspect of the software.

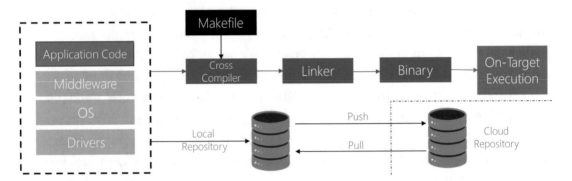

***Figure 1-6.*** *Traditional embedded software development leverages a hand-coded software stack that was only cross-compiled for on-target execution*

Another critical point in the traditional embedded software process is that on-target development is heavily relied upon. From early in the development cycle, teams start development at the lowest firmware levels and then work their way up to the application code. As a result, the application code tends to be tightly coupled to the hardware, making it more difficult to port or reuse application modules. However, that is more an effect of the implementation. In general, the process has nothing wrong with it. The modern approach is more flexible and can dramatically decrease costs, improve quality, and improve the chance to meet deadlines.

# Modern Embedded Software Development

Modern embedded software development can appear complicated, but it offers flexibility and many benefits over the traditional approach. Figure 1-7 provides a brief overview of the modern embedded software build system. There are several essential differences from the traditional approach that is worth discussing.

First, notice that we still have our traditional software stack of drivers, OS, middleware, and application code. The software stack is no longer constrained to only hand-coded modules developed in-house. Most software stacks, starting with low-level drivers through the middleware, can be automatically generated through a platform configuration tool. Platform configuration tools are often provided by the microcontroller vendor and provide development teams with an accelerated path to get them up and running on hardware faster.

While these configuration tools might first appear to violate the principle that there is no hardware only data, at some point we need to provide hooks into the hardware. We maintain our principle by leveraging the platform code behind abstraction layers, APIs. The abstractions allow our application to remain decoupled from the hardware and allow us to use mocks and other mechanisms to test and simulate our application code.

It's important to note that vendor-provided platform configuration tools are often designed for reuse and configuration, not for execution efficiency. They can be a bit more processor cycle hungry, but it's a trade-off between development speed and code execution speed. (Don't forget principle #4! Practical, not perfect! We may do it better by hand, but is there a tangible, value-added benefit?)

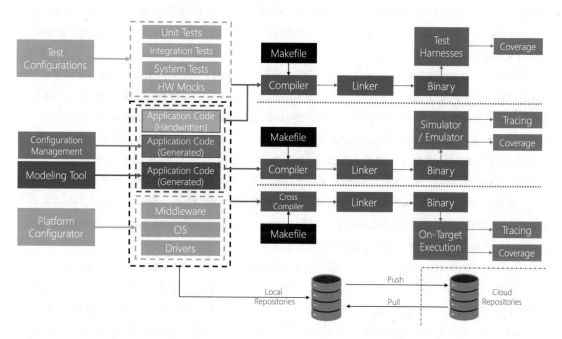

***Figure 1-7.*** *Modern embedded software applies a flexible approach that mixes modeling and configuration tools with hand-coded development. As a result, the code is easily tested and can be compiled for on-target or simulated execution*

Next, the application code is no longer just hand-coded, there are several options to automate code generation through modeling and configuration management. This is a massive advantage because it allows the application business logic to be nearly completely fleshed out and tested in a simulated environment long before the product hardware is available to developers. Business logic can be developed, tested, and adjusted very quickly. The code can then be rapidly autogenerated and deployed to the target for on-target testing. Again, the code generated by a modeling tool is generally bad from a readability standpoint. However, if the modeling tool is used to maintain the business logic, there isn't any harm unless the generated code is grossly inefficient and doesn't meet the systems requirements.

Configuration management is a critical component that is often mismanaged by development teams. It is common to have a core product base that is modified and configured differently based on the mission or customer needs. Embedded software teams can explore general computer science best practices and use hosted configuration files like YAML, XML, etc., to configure and generate flexible components within the application. We'll talk more about this later in the chapter on configuration management.

Another improvement over the traditional approach is that testing is now intricately built into the development process. Embedded software always required testing, but I found that the traditional method used manual test cases almost exclusively which is slow and error prone. Modern techniques leverage a test harness that allows automated unit and regression testing throughout development. This makes it much easier to continuously test the software rather than spot-checking here and there while crossing fingers that everything works as expected. We will be extensively discussing how to automate tests and leverage continuous integration and continuous deployment in the Agile, DevOps, and Processes part of the book.

Finally, perhaps most importantly, the software's output has three separate paths. The first is the traditional path where developers can deploy the compiled code to their target system to run tests and traces and understand test coverage. Developers can also perform on-target debugging if necessary. The second, and more exciting approach, is for developers to run their software in a simulator or emulator instead. The advantage of simulation is that developers can debug and test much faster than on the target. The reason simulation is faster is that it removes the constant cross-compile, flash, and debug process. It can also dramatically remove developers' need to develop and provide prototype hardware, saving cost and time. Finally, the software can also be run within a test harness to verify each unit behaves as expected and that the integration of modules works as expected.

Modern software development is well beyond just the basic build processes we have discussed. If you look again closely at Figure 1-4, you'll notice on the lower right that there is a cloud repository that has been sectioned off from the rest of the diagram. The cloud repository serves a great purpose in software development today. The cloud repository allows developers to implement a continuous integration and continuous deployment (CI/CD) process.

CI/CD is a set of practices that enable developers to frequently test and deploy small changes to their application code.[15] CI/CD is a set of best practices that, when leveraged correctly, can result in highly robust software that is easy to deploy to the devices regularly as new features are released. A general overview of the process for embedded developers can be seen in Figure 1-8.

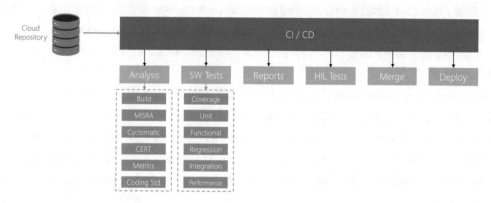

**Figure 1-8.**  *Modern development processes utilize a CI/CD server with various instances to continuously analyze and test the code base*

Every team will have slightly different needs, which allows them to customize their CI/CD process. We will discuss CI/CD in more detail later in the book. For now, it's helpful to understand that it is a great way to automate activities such as

- Software metric analysis

- Software testing

- Reporting activities

---

[15] www.infoworld.com/article/3271126/what-is-cicd-continuous-integration-and-continuous-delivery-explained.html

- Hardware in-loop testing

- Merge requests

- Software deployment

Getting an initial CI/CD system set up can be challenging at first, but the benefits can be fantastic!

# The Age of Modeling, Simulation, and Off-Chip Development

Simulation and off-chip development are by no means something new to embedded software developers. It's been around at least since the 1990s. However, at least 90% of the teams I talk to or interact with have not taken advantage of it! By default, embedded software developers feel like they need to work on the target hardware to develop their applications. In some cases, this is true, but a lot can be done off-chip to accelerate development.

At its core, embedded software handles data that interacts directly with the hardware. But, of course, PC applications also interact with hardware. Still, the hardware is so far removed from the hardware with software layers and abstractions that the developer has no clue what is happening with the hardware. This makes embedded software unique because developers still need to understand the hardware and interact with it directly.

However, embedded software is still just about data, as I've tried to clarify in the design principles we've discussed. Therefore, application development for processing and handling data does not necessarily have to be developed on target. But, on the other hand, developing it on target is probably the least efficient way to develop the software! This is where modeling and simulation tools come into the picture.

A simulation is "a functioning representation of a system or process utilizing another system."[16] Embedded software development has approached a point where it is straightforward for developers to run their application code either in an emulator or natively in a PC environment and observe how the software behaves without the embedded target present. Simulation provides developers with several benefits, such as

---

[16] www.merriam-webster.com/dictionary/simulation

- Reducing debugging time by eliminating the need to program a target

- Ability to examine and explore how their application behaves under various conditions without having to create them in hardware artificially

- Decoupling software development from hardware development

Modeling is very closely tied to simulation. In many cases, developers can use a modeling tool to create a software representation that can run simulations optimally. Simulations exercise the application code in a virtual environment, allowing teams to collect runtime data about their application and system before the hardware is available. Today, there is a big push within embedded communities to move to model and simulation tools to decouple the application from the hardware and speed up development with the hope that it will also decrease costs and time to market. At a minimum, it does improve understanding of the system and can dramatically improve software maintenance.

# Final Thoughts

The challenges that we face as teams and developers will dictate the principles and processes we use to solve those challenges. The challenges that embedded software teams face aren't static. Over the last several decades, they've evolved as systems have become more complex and feature rich. Success requires that we not just define the principles that will help us conquer our challenges, but that we continuously improve how we design and build our systems. We can continue to use traditional ideas and processes, but if you don't evolve with the times, it will become more and more difficult to keep up.

| ACTION ITEMS |
|---|

To put this chapter's concepts into action, here are a few activities the reader can perform to start applying design philosophies:

- First, make a list of the design challenges that you and your team are facing.

- From your list of challenges, what are the top three big ones that require immediate action?

- Next, review the seven modern design philosophy principles and evaluate which ones can help you solve your design problems.

- Identify any additional design philosophies that will guide your designs.

- Identify the behaviors that are decreasing your chances for success. Then, what can you do over the next week, month, or quarter to mitigate these behaviors?

- Review your design and development processes. Are there any areas where you need to add more processes? Any areas where they need to be lightened up?

- Does your development more closely resemble traditional or modern development? What new modern technique can you focus on and add to your toolbox?

- Evaluate whether your entire team, from developers to executives, are on the same page for following best practices. Then, work on getting buy-in and agreement to safeguard using best practices even when the heat is on.

These are just a few ideas to go a little bit further. Carve out time in your schedule each week to apply these action items. Even minor adjustments over a year can result in dramatic changes!

# CHAPTER 2

# Embedded Software Architecture Design

If I could describe my first few years as an embedded software engineer, I would use the words "code first, architect second." I could simplify this statement further to "There is only code!". Creating a software architecture was honestly a foreign concept, and it was an industry-wide issue, not just for me. When a task needed to be done, we would jump into our IDE and start coding. For simple projects, this wasn't an issue. Everything that needed to be done could be kept straight in our minds; however, as projects became more complex, the rework required to add features ballooned out of control.

I recall the first time I went against the wisdom of the time and decided to draw out and design what I was doing before coding it up. It was like I had been coding in the dark my whole life, and suddenly the clouds cleared, the sun shone, and I understood! Architect first and code second became my mantra. Developing a software architecture allowed me to think through what I would code, adjust how it worked, and then code it once. Writing code went from constantly rewriting, modifying, and debugging to designing and implementing. The time required decreased by literally 10x! I had discovered what Ralf Speth had previously stated:

> If you think good design is expensive, you should look at the cost of bad design.

© Jacob Beningo 2022
J. Beningo, *Embedded Software Design*, https://doi.org/10.1007/978-1-4842-8279-3_2

The IEEE 1471 standard defines a software architecture as follows:

---

**Definition**    *A software architecture is the fundamental organization of a system embodied in its components, their relationship to each other and the environment, and the principles **guiding its design and evolution**.*[1]

---

This simple statement tells us much about software architecture and why it is essential to the software development life cycle. Software architecture is a developer's road map, the GPS, on what they are supposed to be building. Without a software architecture, it is straightforward for a developer to get lost and make wrong turns, which forces them to continually retrace their steps, trying to figure out not just how to get where they want to go but where they are trying to go!

A map tells the reader the lay of the land. Landmarks, buildings, and roads provide a general organization of an area and the relationship between those objects. A software architecture does the same thing for software; instead of physical objects and spatial relationships, we are dealing with components, objects, and the relationships between them. The software architecture tells a developer what they are building; it doesn't tell them how to make it!

This brings us to an important point: a software architecture should be hardware independent! The software architecture should guide a developer to successfully write an application, whether it is written for a microcontroller or application processor. The software architecture is also agnostic to the programming language that is used to implement the architecture! For a given architecture, I can write the application in C, C++, Python, Rust, or any other programming language that may suit my fancy.[2]

The software architecture is a map, a blueprint, and a guide for the developer to follow. The developer can decide how they want to implement the architecture; they aren't told by the architecture how to do it. Software architecture can readily be compared to the blueprints used by a builder to construct a building. The blueprints tell

---

[1] https://en.wikipedia.org/wiki/IEEE_1471. IEEE 1471 has been superseded by IEEE 42010. (I just like the IEEE 1471 definition).

[2] In a Q&A session, I was once accused of being a Rust advocate because I mentioned it! As of this writing, I still haven't written a single line of it, although it sounds intriguing.

them the general footprint, how many stories, and where the elevators and stairwells go. However, they don't tell the electrician how to wire a light switch or where the occupant should put the furniture.

---

**Beware**   It is a rare occasion, but sometimes the hardware architecture is so unique that it can influence the software architecture!

---

The software architecture also is not set in stone! Instead, it provides principles that guide the design and implementation of the software. It's so important to realize that software design will evolve. It shouldn't devolve into chaos like so many software systems do but evolve to become more efficient, include additional features, and meet the needs of its stakeholders. I can't tell you a single system I've ever worked on where we got all the details up front, and nothing changed. Software design, by nature, is an exercise in managing the unknown unknowns. To manage them properly, the software architecture must quickly evolve with minimal labor.

Now you might be thinking, I don't need a software architecture. I can design my software on the fly while I'm implementing it. (A lot of open source software is developed this way). The problem, though, is that embedded software today has become so complex that it's nearly impossible to implement it efficiently and timely without an architecture. For example, if you were to design and erect a building, would you just buy concrete and lumber and start building without blueprints? Of course, not! That would be a foolhardy endeavor. Yet, many software products are just as complex as a building, and developers try to construct them without any plan!

Developing a software architecture provides many benefits to developers. Just a few highlights include

- A plan and road map to what is being built

- A software picture that can be used to train engineers and explain the software to management and stakeholders

- Minimizing rework by minimizing code changes

- Decreased development costs

- A clear picture of what is being built

- An evolvable code base that can stand the tests of time (at least product lifetimes)

Now that we understand what software architecture is and what it can do for us, let's discuss an interesting architectural challenge that embedded software developers face.

# A Tale of Two Architectures

Embedded software designs are different when compared to web, mobile, or PC software architectures. General applications have just a single architecture that is required, a hardware-independent business rules–based architecture. These applications are written to be general and can run across multiple hardware architectures and have zero hardware dependence. The architecture focuses on the business logic and the high-level components required to get the job done.

Embedded software designs have a second type of architecture that includes a hardware-dependent real-time architecture. The hardware-dependent architecture requires special knowledge of the hardware the application will be running on. It allows the designer to optimize their architecture for different hardware platforms, if necessary, or just optimize it to run most efficiently on a single architecture. For example, a designer may need to understand the complex behavior of available watchdogs, memory protection units (MPUs), direct memory access (DMA), graphics accelerators, communications buses, etc. This architecture is where the "rubber meets the road" or, in our case, where the application architecture meets real-time hardware.

---

**Best Practice**    *Embedded software architects should separate their architectures into two pieces: business logic and real-time hardware-dependent logic.*

---

I can't stress enough how important it is to recognize that these two different architectures exist in an embedded system. If designers overlook this, they risk developing an architecture that tightly couples the business logic to the hardware. While this may not seem like a problem at first, in today's systems, this can lead to code that is difficult to maintain, port, and scale to meet customer needs. The software architecture should maintain as much independence from the hardware due to the potential for disruptions to occur in the supply chain. If a microcontroller becomes unavailable or obsolete, or suddenly a dramatic increase in lead time occurs, it will be necessary to pivot to other hardware. Software that is tightly coupled to the hardware will make the change expensive and time-consuming.

The separation between the application business architecture and the real-time software architecture can be seen from a 30,000-foot view in Figure 2-1. Note that the two architectures are separated and interact through an abstraction layer.

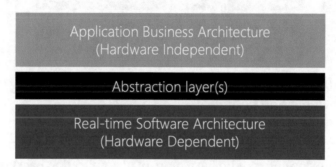

***Figure 2-1.*** *Embedded software architectures are broken into two architectures: a hardware-independent business layer and a hardware-dependent real-time layer*

---

**Definition**    *An abstraction layer is software that translates a high-level request into the low-level commands required to operate.*[3]

---

# Approaches to Architecture Design

Generally, two approaches are typically used to develop an embedded software architecture: the bottom-up approach and the top-down approach. The general difference between these approaches can be seen in Figure 2-2, but let's discuss each and see how they impact the modern designer.

Most embedded software developers I know tend to be naturally inclined to use the bottom-up approach. This approach starts at the hardware and driver layer to identify the software components and behaviors. Once the low-level pieces are placed, the designer works their way up the stack into the application code. The core idea is that

---

[3] https://encyclopedia2.thefreedictionary.com/abstraction+layer

you build up the foundation of the software system first and then work your way up to the product-specific architecture. Figure 2-2 demonstrates how this works with our two different software architectures.

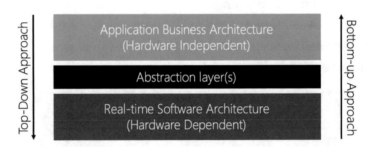

**Figure 2-2.** *Embedded software architectures are developed using two approaches: top-down and bottom-up*

The top-down approach tends to be more comfortable and the starting point for embedded software developers who have more of an application background than a hardware background. The top-down approach starts with the high-level application and the business logic and then works its way down to the abstraction layers and the hardware. The core idea here is that the product differentiator is not in the foundation but the business logic of the device! These designers want to focus on the application code since this is where product differentiation is achieved. The hardware and lower layers are just a means to the end, and it doesn't matter to the product features or the stakeholders (unless it's built upon a poor framework or slow processor).

---

**Beware**  *The goal is to produce a product that generates a profit! Everything else is tertiary (or less).*

---

With two approaches available to us, which one is the right one? Well, it depends on your team and the resources that they have available to them. However, I believe that the right approach for most teams today is to use the top-down approach. There are several reasons why I think this.

First, designers will focus more on their product features and differentiators, the secret sauce that will help sell their products. The more focused I am on my product features, the more I can think through, prioritize, and ensure that my design is going in the right direction. If I save it for the end, I will likely "mess up" on my features and provide a poor experience to the product users.

Second, modern development tools, which we will discuss later in the book, can be used to model and simulate the application's behavior long before the hardware can. This allows us to put the application code to the test early. The earlier we can iterate on our design and adjust before coding on the target, the less expensive development will be. The change cost grows exponentially as a project progresses through the various development stages, as seen in Figure 2-3.[4] The sooner we can catch issues and mistakes, the better it will be from a cost and time-to-market standpoint.

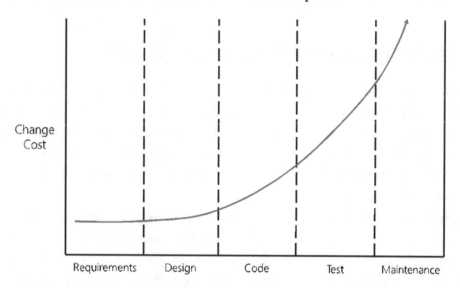

**Figure 2-3.**  *The change cost grows exponentially as a project progresses through the various development stages*

Finally, by designing the top level first, simulating it, and getting feedback, if things don't go well, we can fail quickly. In business, nothing is worse than investing in something that drags on, consumes the budget, and then fails. Companies want projects that are going to fail to fail quickly.[5] If a product isn't going to work, the sooner we know, the better, and the faster we can move on to the next idea. Starting at the top can help us validate our application with the target audience sooner.

---

[4] A great article on change costs can be found at `www.agilemodeling.com/essays/costOfChange.htm`

[5] We'd like to think that businesses don't fail, but 90% of businesses fail within their first ten years of operation!

There is a hybrid approach that I sometimes take with my customers that have a large enough team and resources to make it work. It's where we simultaneously design both architectures! Some team members are assigned to focus on the high-level business logic architecture, while others are given to the low-level real-time architecture. Honestly, this can be a great approach! We essentially have two different application domains that need to interact through a well-defined abstraction layer. There is no reason why these activities cannot be parallelized and meet in the middle!

# Characteristics of a Good Architecture

One concern many developers and software architects have is whether the architecture is good; in other words, does it meet their needs. As strange as it sounds, I see many teams just jump in and start to design their architecture without defining what their end goals are! Of course, every system won't have the same goals; however, some common characteristics that most embedded software architects aim for help us define whether the architecture is good or not.

Many teams aim to ensure that their architecture is reusable, scalable, portable, maintainable, and so forth. Early in the architectural phase, developers should write down their goals for the architecture. If a team is developing a quick rapid prototype, who cares if the architecture is portable or maintainable? However, if the code is the base for multiple products, scalable and maintainable goals are good targets.

No matter what the end goals are for the software, there are two characteristics that every architect needs to carefully manage to reach their end goals: coupling and cohesion.

# Architectural Coupling

Coupling refers to how closely related different modules or classes are to each other and the degree to which they are interdependent.[6] The degree to which the architecture is coupled determines how well a developer can achieve their architectural goals. For example, if I want to develop a portable architecture, I need to ensure that my architecture has low coupling.

---

[6] https://bit.ly/33R6psu

There are several different types and causes for coupling to occur in a software system. First, common coupling occurs when multiple modules have access to the same global variable(s). In this instance, code can't be easily ported to another system without bringing the global variables along. In addition, the global variables become dangerous because they can be accessed by any module in the system. Easy access encourages "quick and dirty" access to the variables from other modules which then increases the coupling even further. The modules have a dependency on those globally shared variables.

Another type of coupling that often occurs is content coupling. Content coupling is when one module accesses another module's functions and APIs. While at first this seems reasonable because data might be encapsulated, developers have to be careful how many function calls the module depends on. It's possible to create not just tightly coupled dependencies but also circular dependencies that can turn the software architecture into a big ball of mud.

Coupling is most easily seen when you try to port a feature from one code base to another. I think we've all gone through the process of grabbing a module, dropping it into our new code base, compiling it, and then discovering a ton of compilation errors. Upon closer examination, there is a module dependency that was overlooked. So, we grab that dependency, put it in the code, and recompile. More compilation errors! Adding the new module quadrupled the number of errors! It made things worse, not better. Weeks later, we finally decided it's faster to just start from scratch.

Software architects must carefully manage their coupling to ensure they can successfully meet their architecture goals. Highly coupled code is always a nightmare to maintain and scale. I would not want to attempt to port highly coupled code, either. Porting tightly coupled code is time-consuming, stressful, and not fun!

## Architectural Cohesion

The coupling is only the first part of the story. Low module coupling doesn't guarantee that the architecture will exhibit good characteristics and meet our goals. Architects ultimately want to have low coupling and high cohesion. Cohesion refers to the degree to which the module or class elements belong together.

In a microcontroller environment, a low cohesion example would be lumping every microcontroller peripheral function into a single module or class. The module would be significant and unwieldy. Instead, a base class could be created that defines the common interface for interacting with peripherals. Each peripheral could then inherit from that interface and implement the peripheral-specific functionality. The result is a highly cohesive architecture, low coupling, and other desirable characteristics like reusable, portable, scalable, and so forth.

Cohesion is really all about putting "things" together that belong together. Code that is highly cohesive is easy to follow because everything needed is in one place. Developers don't have to search and hunt through the code base to find related code. For example, I often see developers using an RTOS (real-time operating system) spread their task creation code throughout the application. The result is low cohesion. Instead, I pull all my task creation code into a single module so that I only have one place to go. The task creation code is highly cohesive, and it's easily ported and configured as well. We'll look at an example later in the Development and Coding Skills part of the book.

Now that we have a fundamental understanding of the characteristics we are interested in as embedded architects and developers, let's examine architectural design patterns that are common in the industry.

# Architectural Design Patterns in Embedded Software

Over the years, several types of architectural patterns have found their way into embedded software. It's not uncommon for systems to have several architectural designs depending on the system and its needs. While there are many types of patterns in software engineering, the most common and exciting for microcontroller-based systems include unstructured monoliths, layered monoliths, event-driven architectures, and microservices. Let's examine each of these in detail and understand the pros and cons of each.

## The Unstructured Monolithic Architecture

The unstructured monolithic software architecture is the bane of a modern developer's existence. Sadly, the unstructured monolith is easy to construct but challenging to maintain scale and port. In fact, given the opportunity, most code bases will try their best to decay into an unstructured monolith if architects and developers don't pay close attention.

The great sage Wikipedia describes a monolithic application as "a single-tiered software application in which the user interface and data access code are combined into a single program from a single platform."[7] An unstructured monolithic architecture might look something like Figure 2-4. The architecture is tightly coupled which makes it extremely difficult to reuse, port, and maintain. Individual modules or classes might have high cohesion, but the coupling is out of control.

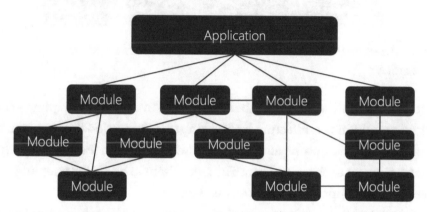

*Figure 2-4.* *Unstructured monolithic architectures are tightly coupled applications that exist in the application layer*

An unstructured monolith was one of the most common architectures in embedded systems when I started back in the late 1990s. A real-time microcontroller-based system was viewed as a "hot rod"; it was designed for a single, one-off purpose, and it needed to be fast and deterministic. For that reason, systems often weren't designed with reuse in mind. Fast forward 20–30 years, microcontrollers are performance powerhouses with lots of memory. Systems are complex, and building that hot rod from scratch in an unstructured monolith doesn't meet today's needs.

## Layered Monolithic Architectures

Although I have no data to prove, I would like to argue that layered monolithic architectures are the most common architecture used in embedded applications today. The layered architecture allows the architect to separate the application into various independent layers and only interact through a well-defined abstraction layer. You may have seen layered monolithic architectures like Figure 2-5.

---

[7]https://en.wikipedia.org/wiki/Monolithic_application

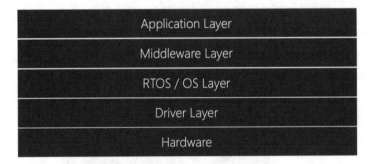

**Figure 2-5.** *An example is layered monolithic architecture for an embedded system*

A layered monolithic application attempts to improve the high coupling of an unstructured monolithic architecture by breaking the application into independent layers. Each layer should only be allowed to communicate with the layer directly above or below it through an abstraction layer. The layers help to break the coupling, which allows layers to be swapped in and out as needed.

The most common example of swapping out a layer is when the architecture will support multiple hardware platforms. For example, I've worked with clients who wanted to use Microchip parts in one product, NXP parts in another, and STM parts in another. The products were all related and needed to run some standard application components. Instead of writing each product application from scratch, we designed the architecture so that the application resembled Figure 2-5. The drivers were placed behind a standard hardware abstraction layer (HAL) that made the application dependent on the HAL, not the underlying hardware.

Leveraging APIs and abstraction layers is one of the most significant benefits of layered monolithic architecture. It breaks the coupling between layers, allows easier portability and reuse, and can improve maintainability. However, one of the biggest gripes one could have with the layered monolithic architecture is that they break the "hot rodness" of the system.

As layers are added to an application, the performance can take a hit since the application can no longer just access hardware directly. Furthermore, clock cycles need to be consumed to circumnavigate the layered architecture. In addition to the performance hit, it also requires more time up front to design the architecture properly. Finally, the code bases do tend to be a little bit bigger. Despite some disadvantages, modern microcontrollers have more than enough processing power in most instances to overcome them.

**Definition**   An OSAL is an operating system abstraction layer. An OSAL can decouple the application dependency on any one specific operating system.

Before we look at the next architecture type, I'd like to point out two critical layered architectures. First, notice that each layer can still be considered an unstructured monolithic architecture. Architects and developers, therefore, need to be careful in each layer that they manage coupling and cohesion carefully. Second, each layer in the layered architecture does not have to extend across the entire previous layer. Like a printed circuit board (PCB) layer, you can create islands and bypasses to improve performance. An example modern layered architecture diagram can be seen in Figure 2-6.

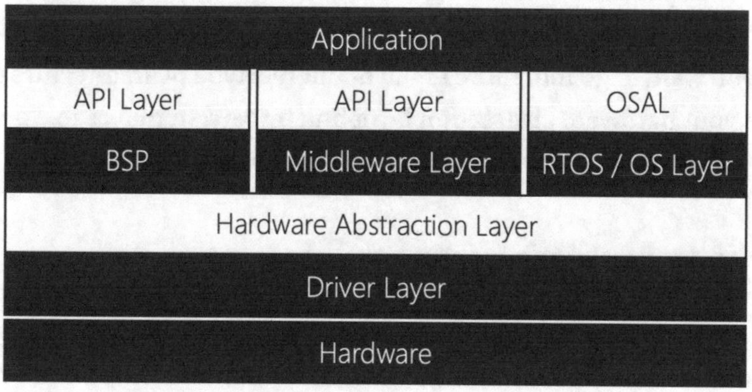

**Figure 2-6.** *An example is modern layered monolithic architecture for an embedded system*

Figure 2-6 has many benefits. First, note how we can exchange the driver layer for working with nearly any hardware by using a hardware abstraction layer. For example, a common HAL today can be found in Arm's CMSIS.[8] The HAL again decouples the hardware drivers from the above code, breaking the dependencies.

Next, notice how we don't even allow the application code to depend on an RTOS or OS. Instead, we use an operating system abstraction layer (OSAL). If the team needs to change RTOSes, which does happen, they can just integrate the new RTOS without

---

[8] CMSIS is Arm's Common Microcontroller Software Interface Standard.

having to change a bunch of application code. I've encountered many teams that directly make calls to their RTOS APIs, only later to decide they need to change RTOSes. What a headache that creates! An example of OSAL can be found in CMSIS-RTOS2.[9]

Next, the board support package exists outside the driver layer! At first, this may seem counterintuitive. Shouldn't hardware like sensors, displays, and so forth be in the driver layer? I view the hardware and driver layer as dedicated to only the microcontroller. Any sensors and so on that are connected to the microcontroller should be communicated through the HAL. For example, a sensor might be on the I2C bus. The sensor would depend on the I2C HAL, not the low-level hardware. The abstraction dependency makes it easier for the BSP to be ported to other applications.

Finally, we can see that even the middleware should be wrapped in an abstraction layer. If someone is using a TLS library or an SD card library, you don't want your application to be dependent on these. Again, I look at this as a way to make code more portable, but it also isolates the application so that it can be simulated and tested off target.

Suppose you want to go into more detail about this type of architecture and learn how to design your hardware abstraction layer and drivers. In that case, I'd recommend reading my other book *Reusable Firmware Development*.

## Event-Driven Architectures

Event-driven architectures make a lot of sense for real-time embedded applications and applications concerned with energy consumption. In an event-driven architecture, the system is generally in an idle state or low-power state unless an event triggers an action to be performed. For example, a widget may be in a low-power idle state until a button is clicked. Clicking the button triggers an event that sends a message to a message processor, which then wakes up the system. Figure 2-7 shows what this might look like architecturally.

*Figure 2-7.* *An example modern layered monolithic architecture for an embedded system*

---

[9] The use of an RTOS with an OSAL will often require an OSAL extension to access unique RTOS features not supported in every RTOS.

Event-driven architectures typically utilize interrupts to respond to the event immediately. However, processing the event is usually offloaded to a central message processor or a task that handles the event. Therefore, event-driven architectures often use message queues, semaphores, and event flags to signal that an event has occurred in the system. Using an RTOS for an event-driven architecture is helpful but not required. (We will discuss more application design using an RTOS in Chapter 4.)

The event-driven architecture has many benefits. First, it is relatively scalable. For example, if the widget in Figure 2-7 needed to filter sensor data when a new sample became available, the architecture could be modified to something like Figure 2-8. New events can easily be added to the software by adding an event handler, an event message, and the function that handles the event.

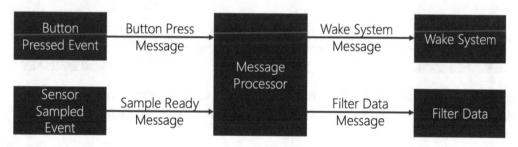

***Figure 2-8.*** *The event-driven architecture is scalable and portable. In this example, a sensor sampled event has been added*

Another benefit to the event-driven architecture is that software modules generally have high cohesion. Each event can be separated and focused on just a single purpose.

One last benefit to consider is that the architecture has low coupling. Each event minimizes dependencies. The event occurs and needs access only to the message queue that is input to the central message processor. Message queue access can be passed in during initialization, decoupling the module from the messaging system. Even the message processor can have low coupling. The message processor needs access to the list of messages it accepts and either the function to execute or a new message to send. Depending on the implementation, the coupling can seem higher. However, the message processor can take during its initialization configuration tables to minimize the coupling.

The disadvantage to using an event-driven architecture with a central message processor is that there is additional overhead and complexity whenever anything needs to be done. For example, instead of a button press just waking the system up, it needs to send a message that then needs to be processed that triggers the event. The result is extra latency, a larger code base, and complexity. However, a trade-off is made to create a more scalable and reusable architecture.

If performance is of concern, architects can use an event-driven architecture that doesn't use or limits the use of the central message processor. For example, the button press could directly wake the system, while other events like a completed sensor sample could be routed through the message processor. Such an architectural example can be seen in Figure 2-9.

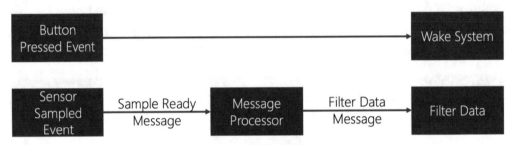

***Figure 2-9.***  *To improve performance, the event-driven architecture allows the button-pressed event to circumvent the message processor and directly perform the desired action*

It's not uncommon for an architectural solution to be tiered. A tiered architecture provides multiple solutions depending on the real-time performance constraints and requirements placed on the system. A tiered architecture helps to balance the need for performance and architectural elegance with reuse and scalability. The tiered architecture also helps to provide several different solutions for different problem domains that exist within a single application. Unfortunately, I often see teams get stuck trying to give a single elegant solution, only to talk themselves in circles. Sometimes, the simplest solution is to tier the architecture into multiple solutions and use the solution that fits the problem for that event.

# Microservice Architectures

Microservice architectures are perhaps the most exciting yet challenging architectures available to embedded architects. A microservice architecture builds applications as a collection of small autonomous services developed for a business domain.[10] Microservices embody many modern software engineering concepts and processes such as Agile, DevOps, continuous integration, and continuous deployment (CI/CD).

Microservices, by nature, have low coupling. Low coupling makes the microservice maintainable and testable. Developers can quickly scale or port the microservice. If a new microservice is developed to replace an old one, it can easily be swapped out without disrupting the system. Engineering literature often cites that microservices are independently deployable, but this may not always be true in constrained embedded systems that use a limited operating system or without a microservice orchestrator to manage the services.

A significant feature of microservices is that they are organized around the system's business logic. Business logic, sometimes referred to as business capabilities, are the business rules and use cases for how the system behaves. I like to think of it as the features that a customer is paying to have in the product. For example, a customer who buys a weather station will be interested in retrieving temperature and humidity data. They don't care the sensor that provides the temperature is a DHT20. The business logic is that a temperature request comes in, and the business layer provides a response with the temperature data.

For embedded systems, microservices can go far beyond just the system's business logic. For example, a weather station might contain microservices for telemetry, sensor interfacing, motor control (to maximize sun illumination for battery charging), and a command service.

The sensor service would be responsible for directly interacting with sensors such as temperature, humidity, motor position, etc. The service would then report this information to the motor and telemetry services. The motor service would drive the motor based on commands from the command service. Status information from the motor service would be reported to the telemetry service. Finally, the command service could command the motor and provide status updates to telemetry. The user could interact with the system, whether connected via a browser, smart app, or whatever. The example architecture can be seen in Figure 2-10.

---

[10] www.guru99.com/microservices-tutorial.html

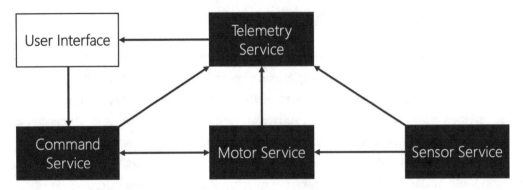

**Figure 2-10.**  *An example weather station microservice architecture. Each service is independent and uses messages to communicate with other services*

At first glance, the microservice architecture may not seem very decoupled. This is because we have not yet discussed what the microservice block looks like. A microservice is made up of five pieces: the microservice, an incoming queue, outbound messages, a log, and microservice status. Figure 2-11 shows each of these pieces.

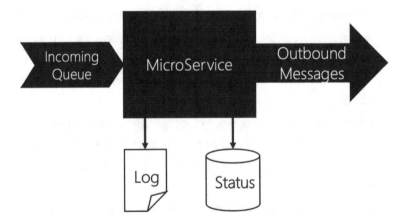

**Figure 2-11.**  *A microservice comprises five standard components, which are graphically represented here*

A microservice is typically a process running on the microprocessor that contains its memory and consists of at least a single task or thread. For systems using an RTOS, a microservice can be just a single task or multiple tasks that use the memory protection unit (MPU) to create a process. The microservice contains all the code and features necessary to fulfill the service's needs.

The microservice contains a message queue used to receive messages from other microservices. In today's systems, these could be messages from other microservices running on the microprocessor or messages obtained from microservices running in the cloud or other systems! The microservice also can send messages to other services.

The important and sometimes overlooked features of a microservice by embedded developers are the logging and status features. The microservice should be able to track its behavior through a logging feature. Information that can be logged varies, but it can include messages received, transmitted, and so forth. In addition, the microservice status can be used to monitor the health and wellness of the microservice.

As we have seen, microservices have a lot of advantages. However, there are also some disadvantages. First, architecturally, they can add complexity to the design. Next, they can add extra overhead and memory needs by having communication features that may not be needed in other architectures. The decentralized nature of the architecture also means that real-time, deterministic behavior may be more challenging. Timing and responses may have additional jitter.

Despite the disadvantages, the microservice architecture can be advantageous if used correctly. However, microservice architectures may not fit your needs or product well. Furthermore, trying to do so could cause increased development times and budgets. Therefore, it's essential to carefully analyze your needs and requirements before committing to any architecture.

Before selecting the best architecture for your application, it's also essential to consider the application domains involved in the product. Architects should decompose their applications into different domains.

# Application Domain Decomposition

A growing number of products require that the system architecture execute across multiple domains. For example, some code may run on an embedded controller while other pieces run in the cloud. Application domain decomposition is when the designer breaks the application into different execution domains. For experienced embedded developers, this might initially seem strange because designers working with microcontroller-based systems lump all their code into a single execution domain. They view all code as running on a single CPU, in a single domain, in one continuous memory map.

It turns out that there are four common application domains that every embedded designer needs to consider. These include

- The Privilege Domain

- The Security Domain

- The Execution Domain

- The Cloud Domain

Let's examine each of these in greater detail and understand the unique insights these domains can bring to the design process.

# The Privilege Domain

Just a moment ago, I mentioned that many embedded software designers lump all their code into a single execution domain. They view all code running on a single CPU in one continuous memory map. While this may be true at the highest levels, developers working with microcontroller will often find that the processor has multiple privilege levels that the code can run in. For example, an Arm Cortex-M processor may have their code running in privileged or unprivileged modes.

In privileged mode, code that is executing has access to all CPU registers and memory locations. In unprivileged mode, code execution is restricted and may not access certain registers or memory locations. Obviously, we don't just want to have all our code running in privileged mode! If the code runs off into the weeds, it would be nice to prevent the code from accessing critical registers.

Within the Cortex-M processor, it's interesting to also note that there are also two modes that code can be running in which affect the privileged mode: thread and handler modes. Thread mode is used for normal code execution, and the code can be in either the privileged or unprivileged mode. Handler mode is specifically for executing exception handlers and only runs in privileged mode. A typical visual summary for these modes can be found in Figure 2-12.

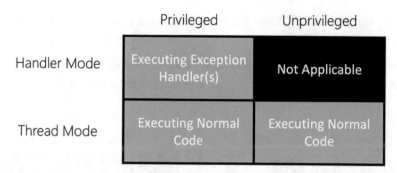

| | Privileged | Unprivileged |
|---|---|---|
| Handler Mode | Executing Exception Handler(s) | Not Applicable |
| Thread Mode | Executing Normal Code | Executing Normal Code |

*Figure 2-12.* *A summary of the privilege states and modes is on Arm Cortex-M microcontrollers[11]*

So, while a developer may think their application is 100% in a single domain, the designer should carefully consider the processor modes in which the application should run.

## The Security Domain

The security domain is an important domain for designers to consider, although unfortunately one that is often overlooked by developers. Stand-alone systems require security to prevent tampering and intellectual property theft as much as a connected device does. However, the attack surface for a connected device is often much larger, which requires the designer to carefully consider the security domain to ensure the product remains secure for its entire useful life.

The security domain is interesting for the designer because security is not something that can be bolted onto a system when it is nearly completed. Instead, security must be thought about as early as possible and built into the design. Plenty of tactical solutions for improving security exist, such as encrypting data and locking down a device's programming port. However, strategically, designers need to create layers of isolation within their applications. Then, each isolated layer in the design can be executed in its own security domain.

---

[11] www.researchgate.net/figure/Armv7-M-operation-modes-and-privileged-levels_fig1_339784921

A designer may find it helpful to decompose their high-level application design into isolated regions that specifically call out the importance of the security context of where the various components are executing. Figure 2-13 shows an example of an updated high-level weather station design to demonstrate security isolation layers.

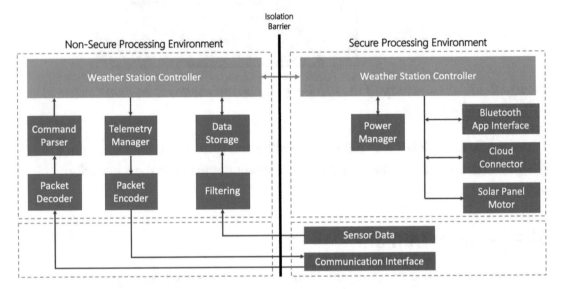

***Figure 2-13.*** *The top-level diagram is now broken up into execution domains in a high-level attempt to demonstrate isolation boundaries*

It's interesting to note that Figure 2-13 doesn't specify which technology should be used for isolation at this point, only that there are secure processing environments (SPE) and nonsecure processing environments (NSPE). The solution could be to use a multicore processor with a dedicated security core or a single-core solution like Arm TrustZone. We will discuss designing secure applications in Chapter 3 and how designers develop their security requirements.

## The Execution Domain

The third domain and the most interesting to most developers is the execution domain. Microcontroller applications have recently reached a point where the needs for performance and energy consumption cannot be achieved in single-core architecture. As a result, multicore microcontrollers are quickly becoming common in many applications. This drives the designer to carefully consider the design's execution domains for various components and features.

Multiple cores naturally force developers to decide which components and tasks run on which cores. For example, developers working with an ESP32, typically a dual-core Tensilica Xtensa LX6 32-bit processor, will often break their application up such that one core runs the Wi-Fi and Bluetooth stack. In contrast, the other core runs the real-time application. This ensures that the application is not choked by the Wi-Fi and Bluetooth stacks and can still achieve real-time requirements. An example domain partition between cores can be seen in Figure 2-14.

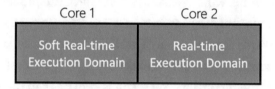

***Figure 2-14.*** *Multicore microcontrollers break an application into multiple execution domains that concurrently run on multiple cores. This is one example of execution domain decomposition*

There are many potential use cases for multicore designers. We will explore the use cases and how to design applications that use multiple cores in Chapter 5.

Execution domain designs today go much further than simply deciding which core to run application components on. Cutting-edge systems and high-compute solutions such as machine learning may also contain several different execution domain types. For example, there may be soft and hard real-time microcontroller cores in addition to application processors, Tensor Processing Units (TPUs),[12] and Graphics Processing Units (GPUs).

Hybrid systems that combine processing elements are becoming much more common. For example, we are familiar with voice assistants such as Alexa, Siri HomePods, etc. These systems often employ a combination of low-power microcontroller cores for initial keyword spotting and then leverage an application processing core for relaying more complex processing to the cloud. These types of hybrid systems will become ubiquitous by the end of the 2020s.

---

[12] https://cloud.google.com/tpu/docs/tpus

# The Cloud Domain

The final application domain to consider is the cloud domain. At first, this probably seems counterintuitive to an embedded designer. After all, doesn't the entire application live on the microcontroller? However, this may not necessarily be the case for a modern embedded system.

Today, designers can potentially offload computing from the embedded controller up to the cloud. For example, voice assistants don't have the processing power to classify all the word combinations that are possible. Instead, these systems are smart enough to recognize a single keyword like "Alexa" or "Hey Siri." Once the keyword is recognized, any remaining audio is recorded and sent to the cloud for processing. Once the cloud processes and classifies the recording, the result is sent back to the embedded system, which performs some action or response.

The cloud provides an application domain that embedded designers might usually overlook and should consider. However, the cloud isn't necessarily a smoking gun to solve all computing problems because it can have a considerable impact in areas such as energy consumption, bandwidth, cost, and latency.

# Final Thoughts

The software architecture is the blueprint, the road map that is used by developers to construct the designers' vision for the embedded software. Designers and developers must carefully decouple the hardware from the application's business logic. Embedded designers have multiple architectures that are available to them today that can help them avoid a giant ball of mud if they minimize coupling and maximize cohesion.

The software architecture in a modern system no longer has just a single domain to consider. Applications have become complex enough that multiple domains must be leveraged and traded against to successfully optimize a design. How the system is decomposed and which architecture will meet your needs depend on your system's end goals and requirements.

## ACTION ITEMS

To put this chapter's concepts into action, here are a few activities the reader can perform to start architecting their embedded software:

- Which approach to software architecture design do you use, top-down or bottom-up? What advantages and disadvantages do you see to the approach that you use? Is now the time to change your approach?

- What benefits can you gain by taking a top-down approach to architecture design? What difference would this make for you as an engineer and the business that employs you?

- Think through some of your recent projects. How would you rate the architecture in the following areas?

  - Coupling

  - Cohesion

  - Portability

  - Scalability

  - Flexibility

  - Reusability

What changes do you need to make to your architecture to improve these characteristics?

- For a current or past project, what architecture did you use? What were the benefits and disadvantages of using that architecture? Are any of the architectures we discussed a better fit for your applications?

- Identify three changes you will make to your current architecture to better fit your team's needs.

- Have you considered the various application domains in your designs? What areas have you overlooked, and how could these have impacted the overall performance, security, and behavior of your applications? Take some time to break down a new application into its various application domains.

These are just a few ideas to go a little bit further. Carve out time in your schedule each week to apply these action items. Even minor adjustments over a year can result in dramatic changes!

# CHAPTER 3

# Secure Application Design

Every embedded system requires some level of security. Yet, security is often the most overlooked aspect in most software designs. Developers may look at their product and say it's not an IoT device, so there is no need to add security. However, they overlook that the same firmware they deploy in their systems may be worth millions of dollars. A competitor interested in reverse-engineering the product or a hacker interested in tampering with the device locally is still a viable threat.

The apparent reason for designing secure embedded software is multifold. First, companies want to protect their intellectual property to prevent competitors from catching up or gaining a competitive edge. Next, companies need to be protecting their users' data, which in itself can be worth a fortune. Finally, a high-profile attack that reaches the media can have catastrophic consequences to a brand that tanks the company stock or causes its users to switch to a competitor. (There are certainly others, but these are probably the biggest.)

Companies are deploying billions of IoT devices yearly. Unfortunately, these connected systems are often ripe for hackers interested in exploiting a device's data for financial gain, pleasure, or to prove their technical skills. There's hardly a day when there is no headline about the latest device hack, malware attack, or successful ransomware.

It's easy to think that your system will not be hacked. There are so many devices on the Internet; what are the chances that someone will find your device and target it? Well, the chances are pretty good! Several years ago, I worked on a project using a custom application processor running a Linux kernel. I was going to be traveling over the weekend and needed remote access to the device to finish my work. The client agreed to provide remote access to the device. It went live on the Internet Friday night; by Sunday morning, the system had become part of a Chinese botnet!

© Jacob Beningo 2022
J. Beningo, *Embedded Software Design*, https://doi.org/10.1007/978-1-4842-8279-3_3

Security is not optional anymore! Developers and companies must protect their intellectual property, customers' data, and privacy and meet the plethora of government regulations steadily being approved (recently, GDPR, NISTIR 8259A, the California SB-327, and the IoT Cybersecurity Improvement Act of 2019). Security must be designed into an embedded system from the very beginning, not added in at the end. Last-minute attempts to secure a device will only leave security vulnerabilities in the system that can be exploited.

In this chapter, we will start to explore how to design secure embedded applications. We'll start by exploring Arm's Platform Security Architecture. By the end of the chapter, you'll understand the fundamentals of how to perform a threat model and security analysis and the options you have to start designing your embedded software with security from the beginning.

# Platform Security Architecture (PSA)

Arm and ecosystem partners developed the Platform Security Architecture (PSA) to "provide the embedded systems industry with a baseline security architecture that could be leveraged to secure an embedded product."[1] PSA is designed to give developers holistic resources to simplify security. PSA consists of

- A set of threat models and security analysis documentation

- Hardware and firmware architecture specifications

- An open source firmware reference implementation[2]

- An independent security evaluation scheme

One way to look at PSA is as an industry best practice guide that allows security to be implemented consistently for hardware and software. Take a moment to absorb that statement; security is not just something done in software but also something in hardware! A secure solution requires both components to work together.

---

[1] https://armkeil.blob.core.windows.net/developer/Files/pdf/white-paper/guide-to-iot-security.pdf

[2] https://developer.arm.com/architectures/security-architectures/platform-security-architecture

> **Beware**    Security requires an interplay between hardware and the software to successfully meet the systems security requirements.

It's important to realize that security is not just something you bolt onto your product when you are ready to ship it out the door. Instead, to get started writing secure software, you need to adopt industry best practices, start thinking about security from the beginning of the project, and ensure that the hardware can support the software needs. On multiple occasions, I've had companies request that I help them secure their products weeks before they officially launched. In several instances, their hardware did not support the necessary features to develop a robust security solution! Their security solution became a "hope and pray" solution.

Developing a secure solution using PSA includes four stages that can help guide a development team to constructing secure firmware. PSA guides teams to identify their security requirements up front so that the right hardware and software solutions can be put in place to minimize the device's attack surface and maximize the device's security. The four PSA stages include

- Analyze

- Architect

- Implement

- Certify

Figure 3-1 shows a complete overview of the PSA stages.

In stage 1, developers analyze their system requirements, identify data assets, and model the threats against their system and assets. In stage 2, Architect, developers select their microcontroller and associated security hardware in addition to architecting their software. In stage 3, Implementation, developers finally get to write their software. Developers can leverage PSA Certified components, processes, and software throughout the process to accelerate and improve software security. PSA Certified is a security certification scheme for Internet of Things (IoT) hardware, software, and devices.[3]

---

[3] https://en.wikipedia.org/wiki/PSA_Certified

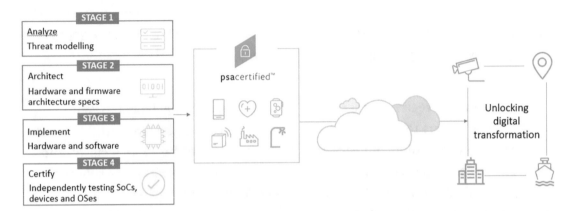

**Figure 3-1.** *The four primary PSA stages and PSA Certified components help unlock a secure digital product transformation[4]*

Since PSA's processes and best practices are essential to designing secure embedded applications, let's take a little more time to examine each PSA stage in a little more detail. Then, we will focus on the areas where I believe embedded developers will benefit the most, although each team may find specific areas more valuable than others.

# PSA Stage 1 – Analyzing a System for Threats and Vulnerabilities

When teams recognize that they need to implement security, they often find implementing security to be a nebulous, undefined activity. Teams know they need to secure the system but don't understand what they are trying to secure. Teams jump straight into concepts like encryption and secure communications in many cases. Encryption on its own is not going to secure a system, it might secure communications, but the system can still be vulnerable to wide variety of attacks.

Designing a secure embedded system starts with identifying the assets that need to be protected and the threats those assets will face. Once the assets and threats are identified, developers can properly define their security requirements to protect them. The security requirements will then dictate the security strategies and tactics used and the hardware and software needed to protect those assets adequately. Formally, the process is known as a threat model and security analysis (TMSA).

---

[4] https://community.arm.com/arm-community-blogs/b/internet-of-things-blog/posts/the-iot-architects-practical-guide-to-security

A TMSA that follows the best practices defined in PSA has five distinct steps that can be seen in Figure 3-2. These include defining assets, identifying adversaries and threats, defining security objectives, writing security requirements, and summarizing the TMSA results.

***Figure 3-2.*** *The five steps developers walk through to perform a TMSA. Image Source: (Arm)[5]*

Thoroughly discussing how to perform a TMSA is beyond the scope of this book,[6] but let's examine a few high-level concepts of what a TMSA might look like for an Internet-connected weather station.

## TMSA Step #1 – Identifying Assets

The first step in the TMSA is to identify assets that might need to be protected. It's interesting to note that most assets that need to be protected in a system are data assets under software control. For example, an IoT device will commonly have data assets such as

- A unique device identification number

- The firmware[7]

- Encryption keys

- Device keys

- Etc.

---

[5] https://community.arm.com/iot/b/blog/posts/five-steps-to-successful-threat-modelling

[6] A good general document to consult is the NIST Guide for Conducting Risk Assessments. https://nvlpubs.nist.gov/nistpubs/Legacy/SP/nistspecialpublication800-30r1.pdf

[7] That's right! Firmware is a system data asset! It's quite common for developers to overlook this fact.

Every product will also have unique data assets that need to be protected. For example, a biomedical device might protect data such as

- A user profile

- Raw sensor data

- Processed sensor data

- System configuration

- Control data

- Etc.

A weather station might protect data such as

- Location

- Local communication interface(s)

- Raw sensor data

- System configuration

- Remote access credentials

- Etc.

This is an excellent time to point out the parallels between identifying data assets from a security perspective with the design philosophies we outlined in Chapter 1. We have already established that data dictates design. In many instances, our data is a data asset that we would identify in this TMSA step. We have outlined the design philosophies that encompass ideas and concepts for secure application development, including that security is king.

Another essential point to consider is that every identified asset in our system doesn't have to be protected. There will be some assets that will be public data that require no security mechanisms. Trying to protect those publicly available assets can decrease the system's security! Imagine taking publicly known data and then encrypting it in a data stream. We can use this information to break the encryption scheme if we know the public data is in the stream and what the data is. Only protect the assets that need to be protected.

**Best Practice**   Only protect data assets that need to be protected. "Overdoing it" can lead to vulnerabilities just as much as not doing anything. Carefully balance based on your system needs.

## TMSA Step #2 – Identifying Adversaries and Threats

Once the data assets that need to be protected have been identified, the developers can specify the adversaries and the threats to the assets. Identifying adversaries can be as simple as creating a list. Still, I often like to include characteristics of the adversary and rank them based on the likelihood and the impact the adversary may have on the product and the business. Table 3-1 shows an example of what an adversary analysis might look like.

***Table 3-1.***  *Example adversary analysis that includes a risk assessment from each*

| Adversary | Impact | Likelihood | Comments |
| --- | --- | --- | --- |
| Competitor | High | High | IP theft |
| End user | Low | Medium | Feature availability |
| Hackers | Medium | Medium | Data theft, malware |
| Nation state | High | Low | Data theft, malware |
| Internal personnel | High | Medium | IP and data theft |

Each adversary will have its purpose for attacking the system. For example, an adversary may be interested in the data on the system, overriding controls, spying, stealing intellectual property, or any number of other reasons. Just because an adversary exists does not necessarily mean that someone will be interested in your system (although assuming no one is interested is super dangerous). Developers should rank each adversary based on the risk (impact) to the business, user, and system and determine the likelihood of an attack. Figure 3-3 shows a matrix often used to evaluate risk.

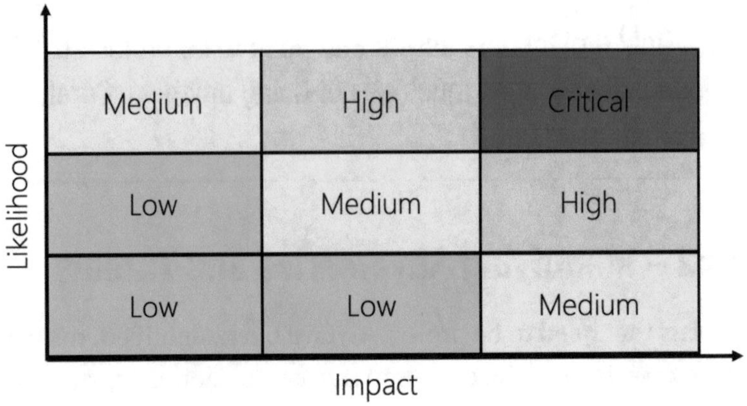

| Medium | High | Critical |
|---|---|---|
| Low | Medium | High |
| Low | Low | Medium |

Likelihood / Impact

*Figure 3-3.* *A standard risk matrix trades off impact and the likelihood of an event[8]*

From a quick glance, Figure 3-3 gives us the foundation for developing our own risk matrix. Unfortunately, the generalized matrix doesn't really tell you much. Instead, I would recommend that you take the format and then update it for your specific situation. For example, if I take Table 3-1 and overlay it on the risk matrix, the result will be something like Figure 3-4. Figure 3-4 helps to create a better picture of what we are facing and hopefully allows us to generate some actionable behaviors from it.

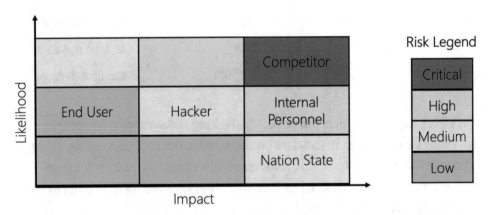

*Figure 3-4.* *An actionable risk table that demonstrates the risk for potential adversaries*

---

[8]www.alertmedia.com/blog/business-threat-assessment/

We can see that for our hypothetical system, the highest risk is from our competitors followed by our internal personnel. Now that we understand who might be attacking our system, we can decide where to focus our security efforts. If the end user will have minimal impact and likelihood, we may just put enough security in place to make it not worth their time. Instead, we focus on making sure our competitors can't steal our intellectual property or gain a competitive advantage.

Adversaries will target and attack a system's assets in various ways; these attacks are often referred to as threats. Every embedded system will face different and unique threats, but there will be a standard set of threats. For example, an adversary interested in gaining access to the system might attack the system's credentials through an impersonation attack. Another adversary may attack the firmware to inject code or modify its functions.

TMSA recommends developers identify the various threats that each asset will face. Teams often find it helpful to build a table that lists each asset and includes columns for various analytical components. For example, Table 3-2 shows several assets and the expected threats for a connected weather station.

***Table 3-2.*** *A list of assets and their threats for a fictitious connected weather station*[9]

| Data Asset | Threats |
| --- | --- |
| Location | Man-in-the-middle, repudiation, impersonation, disclosure |
| Raw sensor data | Tamper (disclosure) |
| System configuration | Tamper (disclosure) |
| Credentials | Impersonation, tamper (disclosure), man-in-the-middle, escalation of privilege (firmware abuse) |
| Firmware | Tamper, escalation of privilege, denial of service, malware |

---

[9] See Appendix A for the definition of "important terms."

As the reader may imagine, it can also be advantageous to reverse this table and list threats that list the data assets. I won't display such a table for brevity, but you can see an expanded version in Table 3-3. I'll only recommend that both tables be created and maintained.

## TMSA Step #3 – Defining Security Objectives

The TMSA helps developers and teams define the security objects for their systems. There are typically two parts to the security objective definitions. First is identifying which combination of confidentiality, integrity, and authenticity will protect the assets. Second is the security objective that will be used to secure the asset. Let's examine each of these two areas separately.

Confidentiality, integrity, and authenticity are often called the CIA model.[10] Every data asset that a team is interested in protecting will require some combination of CIA to protect the asset. For example, credentials would require confidentiality. The system configuration might require confidentiality and integrity. The product's firmware would require confidentiality, integrity, and authorization. Before we can define the combination that fits an asset, it's helpful to understand what each element of the CIA does.

Confidentiality indicates that an asset needs to be kept private or secret. For example, data assets such as passwords, user data, and so forth would be confidential. Access to the data asset would be limited to only authorized users. Confidentiality often indicates that encryption needs to be used. However, that encryption often goes beyond just point-to-point encryption that we find in the connection between the device and the cloud. Confidential assets often remain encrypted on the device.

Integrity indicates that the data asset needs to remain whole or unchanged. For example, a team would want to verify the integrity of their firmware during boot to ensure that it has not been changed. The transfer of a firmware update would also need to use integrity. We don't want the firmware to be changed in-flight. Data integrity is often used to protect the system or a data asset against malware attacks.

Authenticity indicates that we can verify where and from whom the data asset came. For example, when performing an over-the-air (OTA) update, we want the system to authenticate the firmware update. Authentication involves using a digital signature to verify the integrity of the asset and verify the origin of the asset. Authenticity is all about confirming the trustworthiness of the data asset.

---

[10] Sometimes, authenticity is replaced with availability when we are considering robust software models.

When performing a TMSA, it is relatively common to analyze each data asset and assign which elements of the CIA model will be used to protect it. Of course, the exact methods will often vary, but Table 3-3 shows a simple way to do this using the assets we have already defined using a weather station as an example.

***Table 3-3.***  *A list of assets, threats, and the CIA elements used to protect the asset. C = Confidentiality, I = Integrity, and A = Authenticity*

| Data Asset | CIA | Threat |
|---|---|---|
| Location | C, I | Man-in-the-middle, Repudiation, Impersonation, Disclosure |
| Raw Sensor Data | C, I | Tamper (Disclosure) |
| System Configuration | C, I | Tamper (Disclosure) |
| Credentials | C, I | Impersonation, Tamper (Disclosure), Man-in-the-middle, Escalation of Privilege (firmware abuse) |
| Firmware | C, I, A | Tamper, Escalation of Privilege, Denial of Service, Malware |

Once we understand the data assets and the CIA model elements we need to use to protect them, we can define the security objectives for the data asset. Of course, we may want to implement many security objectives. Still, the most common in an embedded system are access control, secure storage, firmware authenticity, communication, and secure storage. The following is a list of definitions for each:[11]

**Access control** – The device authenticates all actors (human or machine) attempting to access data assets. Access control prevents unauthorized access to data assets. It counters spoofing and malware threats where the attacker modifies firmware or installs an outdated flawed version.

**Secure storage** – The device maintains confidentiality (as required) and integrity of data assets. This counters tamper threats.

**Firmware authenticity** – The device verifies firmware authenticity before boot and upgrade. This counters malware threats.

---

[11] www.mouser.com/pdfDocs/threat-based_analysis_method_for_iot_devices_-_cypress_and_arm.pdf

**Communication** – The device authenticates remote servers, provides confidentiality (as required), and maintains the integrity of exchanged data. This counters man-in-the-middle (MitM) threats.

**Secure state** – This ensures that the device maintains a secure state even in case of failure to verify firmware integrity and authenticity. This counters malware and tamper threats.

Usually, once the security objectives are outlined, the team will discover that a single security objective will often cover multiple data assets. Therefore, it can be helpful to list the security objectives, threats, and data assets covered by the objective.

With the objectives outlined, the next step in the TMSA is to define the security requirements.

# TMSA Step #4 – Security Requirements

Once a team understands their data assets, threats, and the objectives to protect those assets, they can then define their security requirements. Security requirements are more interesting than standard software requirements because security is not static throughout the product life cycle. The system may have different security requirements at various stages of development, such as during design, manufacturing, inventory, normal operation, and end of life.

As part of the security requirements, teams should outline how their requirements may change with time. You might think I just want to set my requirements and never change them. However, this problem can add a lot of extra heartache at specific points during development. For example, a team will not want to disable the SWD/JTAG interface during design and implementation. If they do, debugging will become a slow, hellish nightmare. Instead, they want to ensure the SWD/JTAG port is disabled when the devices enter the field.

A simple example to consider is the configuration data asset. We want to protect the asset from spoofing using access control and confidentiality security property. We don't have a security objective or requirement during design, manufacturing, and while the product sits in the inventory. However, once the product is in a user's hands, we want the system configuration to be encrypted. At the end of life, we want the configuration scrubbed and the microcontroller placed into a non-operational state.

I highly recommend taking the time to think through each asset at each stage of development to properly define the security requirements. These requirements will be used to identify hardware and software assets necessary to fulfill and secure the system.

# TMSA Step #5 – Summary

The last step in the TMSA process is to summarize the results. A good TMSA will be summarized using several documents, such as a detailed TMSA analysis document and a TMSA summary.

The detailed TMSA analysis document will walk through the following:

- Explain the target of evaluation

- Define the security problem

- Identify the threats

- Define organizational security policies

- Define security objectives

- Define security requirements

The TMSA overview is often just a spreadsheet that summarizes the assets, threats, and how the assets will be protected. An example summary from the Arm TMSA for the firmware data asset for an asset tracker can be found in Figure 3-5.

This spreadsheet is an "at a glance" summary of a Threat Model and Security Analysis (TMSA) document (DEN0075). It is a quick-reference summary of assets, threats, impact and security requirements that are provided in Excel format to allow you to easily edit and extend. We hope that you find this document useful as a starting point.

**arm**

© Copyright Arm Limited 2018. All rights reserved.                    CI=Arm CryptoIsland, PSA=Arm Platform Security Architecture

| Asset | Security Property | Threat | Entry Point of Threat (where the attack is launched from ex: malware, network, JTAG, etc) | Impact of Vulnerability | Severity (CVSS Rating) | Mitigation/Security Requirement | Arm's Technology |
|---|---|---|---|---|---|---|---|
| Firmware | Integrity | Tamper | Malware, bug, mass storage access, JTAG, network, update | Install malware | Critical: 9 CVSS:3.0/AV:N/AC:H/ PR:N/UI:N/S:C/C:H/I: H/A:H | Support secure boot flows (authenticate firmware) | [CI] Loaded SW validation functionality |
| | | Escalation of privilege (Firmware Abuse) | | Launch DDoS | | Enforce principle of least privilege | |
| | | | | | | Support secure firmware update (authenticate update) | [CI] SW update validation |
| | | | | | | | [PSA] Trusted Boot features |
| | | DoS (Firmware Abuse) | | Tamper/Steal location | | Support anti-rollback of firmware | [CI] SW update validation |
| | | | | Permanent bricking of device | | | [PSA] Firmware Update features |

***Figure 3-5.*** *A TMSA summary for a firmware data asset (Source: psacertified.org)*

There are several TMSA examples that Arm has put together that can be found here.[12] The PSA website also has many resources for getting started with a TMSA.[13] If you plan to perform your TMSA, I would highly recommend that a professional perform your TMSA or have a security professional review your results. Several times, I've gone to review a client's TMSA and found some wide gaps, holes, and oversights in their analysis.

# PSA Stage 2 – Architect

Once we understand our data assets, threats, security properties, objectives, and requirements, we are ready to move to the second PSA stage, architecting our software. The key to securing an embedded system is through isolation. Isolation is the foundation on which security is built into an application. It separates software modules, processes, and memory into separate regions with minimal or no interaction with each other. The isolation acts as a barrier limiting the attack surface that an attacker can leverage to access an asset of interest on the device.

## Security Through Isolation

There is an easy way to think about the benefits of isolation in a secure application. Consider an office building where the office is organized as an open floor plan. Everyone's desks are out in the open, and everyone can get access to everyone else's workspace. The building has two security access points: a front and a back door. An attacker needs to circumvent the front or back door security to gain access to the entire building. They have access to the whole building to accomplish this, as shown in Figure 3-6.

---

[12] www.psacertified.org/development-resources/building-in-security/threat-models/

[13] psacertified.org

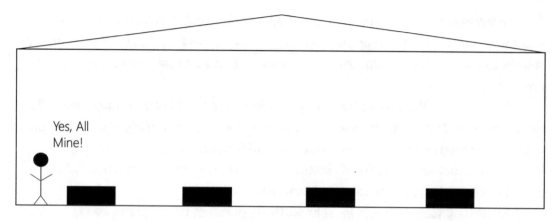

**Figure 3-6.** *An adversary can bypass security at the front and access the entire building*

Now, consider an office building where the office is organized into private offices for every employee. Each office has an automatic lock that requires a keyed entry. On top of this, building security is still at the front and back doors. If an attacker can breach the outer security, they have access to the building but don't have access to everyone's workspace! The doors isolate each office by providing a barrier to entry, as shown in Figure 3-7. The attacker would now need to work harder to access one of the offices, which only gives them access to one office, not all of them.

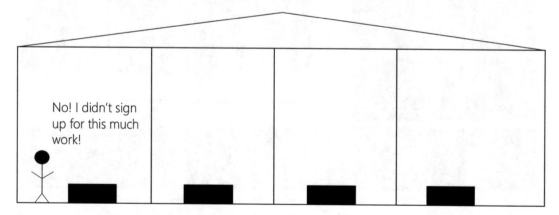

**Figure 3-7.** *An adversary can bypass security at the front or back, but they don't have free reign and have additional work to access critical assets*

The key to designing secure embedded software is to build isolation into the design that erects barriers to resources and data assets. Then, if hackers find their way into the system, they are provided minimal access to the system with essential assets, hopefully further isolated.

Take a moment to consider how designers currently architect most microcontroller-based embedded systems. For example, it's not uncommon to find applications running in privileged mode with the entire application written into one giant memory space where every function can access all memory and resources without constraint. As you can imagine, this results in a not very secure system.

A secure application design must break the application into separate tasks, processes, and memory locations where each has limited access to the data assets and system resources. In addition, a successful, secure application design requires that the underlying hardware support hardware-based isolation. There are three levels of isolation that designers need to consider: processing environments, Root-of-Trust with trusted services, and trusted applications. An overview of these three levels can be seen in Figure 3-8.

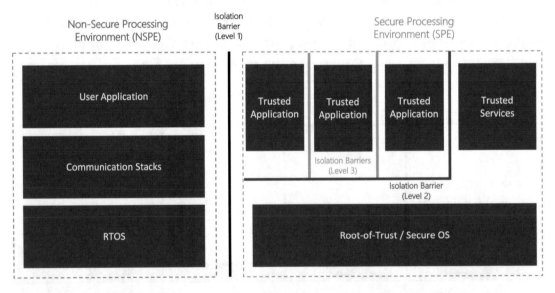

***Figure 3-8.*** *Securing an embedded application requires three levels of isolation: isolated processing environments, an isolated Root-of-Trust with trusted services, and application isolation*

# Isolation Level #1 – Processing Environments

The first layer of isolation in a secure application separates the execution environments into two isolated hardware environments: the secure processing environment (SPE) and the nonsecure processing environment (NSPE). The SPE, also known as the trusted execution environment (TEE), contains the Root-of-Trust (RoT), trusted services like cryptographic services, and trusted application components. The NSPE, also known as the rich execution environment, includes the typical software developers' use, like an RTOS, and most of their general application code.

A designer can use two mechanisms to create the first level of isolation: multicore processors and Arm TrustZone. The designer sets one of the processing cores in the multicore solution to act as the NSPE and the second core to serve as the SPE. The two cores would be isolated with some shared memory and interprocessor communications (IPC). The secure core would only allow a minimal set of operations publicly available to the NSPE. One advantage to using multiple processors is that the NSPE and the SPE can execute code simultaneously.

Arm TrustZone is a solution that provides hardware-based isolation between the NSPE and the SPE, but it is done in a single-core solution. When an SPE function is called, the processor switches from the NSPE environment to the SPE environment in deterministic three clock cycles. When the SPE is complete, the processor switches back to the NSPE. We will explore a little bit more about these environments later.

# Isolation Level #2 – Root-of-Trust and Trusted Services

The second level of isolation in a secure application is established in the SPE through a Root-of-Trust (RoT) and trusted services.

A Root-of-Trust provides the trust anchor in a system to support the secure boot process and services.[14] A Root-of-Trust should be immutable and hardware based to make it immune to attack. The Root-of-Trust often includes hardware-accelerated cryptography, true random number generation (TRNG), and secure storage.

A good Root-of-Trust will start with the chip manufacturer. For example, the Cypress PSoC 64, now part of Infineon, ships with a hardware-based Root-of-Trust from the factory, with Cypress owning the Root-of-Trust. As part of the Root-of-Trust, Cypress has a unique identifier for each chip with Cypress keys that allow a user to verify the

---

[14] https://bit.ly/3GPlIj3

authenticity and integrity of the hardware. Developers can then verify the chip came from Cypress and transfer ownership of the Root-of-Trust to their company. From there, they can inject their user assets, provision the device, and so forth.

Trusted services also reside in the SPE. There are many types of services that can be supported. For example, it is not uncommon for trusted services to include a secure boot process through a chain of trust, device provisioning, attestation, secure storage, Transport Layer Security (TLS), and even firmware over-the-air (FOTA) updates. These are all standard services that nearly every connected needs to build a secure application successfully.

## Isolation Level #3 – Trusted Applications

Trusted applications are the developer's applications that run in the SPE in their own isolated memory space. Trusted applications provide services to the system that require secure access to data and hardware resources.

A trusted application is built upon the Root-of-Trust and other trusted services. In addition, trusted applications are often implemented by leveraging the capabilities provided by an open source framework called Trusted Firmware M (TF-M). TF-M is a reference software platform for Arm Cortex-M processors that implements SPE and provides baseline services that trusted applications are built upon, such as

- Secure boot

- Managing isolation

- Secure storage and attestation services

One of the goals of TF-M is to provide a reference code base that adheres to the best practices for secure systems defined by the Platform Security Architecture (PSA).

## Architecting a Secure Application

Architecting a secure application requires the developer to create layers of isolation. The process starts with separating the processing environments, isolating the RoT and trusted services, and then the trusted applications. What I often find interesting is that developers architecting a solution try to jump straight into microcontroller or processor selection! They act as if the microcontroller dictates the architecture when it's the other way around! So at the architectural stage, we architect the best security solution for our product and let the implementers figure out the details.[15]

I highly recommend that embedded software architects create a model of their secure application. The model doesn't have to be a functional model per se, but something that is thought through carefully and diagrammed. The model should

- Identify what goes in each processing environment

- Break up the system into trusted applications

- Define the RoT and trusted services

- Identify general mechanisms for protection such as MPUs, SMPUs, etc.

- Define general processes and memory organization (in a generic way)

A simplified example of what this might look like can be found in Figure 3-9. Usually, this would also break out references to the TMSA and other services, components, and more. Unfortunately, such a diagram would be unreadable in book format. However, the reader gets a general idea and should extrapolate for their application design.

---

[15] If microcontroller selection dictated the architecture, then microcontroller selection would be Chapter 1, not Chapter 11!

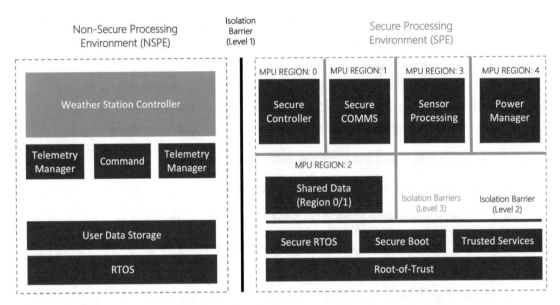

**Figure 3-9.** *A secure application model lays the foundation for the application developers to understand the security architecture and designers' intentions*

When you architect your security solution, it's essential to realize that you aren't just looking at where different components and pieces of code and data should be. The end solution may be using a real-time operating system (RTOS) which will also require developers to carefully think through where the various application threads should be located and which state the processor will be in to execute the functions within those threads. There are three different types of threads that architects and developers should consider:

- **Nonsecure threads** only call functions that exist within the NSPE.

- **Secure threads** which only call functions that exist within the SPE.

- **Hybrid threads** execute functions within both the SPE and the NSPE. Hybrid threads often need access to an asset the architect is trying to protect.

A simple UML sequence diagram can be architected how the application will behave from a thread execution standpoint. The sequence diagram can show the transition points from the NSPE to the SPE and vice versa. The sequence diagram can dramatically clarify a developer's understanding of the implementation and how hybrid threads will transition between the NSPE and the SPE. Figure 3-10 demonstrates what an example hybrid thread sequence diagram might look like.

The diagram contains two threads, a sensor, and a communications (comm) thread. Certain operations for each thread are executed within the NSPE and the SPE. We can see the general operations blocks at specific points along the thread's lifeline. Again, this is just another example of what an architect might do to help clarify the intent of the design so that developers can implement the solution properly.

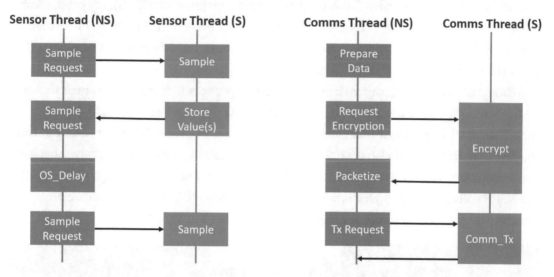

***Figure 3-10.*** *A sequence diagram can demonstrate transition points in a thread-based design between NSPE and SPE execution*[16]

# PSA Stage 3 – Implementation

The third stage of PSA is all about implementation. Implementation is where most developers' minds go first and are most tempted to begin. However, as stated before, we need to have our architecture in place first to implement a secure application successfully.[17] In general, two methods are used in microcontroller-based systems to implement and isolate the runtime environments that don't require an external security processor: multicore processors and Arm TrustZone.

---

[16] www.beningo.com/insights/software-developers-guide-to-iot-security/

[17] Just because we architect first doesn't mean that we can't in parallel experiment and evaluate different hardware solutions. Implementation feedback is almost always needed to evolve and tweak the architecture.

# Multicore Processors

Multicore processors provide developers unique and interesting solutions to create an SPE and NSPE. Multicore solutions are typically dual-core solutions with one core dedicated to the nonsecure processing environment (NSPE). The second is dedicated as a security processor that runs the secure processing environment (SPE). My prediction is that over the next decade, or so, we will start to see microcontrollers with more than two cores. Multiple cores will provide multiple hardware-based execution partitions that would increase the security capabilities of processors.

Today, a typical multicore architecture has two processing cores that may or may not be the same processor architecture. For example, the Cypress PSoC 64 series have an Arm Cortex-M0+ processor that runs the SPE and an Arm Cortex-M4 processor that runs the NSPE. Each core will also access its isolated memory regions and an interprocessor communication (IPC) bus. The IPC is used to communicate between the two. Generally, the NSPE will request an operation from the SPE, which the SPE will perform. The general architecture for a multicore solution can be seen in Figure 3-11.

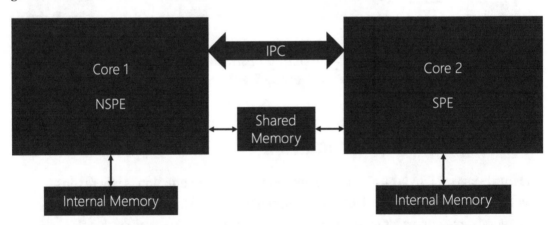

***Figure 3-11.*** *A multicore microcontroller can set up an SPE in one core while running the NSPE in the other. Each core has dedicated isolated memory, shared memory for data transfer, and interprocessor communications (IPC) to execute secure functions*

The multicore solution does have several advantages and disadvantages, just like any engineering solution. One advantage is that each core can execute application instructions in parallel, improving the application's responsiveness and performance. For example, if data needs to be encrypted, the cryptographic operation can be done in the SPE while the NSPE is still processing a touch screen and updating the display.

However, it's not uncommon for developers to struggle with performance issues when implementing their security solutions, especially if they did not select their microcontroller hardware properly (see Chapter 11).

There are several other advantages to using a multicore processor as well. For example, having both cores on a single die can decrease costs and limit the attack surface provided by an external security processor. Each core is also isolated, which limits access and adds additional complexity to trying to exploit the system.

Multicore solutions also have some disadvantages. First, the cost of a multicore microcontroller tends to be higher than single-core solutions. However, this may not always be the case, as economics drives these prices, such as volumes, demand, and so forth. Writing multicore applications can also be more complex and costly and require more debug time. However, developers should minimize what they put into the SPE to reduce their attack surface, limiting how much extra complexity is added.

## Single-Core Processors with TrustZone

Starting with the Armv8-M architecture, silicon providers have the opportunity to include hardware-based isolation within a single-core processor through the use of TrustZone technology. TrustZone offers an efficient, system-wide approach to security with hardware-enforced isolation built into the CPU.[18] I mention "have the opportunity to" because not all Armv8-M processors and beyond have TrustZone. Furthermore, it's optional, so if you want the technology, you must be careful about the microcontroller you select to ensure it has it.

The idea behind TrustZone is that it allows the developer to break their processors' memory up into an NSPE and SPE. Flash, RAM, and peripheral memory can all be assigned to either NSPE or SPE. In addition, developers can leverage CMSIS-Zone,[19] which creates a configuration file that allows developers to partition their system resources across multiple projects and processors. (Yes, for secure applications, you will have to manage one project for the NSPE and one for the SPE!) CMSIS-Zone includes an Eclipse-based utility that provides a simple GUI to[20]

---

[18] Definition is defined at https://bit.ly/3F4k5N2

[19] https://arm-software.github.io/CMSIS_5/Zone/html/index.html

[20] www.beningo.com/insights/software-developers-guide-to-iot-security/

- Assign resources

- Display all available system resources, including memory and peripherals

- Allow the developer to partition memory and assign resources to subsystems

- Support the setup of secure, nonsecure, and MPU-protected execution zones with the assignment of memory, peripherals, and interrupts

- Provide a data model to generate configuration files for tool and hardware setup

Developers should keep in mind that TrustZone is a single-core solution. Unlike the multicore solution, it can only execute the SPE or the NSPE simultaneously. For example, if the NSPE is running and an operation is requested that runs in the SPE, deterministic three clock cycles occur to transition the hardware from NSPE to SPE. Once in the SPE, another setup may occur based on the application before the SPE runs the operation. When the procedure is complete, the processor scrubs CPU registers (not all of them, so see your documentation) and then transitions back to the NSPE.

The boot sequence for a TrustZone application is also slightly different from multicore. In multicore, the developer can bring both cores up simultaneously or bring the security core and validate the NSPE first. Figure 3-12 shows the typical TrustZone application boot and execution process. The boot process starts in the SPE, allowing the secure boot and bootloader to execute and validate the NSPE before jumping to the NSPE for execution.

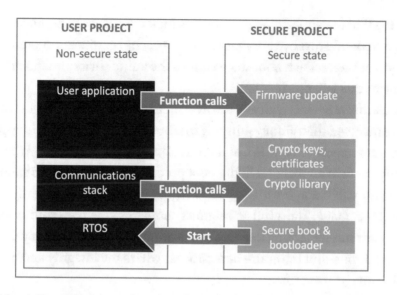

*Figure 3-12.* *A TrustZone application boots into the SPE, validates the NSPE, and executes the user project*

While the user project executes its application, it will make function calls into the SPE through a secure gateway. A secure gateway is an opening in the SPE that only allows access to function calls. For example, the secure state project can expose the function calls for the cryptographic library. The user project can then make calls to those functions, but only those functions! An exception will be thrown if a developer tries to create a pointer and access memory that is not exposed by a secure gateway. The only area of the SPE that the NSPE can access is the functions exposed by the SPE. However, the SPE can access all memory and functions, including the NSPE! Therefore, developers must be cautious with what they access and where those accesses are.

Single-core processors do have the advantage of being less expensive than multicore processors. However, they can also access the entire memory map, which is a disadvantage. One drawback is that clock cycles are wasted when transitioning from NSPE to SPE and SPE to NSPE. Admittedly, it's not very many clock cycles, but those few clocks could add up over time if the application is concerned with performance or energy consumption.

## The Secure Boot Process

The boot process I see companies using in the microcontroller space generally leaves much, much desired. The typical boot process I encounter verifies an application checksum, and that's about it. However, a secure boot process needs must more than just confirming that the application's integrity is intact. A secure boot process involves establishing a Root-of-Trust by verifying the onboard chip certificate and then booting the application in stages. Each boot stage verifies the authenticity and integrity of the firmware images that will be executed.

Figure 3-13 shows what is typically involved in the boot process. As you can see, this is much more involved than simply jumping to a bootloader, verifying the application is intact through a checksum or CRC, and then running the application. At each step, before loading the next application in the boot process, the new image is validated. For our purposes, validation is verifying the integrity of the image and verifying the authenticity of the image. The application must come from the right source and also be intact. Developers implementing a secure application need to ensure they develop a Chain-of-Trust throughout the entire boot process where the integrity and authenticity of each loaded image are verified.

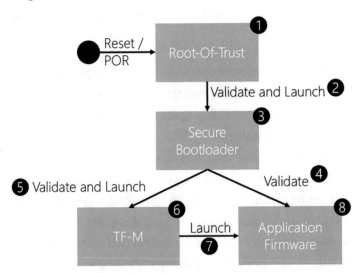

***Figure 3-13.*** *A Chain-of-Trust is often used to securely boot a system where each stage is verified through authenticity and validated for integrity before loading and executing the image*

Developing a Chain-of-Trust involves several steps for an embedded application. First, the Root-of-Trust is a trusted anchor from which all other trusts in the system originate. I often recommend that developers find microcontrollers that have a hardware-based Root-of-Trust. In these cases, the microcontroller manufacturer often creates a Root-of-Trust and transfers it to the developing company.[21] The developing company can then add their credentials and build their system.

Second, the RoT during boot will authenticate and validate the next application image. In most cases, the following application is a secure bootloader. The RoT will use a hash to validate the secure bootloader. Once validating, the RoT will launch the bootloader. The bootloader will then check for any new firmware that is being requested to load. If so, the firmware update process will occur (which is beyond our scope now); otherwise, the existing application images will be used.

When not updating firmware, the secure bootloader has two jobs that it must complete: validate the existing application firmware (NSPE user application) and then validate and launch the secure firmware (SPE). The secure bootloader does not launch the user application that exists in the NSPE. Instead, the SPE, which contains the secure runtime, will launch the NSPE once it has been deemed that the system is secure and the NSPE is ready to launch.

The secure firmware in Arm Cortex-M processors is often based on Trust Firmware for Cortex-M (TF-M) processors. TF-M is an open source reference implementation of the PSA IoT Security Framework.[22] TF-M contains common resources across security application implementations, such as secure storage, attestation, and cryptographic operations. If the user application that runs in the NSPE is validated successfully, it will launch and allow the system to operate.

# PSA Stage 4 – Certify

How do you know that the components you are using in implementation and your system meet its security requirements? One potential answer is that you certify your device. Certification is the last stage in PSA, and it's also the one stage that is potentially optional (although I always recommend that even if you don't plan to certify your device or software, you at least hire an expert to perform penetration testing on it).

---

[21] There are also some interesting mechanisms for creating a hardware-based Root-of-Trust using a microcontroller SRAM known as SRAM PUF (physical unclonable function). SRAM PUF can be applied to any microcontrollers without the need for the microcontroller vendor to be involved. An introduction is beyond our scope, but you can learn more at www.embedded.com/basics-of-sram-puf-and-how-to-deploy-it-for-iot-security/

[22] https://bit.ly/335RcTV

Certification can be performed on both hardware and software. Usually, PSA certification is focused on component or hardware providers who want to certify their components for companies that will develop products. However, consumer-focused devices can also be certified. Certification guarantees that the developers considered and implemented industry-recommended best practices when they designed and implemented their system.

PSA Certified is an independent security evaluation scheme for IoT chips, operating systems, and devices. There are several lab companies that can perform the certification process.[23] Each lab is overseen by TrustCB, which is the company that certifies the labs and was involved in the creation of PSA. The time and costs required to certify will vary depending on the level of certification sought and what is being certified. For example, companies can certify for their hardware device, silicon chips, system software, and IP providers.

PSA Certified provides a much-needed scheme for microcontroller-based devices, plus a common language that the whole industry can use and understand. PSA Certified contains three different levels, which increase the security robustness at each level. Figure 3-14 shows the three different general ideas behind each.

*Figure 3-14.* *PSA Certified provides three certification levels that increase the robustness of hardware or software at each level*

---

[23] A list of labs can be found at `www.psacertified.org/getting-certified/evaluation-labs/`

# Final Thoughts

We are entering an age for embedded software where developers can no longer ignore security. If a device connects to the Internet, then the system must take security into account. If the system is not connected, well, you still might want to add basic security like tamper protection to protect your intellectual property. Security can be a touchy subject because the threats are always evolving, and many embedded software developers don't have the knowledge or experience with designing and building secure software.

Developers can leverage existing frameworks like PSA to help guide them through the process. There is a big push right now to simplify security for embedded systems. Software frameworks like TF-M can help teams get a jump-start in deploying trusted services to their devices. If there is anything that this chapter has taught you, I hope that it is that you can't bolt security onto your system at the end. Security must be designed into the system from the very beginning, or else the system may be left vulnerable to attack.

Don't wait until your system and company make headlines news! Start thinking about security today and with every line of code you write!

---

## ACTION ITEMS

To put this chapter's concepts into action, here are a few activities the reader can perform to start applying secure application design principles to their application(s):

- Explore the many articles and white papers that explain PSA. A good starting point is

    - `https://developer.arm.com/architectures/architecture-security-features/platform-security`

- Perform a TMSA on an existing product or a fictitious product for practice. Leverage the PSA TMSA examples to get started.

- Investigate TF-M or a similar platform that is used in a security application. First, review the APIs, functions, and capabilities. Then, how can you use these in your secure applications?

- Practice developing a secure architecture. Either use an existing product or a fictitious product to practice decomposing an application into NSPE and SPE.

  - How can the application be broken up into components?

  - Where should the application components live?

- Download and install CMSIS-Zone. Explore and build an example system to understand how to partition applications and memory regions.

- Review your current secure application development processes. Ask yourself the following questions:

  - Are these processes sufficient to protect our data assets?

  - Are there any new threats present that were previously overlooked?

  - Are the proper hardware and software resources in place to protect our data assets?

- If you are activating and building a secure application for the first time, reach out to an expert(s) who will help guide you and ensure that it is done correctly.

These are just a few ideas to go a little bit further. Carve out time in your schedule each week to apply these action items. Even minor adjustments over a year can result in dramatic changes!

# CHAPTER 4

# RTOS Application Design

Embedded software has steadily become more and more complex. As businesses focus on joining the IoT, the need for an operating system to manage low-level hardware, memory, and time has steadily increased. Embedded systems implement a real-time operating system in approximately 65% of systems.[1] The remaining systems are simple enough for bare-metal scheduling techniques to achieve the systems requirements.

Real-time systems require the correctness of the computations, the computation's logical correctness, and timely responses.[2] There are many scheduling algorithms that developers can use to get real-time responses, such as

- Run to completion schedulers
- Round-robin schedulers
- Time slicing
- Priority-based scheduling

---

**Definition**    A real-time operating system (RTOS) is an operating system designed to manage hardware resources of an embedded system with very precise timing and a high degree of reliability.[3]

---

It is not uncommon for a real-time operating system to allow developers to simultaneously access several of these algorithms to provide flexible scheduling options.

---

The original version of this chapter was revised. A correction to this chapter is available at https://doi.org/10.1007/978-1-4842-8279-3_17

---

[1] www.embedded.com/wp-content/uploads/2019/11/EETimes_Embedded_2019_Embedded_Markets_Study.pdf

[2] www.embedded.com/program-structure-and-real-time/

[3] www.embeddedrelated.com/thread/5762/rtos-vs-bare-metal

© Jacob Beningo 2022, corrected publication 2023
J. Beningo, *Embedded Software Design*, https://doi.org/10.1007/978-1-4842-8279-3_4

RTOSes are much more compact than general-purpose operating systems like Android or Windows, which can require gigabytes of storage space to hold the operating system. A good RTOS typically requires a few kilobytes of storage space, depending on the specific application needs. (Many RTOSes are configurable, and the exact settings determine how large the build gets.)

An RTOS provides developers with several key capabilities that can be time-consuming and costly to develop and test from scratch. For example, an RTOS will provide

- A multithreading environment

- At least one scheduling algorithm

- Mutexes, semaphores, queues, and event flags

- Middleware components (generally optional)

The RTOS typically fits into the software stack above the board support package but under the application code, as shown in Figure 4-1. Thus, the RTOS is typically considered part of the middleware, even though middleware components may be called directly by the RTOS.

While an RTOS can provide developers with a great starting point and several tools to jump-start development, designing an RTOS-based application can be challenging the first few times they use an RTOS. There are common questions that developers encounter, such as

- How do I figure out how many tasks to have in my application?

- How much should a single task do?

- Can I have too many tasks?

- How do I set my task priorities?

This chapter will explore the answers to these questions by looking at how we design an application that uses an RTOS. However, before we dig into the design, we first need to examine the similarities and differences between tasks, threads, and processes.

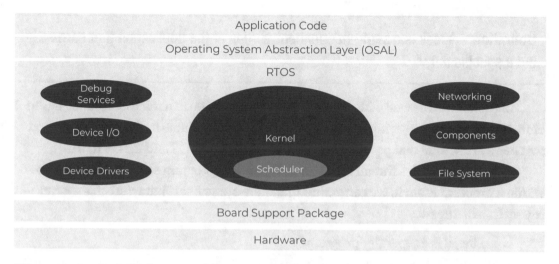

*Figure 4-1.* *An RTOS is a middleware library that, at a minimum, includes a scheduler and a kernel. In addition, an RTOS will typically also provide additional stacks such as networking stacks, device input/output, and debug services*

# Tasks, Threads, and Processes

An RTOS application is typically broken up into tasks, threads, and processes. These are the primary building blocks available to developers; therefore, we must understand their differences.

A task has several definitions that are worth discussing. First, a task is a concurrent and independent program that competes for execution time on a CPU.[4] This definition tells us that tasks are isolated applications without interactions with other tasks in the system but may compete with them for CPU time. They also need to appear like they are the only program running on the processor. This definition is helpful, but it doesn't represent what a task is on an embedded system.

The second definition, I think, is a bit more accurate. A task is a semi-independent portion of the application that carries out a specific duty.[5] This definition of a task fits well. From it, we can gather that there are several characteristics we can expect from a task:

- It is a separate "program."

- It may interact with other tasks (programs) running on the system.

- It has a dedicated function or purpose.

---

[4] Unknown reference.

[5] https://docs.microsoft.com/en-us/azure/rtos/threadx/chapter1

---

**Definition**    A task is a semi-independent portion of the application that carries out a specific duty.[5]

---

This definition fits well with what we expect a task to be in a microcontroller-based embedded system. Surveying several different RTOSes available in the wild, you'll find that there are several that provide task APIs, such as FreeRTOS and uC OS II/III.

On the other hand, a thread is a semi-independent program segment that executes within a process.[6] From it, we can gather that there are several characteristics we can expect from a thread:

- First, it is a separate "program."

- It may interact with other tasks (programs) running on the system.

- It has a dedicated function or purpose.

For most developers working with an RTOS, a thread and a task are synonyms! Surveying several different RTOSes available in the wild, you'll find that there are several that provide thread APIs, such as Azure RTOS, Keil RTX, and Zephyr. These operating systems provide similar capabilities that compete with RTOSes that use task terminology.

A **process** is a collection of tasks or threads and associated memory that runs in an independent memory location.[7] A process will often leverage a memory protection unit (MPU) to collect the various elements part of the process. These elements can consist of

- Flash memory locations that contain executable instructions or data

- RAM locations that include executable instructions or data

- Peripheral memory locations

- Shared RAM, where data is stored for interprocess communication

A process groups resources in a system that work together to achieve the application's goal. Processes have the added benefits of improving application robustness and security because it limits what each process can access. A typical multithreaded application has all the application tasks, input/output, interrupts, and other RTOS objects in a single address space, as shown in Figure 4-2.

---

[6] https://docs.microsoft.com/en-us/azure/rtos/threadx/chapter1
[7] https://docs.microsoft.com/en-us/azure/rtos/threadx/chapter1

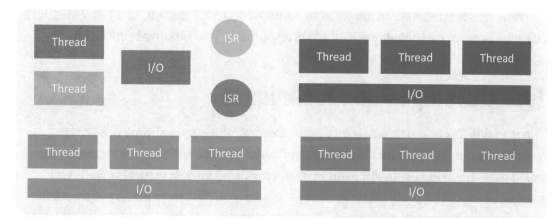

***Figure 4-2.*** *A typical multithreaded application has all the application tasks, input/output, interrupts, and other RTOS objects in a single address space[8]*

This approach is acceptable and used in many applications, but it does not isolate or protect the various elements. For example, if one of the green threads goes off the rails and starts overwriting memory used by the blue thread, there is nothing in place to detect or protect the blue thread. Furthermore, since everything is in one memory space, everyone can access everyone else's data! Obviously, this may be unwanted behavior in many applications, which is why the MPU could be used to create processes, resulting in multiprocess applications like that shown in Figure 4-3.

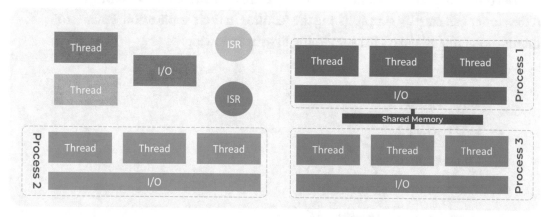

***Figure 4-3.*** *A multiprocess application can leverage an MPU to group threads, input/output, interrupts, and other RTOS objects into multiple, independent address spaces[9]*

---

[8] This diagram is inspired by and borrowed from Jean Labrosse's "Using a MPU with an RTOS" blog series.

[9] This diagram is inspired by and borrowed from Jean Labrosse's "Using a MPU with an RTOS" blog series.

Now that we understand the differences between tasks, threads, and processes, let's examine how we can decompose an RTOS application into tasks and processes.

# Task Decomposition Techniques

The question I'm asked the most by developers attending my real-time operating systems courses is, "How do I break my application up into tasks?". At first glance, one might think breaking up an application into semi-independent programs would be straightforward. The problem is that there are nearly infinite ways that a program can be broken up, but not all of them will be efficient or result in good software architecture. We will talk about two primary task decomposition techniques: feature-based and the outside-in approach.

## Feature-Based Decomposition

Feature-based decomposition is the process of breaking an application into tasks based on the application features. A feature is a unique property of the system or application.[10] For example, the display or the touch screen would be a feature of an IoT thermostat. However, these could very quickly be their own task within the application.

Decomposing an application based on features is a straightforward process. A developer can start by simply listing the features in their application. For an IoT thermostat, I might make a list something like the following:

- Display
- Touch screen
- LED backlight
- Cloud connectivity
- Temperature measurement
- Humidity measurement
- HVAC controller

---

[10] www.webopedia.com/definitions/feature/#:~:text=(n.),was%20once%20a%20simple%20 application

Most teams will create a list of features the system must support when they develop their stakeholder diagrams and identify the system requirements. This effort can also be used to determine the tasks that make up the application software.

Feature-based task decomposition can be very useful, but sometimes it can result in an overly complex system. For example, if we create tasks based on all the system features, it would not be uncommon to identify upward of a hundred tasks in the system quickly! This isn't necessarily wrong, but it could result in an overly complex system with more memory and RAM than required.

When using the feature-based approach, it's critical that developers also go through an optimization phase to see where identified tasks can be combined based on common functionality. For example, tasks may be specified for measuring temperature, pressure, humidity, etc. However, having a task for each individual task will overcomplicate the design. Instead, these measurements could all be combined into a sensor task. Figure 4-4 provides an example of a system's appearance when feature-based task decomposition is used.

**Figure 4-4.** *An example is feature-based task decomposition for an IoT thermostat that shows all the tasks in the software*

Using features is not the only way to decompose tasks. One of my favorite methods to use is the outside-in approach.

# The Outside-In Approach to Task Decomposition

The outside-in approach[11] to task decomposition is my preferred technique for decomposing an application. The primary reason is that the approach is data-centric and adheres to the design philosophy principles identified in Chapter 1. Thus, rather than looking at the application features, which will change and morph over time, we focus on the data and how it flows through the application. Let's now look at the steps necessary to apply this approach and a simple example for an IoT-based device.

## The Seven-Step Process

Developers can follow a simple process to decompose their applications using the outside-in approach. The outside-in process helps developers think through the application, examine the data elements, and ensure a consistent process for developing applications. The steps are relatively straightforward and include

1. Identify the major components

2. Draw a high-level block diagram

3. Label the inputs

4. Label the outputs

5. Identify first-tier tasks

6. Determine concurrency, dependencies, and data flow

7. Identify second-tier tasks

The easiest way to examine and explain each step is to look at an example. One of my favorite teaching examples is to use an Internet-connected thermostat. So let's discuss how we can decompose an IoT thermostat into tasks.

## Decomposing an IoT Thermostat

I like to use an IoT thermostat as an example because they can be sophisticated devices and they have become nearly ubiquitous in the market. An IoT thermostat has

---

[11] An abbreviated version of this approach is introduced in "Real-Time Concepts for Embedded Systems" by Qing Li with Caroline Yao in Chapter 14, Section 2, pages 214–216.

- A wide range of sensors.

- Sensors with various filtering and sample requirements.

- Power management requirements, including battery backup management.

- Connectivity stacks like Wi-Fi and Bluetooth for reporting and configuration management.

- Similarities to many IoT control and connectivity applications make the example scalability to many industries.

There are also usually development boards available that contain many sensors that can be used to create examples.

Like any design, we need to understand the hardware before designing the software architecture. Figure 4-5 shows an example thermostat hardware block diagram. We can use it to follow then the steps we have defined to decompose our application into tasks.

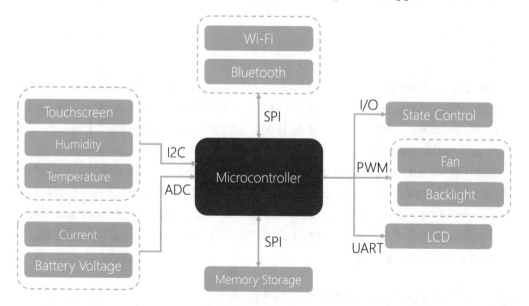

***Figure 4-5.***  *The hardware block diagram for an IoT thermostat. Devices are grouped based on the hardware interface used to interface with them*

## Step #1 – Identify the Major Components

The first step to decomposing an application into tasks is identifying the major components that make up the system. These are going to be components that influence the software system. For example, the IoT thermostat would have major components such as

- Humidity/temperature sensor

- Gesture sensor

- Touch screen

- Analog sensors

- Connectivity devices (Wi-Fi/Bluetooth)

- LCD/display

- Fan/motor control

- Backlight

- Etc.

I find it helpful to review the schematics and hardware block diagrams to identify the major components contributing to the software architecture. This helps me start thinking about how the system might be broken up, but at a minimum, it lets me understand the bigger picture.

One component that I find people often overlook is the device itself! Make sure you add that to the list as well. For example, add an IoT device to your list if you are building an IoT device. If you are making a propulsion controller, add a propulsion controller. The reason to do this is that this block acts as a placeholder for our second-tier tasks that we will decompose in the last step.

## Step #2 – Draw a High-Level Block Diagram

At this point, we now know the major components that we will be trying to integrate into our software architecture. We don't know what tasks we will have yet, though. That will depend on how these different components interact with each other and how they produce data. So, the next logical step is to take our list of components and build a block diagram with them. A block diagram will help us visualize the system, which I find helpful. We could, in theory, create a table for the decomposition steps, but I've always found that a picture is worth a thousand words and much more succinctly gets the point across.

Figure 4-6 demonstrates how one might start to develop a block diagram for the IoT thermostat. Notice that I've grouped the analog and digital sensors into a single sensor block to simplify the diagram and grouped them with the fan/state control components.

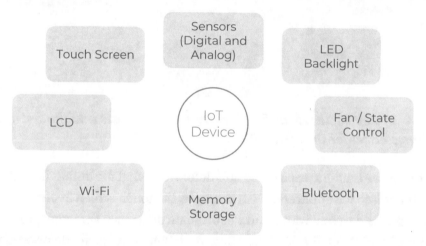

***Figure 4-6.*** *The major components are arranged in a way that allows us to visualize the system*

## Step #3 – Label the Inputs

We don't want to lose sight of our design principles when designing embedded software. We've discussed how important it is to let data dictate the design. When we are working on decomposing our application into tasks, the story is not any different. Therefore, we want to examine each major component and identify which blocks generate input data into the IoT device block, as shown in Figure 4-7.

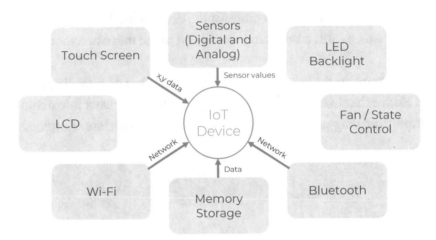

***Figure 4-7.*** *In step #3, we identify where input data comes into the system. These are the data sources for the application*

In this step, we examine each block and determine what the output from the block is and the input into the application. For example, the touch screen in our application will likely generate an event with x and y coordinate data when someone presses the touch screen. In addition, the Wi-Fi and Bluetooth blocks will generate network data, which we are leaving undefined now. Finally, the sensor block we mark as generating sensor values.

At this stage, I've left the data input into the IoT device block as descriptive. However, it would not hurt to add additional detail if it is known. For example, I might know that the touch screen data will have the following characteristics:

- X and y coordinates.

- Event driven (data only present on touch).

- Data generation is every 50 milliseconds when events occur.

- Coordinate data may be used to detect gestures.

It would be a good idea to either list this information on the diagram or, better yet, add a reference number that can then be looked up in a table with the additional information about the data.

This additional information will help understand the rates at which each task will need to run and even how much work each task must do. In addition, the details will be used later when discussing how to set task priorities and perform a rate monotonic analysis.

## Step #4 – Label the Outputs

After identifying all the inputs into the IoT device block, we naturally want to look at all the outputs. Outputs are things that our device is going to control in the "real world."[12] An example of the outputs in the IoT thermostat example can be found in Figure 4-8.

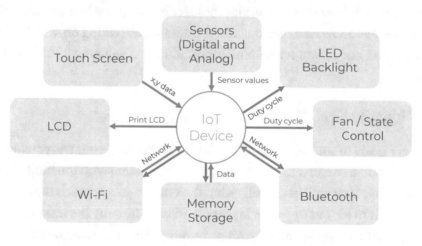

***Figure 4-8.*** *In step #4, we identify the system's outputs. These are the data products that are generated by the system*

We can see from the diagram that this system has quite a few outputs. First, we have an output to the LCD, which is character data or commands to the LCD. The LED backlight and fan are each fed duty cycle information. There is then data associated with several other blocks as well.

Just like with the input labeling, it would not hurt to add a reference label to the block diagram and generate a table or description that gives more information about each. For example, I might expand on the LED backlight duty cycle output data to include

- Duty cycle 0.0–100.0%.

- The update rate is 100 milliseconds.

---

[12] I mention "real world" here because there will be inputs and outputs in the application layer that don't physically interact with the world but are just software constructs.

For the fan data, I might include something like

- Duty cycle 0.0–100.0%.

- The update rate is 50 milliseconds.

- Controlled by a state machine.

The additional details are used as placeholders for when we identify our tasks. In addition, the details allow us to understand the data we are acting on, the rate at which we are acting on it, and any other helpful information that will help us fully architect the system.

## Step #5 – Identify First-Tier Tasks

Finally, we are ready to start identifying the first tasks in our system! As you may have already figured out, these first tasks are specific to interacting with the hardware components in our system! That's really what the outside-in approach is doing. The application needs to look at the outer layer of hardware, identify the tasks necessary to interact with it, and finally identify the inner, purely software-related tasks.

Typically, I like to take the block diagram I was working with and transform it during this step. I want to look at the inputs and outputs to then rearrange the blocks and data into elements that can be grouped. For example, I may take the sensor and touch screen blocks and put them near each other because they can be looked at as sensor inputs to the application. I might look at the Wi-Fi and Bluetooth blocks and say these are connectivity blocks and should also be placed near each other. This helps to organize the block diagram a bit before we identify our first tasks.

I make one critical adjustment to the block diagram; I expand the IoT device block from a small text block to a much larger one. I do this so that the IoT device block can encompass all the tasks I am about to define! I then go through the system inputs and outputs and generate my first-tier tasks when I do this. The result can be seen in Figure 4-9.

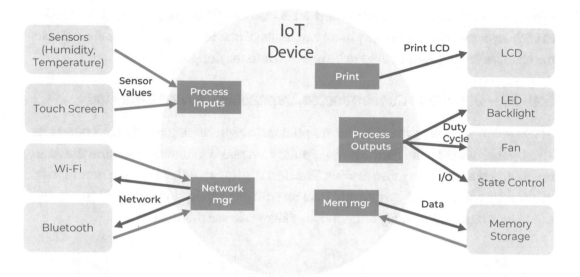

**Figure 4-9.** *In step #5, we group data inputs and outputs and identify the first tasks in the outer layer of the application*

Five tasks have been identified in this initial design which include

- Process inputs

- Network manager

- Print

- Process outputs

- Memory manager

What can be seen here is that we are trying to minimize the number of tasks included in this design. We've grouped all the sensor inputs into a single process input task. All connectivity has been put into a network management task. The nonvolatile memory source is accessed through a memory manager gatekeeping task rather than using a mutex to protect access.

This design shows a design philosophy to keep things simple and minimize complexity. I just as quickly could have created separate tasks for all sensors, individual tasks for Wi-Fi and Bluetooth, and so forth. How I break my initial tasks up will depend on my design philosophy and a preference for how many tasks I want in my system. I try to break my system into layers; each layer does not contain more than a dozen tasks. This is because the human mind generally can only track seven plus or minus two pieces of information at any given time to short-term memory.[13] Limiting the number of tasks in each layer makes it more likely that I can manage that information mentally, reducing the chances for mistakes or bugs to be injected into the design.

## Step #6 – Determine Concurrencies, Dependencies, and Data Flow

At this point, we now have our initial, first-tier tasks representing our design's outer layer. This layer is based on the hardware our application needs to interact with and the data inputs and outputs from those devices. The next step is carefully reviewing how that data will flow through the task system and interact with our application task(s). Figure 4-10 shows how one might modify our previous diagram to see this data flow.

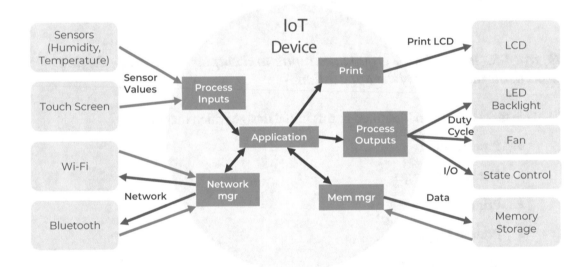

***Figure 4-10.*** *Carefully identify the high-level dependencies between the various tasks. This diagram shows arrows between the application task and the input/ output tasks*

In Figure 4-10, we have added an application task that encompasses all the application's activities. We have not yet decomposed this block but will do that once we understand the surrounding software architecture. The diagram allows us to think through how data will flow from the existing tasks to other tasks in the system. For

---

[13] Miller, G. (1956), The psychological review, 63, 81–97.

example, we can decide whether the network manager can directly interact with the memory manager to retrieve and store settings or whether it must go first through the application. This helps us start understanding how data will flow through our design and helps us understand how tasks might interact with each other.

The diagram also allows us to start considering which tasks are concurrent, that is, which ones may need to appear to run simultaneously and which are feeding other tasks. For example, if we have an output that we want to generate, we may want the input task and the application task to run first to act on the latest data. (We may also want to act on the output data from the last cycle first to minimize jitter in the output!)

Once we have identified our initial tasks, it's the perfect time to develop a data flow diagram. A data flow diagram shows how data is input, transferred through the system, acted on, and output from the system. In addition, the data flow diagram allows the designer to identify event-driven mechanisms and RTOS objects needed to build the application. These will include identifying

- Interrupts and direct memory access (DMA) accesses

- Queues to move data

- Shared memory locations to move data

- Mutexes to protect shared data

- Semaphores to synchronize and coordinate task execution

- Event flags to signal events and coordinate task execution

Figure 4-11 provides an example data flow diagram for our IoT thermostat.

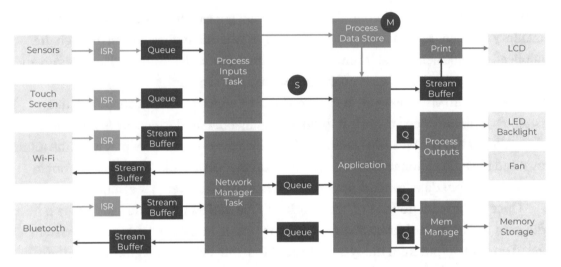

***Figure 4-11.*** *Developing a data flow diagram is critical in determining how data moves through the application and identifying shared resources, task synchronization, and resources needed for the application*

We now have gone from a hardware diagram to identifying our first-tier tasks and generating a data flow diagram that identifies the design patterns we will use to move data throughout the application. However, the critical piece still missing in our design is how to break up our application task into second-tier tasks focusing specifically on our application. (After all, it's improbable that having a single task for our application will be a robust, scalable design.)

## Step #7 – Identify Second-Tier Tasks

The data flow diagram we created in the last step can be beneficial in forming the core for our second-tier task diagram. In this step, we want to look at our application, its features, processes, etc., and break it into more manageable tasks. It wouldn't be uncommon for this generic block we have been looking at to break up into a dozen or more tasks suddenly! We've essentially been slowly designing our system in layers, and the application can be considered the inner layer in the design.

To identify our second-tier or application tasks, I start with my data flow diagram and strip out everything in the data flow outside the tasks that directly interact with the application. This essentially takes Figure 4-11 and strips it down into Figure 4-12. This diagram focuses on the application block's inputs, outputs, and processing.

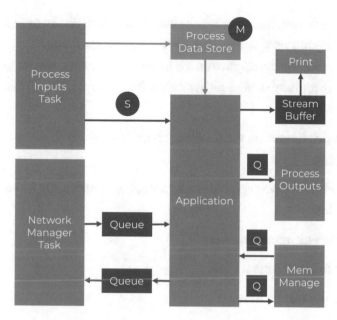

**Figure 4-12.** *Identifying the second-tier application tasks involves focusing on the application task block and all the input and output data from that block*

We now remove the application block and try to identify the tasks and activities performed by the application. The approach should look at each input and determine what application task would handle that input. Each output can be looked at to determine what application task would take that output. The designer can be methodical about this.

Let's start by looking at the process input task. Process inputs execute and receive new sensor data. The sensor data is passed to the application through a data store protected by a mutex. The mutex ensures mutually exclusive access to the data so that we don't end up with corrupted data. How process inputs execute would be as follows:

1. Process inputs acquire new sensor data.

2. Process inputs lock the mutex (preventing the application from accessing the data).

3. New sensor data is stored in the data store.

4. Process inputs unlock the mutex (allowing the application to access the data).

5. Process inputs signal the application that new data is available through a semaphore.

The process input task behavior is now clearly defined. So the question becomes, how will the application receive and process the new sensor data?

One design option for the application is to have a new task, sensor DSP, block on a semaphore waiting for new data to be available. For example, the sensor DSP task would behave as follows:

1. First, block execution until a semaphore is received (no new data available).

2. Then, when the semaphore is acquired, begin executing (data available).

3. Lock the sensor data mutex.

4. Access the new sensor data.

5. Unlock the sensor mutex.

6. Perform digital signal processing on the data, such as averaging or filtering.

7. Pass the processed data or results to the application task and pass the data to a gesture detection task.

This process has allowed us to identify those three tasks that may be needed to handle the process input data: a sensor DSP task, a controller task (formerly the application), and a gesture detection task. Figure 4-13 demonstrates what this would look like in addition to filling in the potential second-tier task for other areas in the design.

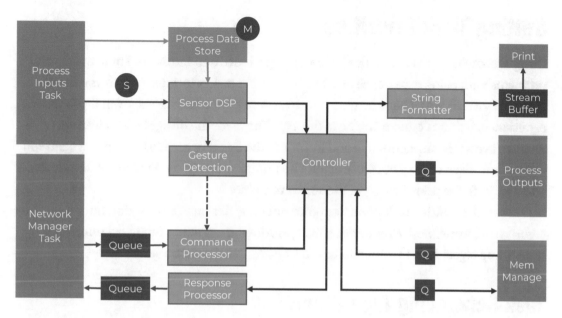

**Figure 4-13.**  *Second-tier tasks are created by breaking down the application task and identifying semi-independent program segments that assist in driving the controller and its output*

A designer can, at this point, decompose the application further by looking at each block and identifying tasks. Once those tasks are identified, the designer can identify the RTOS objects needed to transfer data and coordinate task behavior. Figure 4-13 doesn't show these additional items to keep the diagram readable and straightforward. For example, the Sensor DSP task has an arrow going into the controller task. How is the result being transferred? Is it through a queue? Is it through a shared data store? At this point, I don't know. Again, it's up to the designer to look at the timing, the data rates, and so forth to determine which design pattern makes the most sense.

Now that we have identified our tasks, several questions should be on every designer's mind:

- How am I going to set my task priorities?

- Can the RTOS even schedule all these tasks?

Let's discuss how to set task priorities and perform a schedulable task analysis to answer these questions.

# Setting Task Priorities

Designers often set task priorities based on experience and intuition. There are several problems with using experience and intuition to set task priorities. First, if you don't have experience, you'll have no clue how to set the preferences! Next, even if you have experience, it doesn't mean you have the experience for the designed application. Finally, if you rely on experience and intuition, the chances are high that the system will not be optimized or may not work at all! The implementers may need to constantly fiddle and play with the priorities to get the software stable.

The trick to setting task priorities relies not on experience or intuition but on engineering principles! When setting task priorities, designers want to examine task scheduling algorithms that dictate how task priorities should be set.

## Task Scheduling Algorithms[14]

There are typically three different algorithms that designers can use to set task priorities:

- Shortest job first (SJF)

- Shortest response time (SRT)

- Periodic execution time (RMS)

To understand the differences between these three different scheduling paradigms, let's examine a fictitious system with the characteristics shown in Table 4-1.

*Table 4-1.  An example system has four tasks: two periodic and two aperiodic*

| Task Number | Task Type | Response Time (ms) | Execution Time (ms) | Period (ms) |
|---|---|---|---|---|
| 1 | Periodic | 30 | 20 | 100 |
| 2 | Periodic | 15 | 5 | 150 |
| 3 | Aperiodic | 100 | 15 | – |
| 4 | Aperiodic | 20 | 2 | – |

---

[14] The concepts, definitions, and examples in this section are taken from *Real-Time Operating Systems Book 1 – The Theory* by Jim Cooling, Chapter 9, page 6.

In shortest job first scheduling, the task with the shortest execution time is given the highest priority. For example, for the tasks in Table 4-1, task 4 has a two-millisecond execution time, which makes it the highest priority task. The complete task priority for the shortest job first can be seen in Table 4-2.

It is helpful to note that exact measurements for the task execution times will not be available during the design phase. It is up to the designer to estimate, based on experience and intuition, what these values will be. Once the developers begin implementation, they can measure the execution times and feed them back into the design to ensure that the task priorities are correct.

Shortest response time scheduling sets the task with the shortest response time as the highest priority. Each task has a requirement for its real-time response. For example, a task that receives a command packet every 100 milliseconds may need to respond to a completed packet within five milliseconds. For the tasks in Table 4-1, task 2 has the shortest response time, making it the highest priority.

***Table 4-2.*** *Priority results for the system are defined in Table 4-1 based on the selected scheduling algorithm*

| Scheduling Policy | Shortest Response Time | Shortest Execution Time | Rate Monotonic Scheduling |
|---|---|---|---|
| Assigned Task Priority | Task 2 – Highest | Task 4 – Highest | Task 1 – Highest |
| | Task 4 | Task 2 | Task 2 |
| | Task 1 | Task 3 | Undefined |
| | Task 3 | Task 1 | Undefined |

Finally, the periodic execution time scheduling sets the task priority with the shortest period. Therefore, the task must be a periodic task. For example, in Table 4-1, tasks 3 and 4 are aperiodic (event-driven) tasks that do not occur at regular intervals. Therefore, they have an undefined task priority based on the periodic execution time policy.

Table 4-2 summarizes the results for setting task priorities for the tasks defined in Table 4-1 based on the selected scheduling policy. The reader can see that the chosen scheduling policy can dramatically impact the priority settings. Conversely, selecting an inappropriate scheduling policy could affect how the system performs and even determine whether deadlines will be met successfully or not.

For many real-time systems, the go-to scheduling policy starts with the periodic execution time, also known as rate monotonic scheduling (RMS). RMS allows developers to set task priorities and verify that all the tasks in a system can be scheduled to meet their deadlines! Calculating CPU utilization for a system based on its tasks is a powerful tool to make sure that the design is on the right track. Let's now look at how verifying our design is possible by using rate monotonic analysis (RMA).

---

**Best Practice**    While we often default to rate monotonic scheduling, multiple policies can be used simultaneously to handle aperiodic tasks and interrupts.

---

# Verifying CPU Utilization Using Rate Monotonic Analysis (RMA)

Rate monotonic analysis (RMA) is an analysis technique to determine if all tasks can be scheduled to run and meet their deadlines.[15] It relies on calculating the CPU utilization for each task over a defined time frame. RMA comes in several different flavors based on how complex and accurate an analysis the designer wants to perform. However, the basic version has several critical assumptions that developers need to be aware which include

- Tasks are periodic.

- Tasks are independent.

- Preemptive scheduling is used.

- Each task has a constant execution time (can use worst case).

- Aperiodic tasks are limited to start-up and failure recovery.

- All tasks are equally critical.

- Worst-case execution time is constant.

---

[15] https://en.wikipedia.org/wiki/Rate-monotonic_scheduling

At first glance, quite a few of these assumptions seem unrealistic for the real world! For example, all tasks are periodic except for start-up and failure recovery tasks. This is an excellent idea, but what about a task kicked off by a button press or other event-driven activities? In a case like this, we either must ignore the event-driven task from our analysis or assume a worst-case periodic execution time so that the task is molded into the RMA framework.

The other big assumption in this list is that all tasks are independent. If you have ever worked on an RTOS-based system, it's evident that this is rarely the case. Sometimes, tasks will protect a shared resource using a mutex or synchronize task execution using a semaphore. These interactions make it so that the task execution time may be affected by other tasks and not wholly independent. The basic RMA assumptions don't allow for these cases, although there are modifications to RMA that cover these calculations beyond our discussion's scope.[16]

The primary test to determine if all system tasks can be scheduled successfully is to use the basic RMA equation:

$$\sum_{k=1}^{n} \frac{E_k}{T_k} \leq n \left( 2^{\frac{1}{n}} - 1 \right)$$

In this equation, the variables are defined as follows:

- $E_k$ is the worst-case execution time for the task.

- $T_k$ is the period that the task will run at.

- n is the number of tasks in the design.

At its core, this equation is calculating CPU utilization. We calculate the CPU utilization for each task and then add up all the utilizations. The total CPU utilization, however, doesn't tell the whole story. Just because we are less than 100% does not mean we can schedule everything!

The right side of the equation provides us with an inequality that we must compare the CPU utilization. Table 4-3 shows the upper bound on the CPU utilization that is allowed to schedule all tasks in a system successfully. Notice that the inequality quickly bounds itself to 69.3% CPU utilization for an infinite number of tasks.

---

[16] A basic reference for additional techniques is *Real-Time Concepts for Embedded Systems* by Qing Li with Caroline Yao, Chapter 14, and *A Practitioner's Handbook for Real-Time Analysis*.

**Table 4-3.** *Scheduling all tasks in a system and ensuring they meet their deadlines depends on the number of tasks in the system and the upper bound of the CPU utilization*

| Number of Tasks | CPU Utilization (%) |
|---|---|
| 1 | 100% |
| 2 | 82.8% |
| 3 | 77.9% |
| 4 | 75.6% |
| ∞ | 69.3% |

RMA allows the designer to use their assumptions about their tasks, and then based on the number of tasks, they can calculate whether the system can successfully schedule all the tasks or not. It is essential to recognize that the basic analysis is a sanity check. In many cases, I view RMA as a model that isn't calculated once but tested using our initial assumptions and then periodically updated based on real-world measurements and refined assumptions as the system is built.

The question that may be on some readers' minds is whether the example system we were looking at in Table 4-1 can be scheduled successfully or not. Using RMA, we can calculate the individual CPU utilization as shown in Table 4-4.

**Table 4-4.** *An example system has four tasks: two periodic and two aperiodic*

| Task Number | Task Type | Response Time (ms) | Execution Time (ms) | Period (ms) | CPU Utilization (%) |
|---|---|---|---|---|---|
| 1 | Periodic | 30 | 20 | 100 | 20.0 |
| 2 | Periodic | 15 | 5 | 150 | 3.33 |
| 3 | Aperiodic | 100 | 15 | 100[17] | 15.0 |
| 4 | Aperiodic | 20 | 2 | 50[18] | 4.0 |

---

[17] This is the worst-case periodic rate for this aperiodic, event-driven task.

[18] This is the worst-case periodic rate for this aperiodic, event-driven task.

The updated table now contains a CPU utilization column. When we sum the CPU utilization column, the result is 42.33%. Can the tasks in this system be scheduled successfully? We have two options to test. First, we can plug n = 4 into the basic RMA equation and find that the result is 75.6%. Since 42.33% < 75.6% is valid, a system with tasks with these characteristics can be scheduled successfully. Second, we could just use the infinite task number value of 69.3% and test the inequality. Either way, the result will show us that we should be able to schedule this system without any issues.

# Measuring Execution Time

The discussion so far in this chapter has been focused on designing RTOS applications. Everything that we have discussed so far is important, but the reader might be wondering how you can measure execution time. Our rate monotonic analysis, the priorities we assign our tasks, and so forth, are very dependent upon the execution time of our tasks. For just a few moments, I want to discuss how we can make these measurements in the real world and use them to feed back into our design.

There are several different techniques that can be used to measure task execution time. First, developers can manually make the measurements. Second, developers can rely on a trace tool to perform the measurements for them. Let's look at both approaches in a little more detail.

## Measuring Task Execution Time Manually

Measuring task execution time manually can provide developers with fairly accurate timing results. The problem that I often see with manual measurements is that they are performed inconsistently. Developers often must instrument their code in some manner and then purposely make the measurements. The result is that they are performed at inconsistent times, and it's easy for system timing to get "out of whack."

In many systems, the manual measurements are performed either by using GPIO pins or using an internal timer to track the time. The problem with the manual measurements becomes that it may not be possible to instrument all the tasks simultaneously, making it an even more manual and time-consuming process to make the measurements. For these reasons, I often don't recommend that teams working with an RTOS perform manual measurements on their timing. Instead, tracing is a much better approach.

# Measuring Task Execution Time Using Trace Tools

Most RTOS kernels have a trace capability built into them. The trace capability is a setting in the RTOS that allows the kernel to track what is happening in the kernel so that developers can understand the application's performance and even debug it if necessary. The trace capability is an excellent, automated, instrumentation tool for designers to leverage to understand how the implemented application is performing compared to the design.

Within the RTOS, whenever an event occurs such as a task is switched, a semaphore is given or taken, a system tick occurs, a mutex is locked or unlocked, and so forth, the event is recorded in the kernel. As each event occurs, the kernel tracks the time at which the event occurs. If these events are stored and then pushed to a host environment, it's possible to take the event data and reconstruct not just the ordering of the events but also the timing of the events!

Designers need two things to retrieve the event data from the RTOS kernel. First, they need a recording library that can take the events from the kernel and store them in a RAM buffer. The RAM buffer can then either be read out when it is full (snapshot mode) or it can be emptied at periodic intervals and streamed to the host (streaming mode) in real time. Second, there needs to be a host-side application that can retrieve the event data and then reconstruct it into a visualization that can be reviewed by the developer.

We will discuss tools a bit more in Chapter 15, but I don't want to move on without giving you a few examples of the tools I use to perform RTOS measurements. The first tool that can be used is SEGGER SystemView. An example trace from SystemView can be seen in Figure 4-14. The RTOS generates the events which are then streamed using the SEGGER RTT library through SWD into a SEGGER J-Link which then distributes it through the J-Link server to SystemView. SystemView then records the events and generates

- An event log

- A task context switch diagram in a "chart strip" visualization

- The total CPU utilization over time

- A context summary that lists the system tasks along with statistical data about them such as total runtime, minimum execution time, average execution time, maximum execution time, and so forth.

SystemView is a free tool that can be downloaded by any team that is using a J-Link. SystemView can provide some amazing insights, but as a free tool, it does have limitations as to what information it reports and its capabilities. However, for teams looking for quick and easy measurements without spending any money, this is a good first tool to investigate.

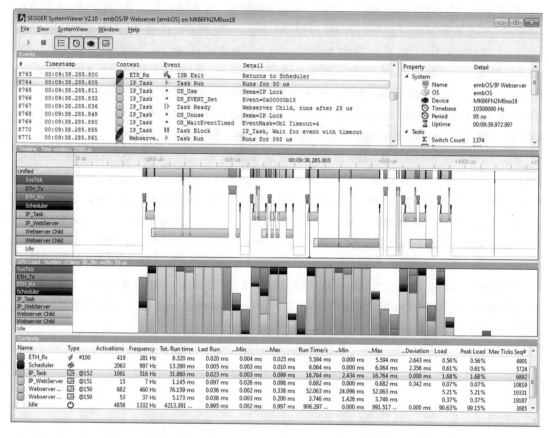

**Figure 4-14.** *An example RTOS trace taken using SEGGER SystemView. Notice the running trace data that shows task context switches and the event log with all the events generated in the system. (Image Source: SEGGER[19])*

The second tool that I would recommend and the tool that I use for all my RTOS measurements and analysis is Percepio's Tracealyzer. Tracealyzer is the "hot rod" of low-cost RTOS visualization tools. Tracealyzer provides the same measurements and

---

[19] https://c.a.segger.com/fileadmin/documents/Press_Releases/PR_151106_SEGGER_SystemView.pdf

visualizations as SystemView but takes things much further. The event library can record the standard RTOS events but also allows users to create custom events as well. That means if a developer wanted to track the state of a state machine or the status of a network stack and so forth, they can create custom events that are recorded and reported.

The Tracealyzer library also allows developers to use a J-Link, serial interface, TCP/IP, and other interfaces for retrieving the trace data. I typically pull the data over a J-Link, but the option for the other interfaces can be quite useful. The real power within Tracealyzer is the reporting capabilities and the linking of events within the interface. It's very easy to browse through the events to discover potential problems. For example, a developer can monitor their heap memory usage and monitor stacks for overflows. An example of the Tracealyzer user interface can be seen in Figure 4-15.

Okay, so what does this have to do with measuring execution times? As I mentioned, both tools will report for each task in your system, the minimum, average, and maximum execution times. In fact, if you are using Tracealyzer, you can generate a report that will also tell you about the periodicity for each task and provide minimum, average, and maximum times between when the task is ready to execute and when it executes! There's no need to manually instrument your code, you rely on the RTOS kernel and the event recorder which can then provide you with more insights into your system execution and performance than you will probably ever imagine. (You can even trace which tasks are interacting with semaphores, mutexes, and message queues.)

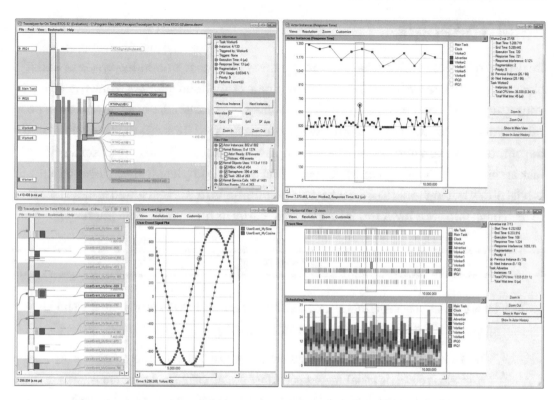

***Figure 4-15.*** *An example RTOS trace taken using Percepio Tracealyzer. Notice the multiple views and the time synchronization between them to visualize what is happening in the system at a given point in time. (Image Source: Percepio[20])*

I'm probably a little bit biased when it comes to using these tools, so I would recommend that you evaluate the various tools yourself and use the one that best fits your own needs and budget. There are certainly other tools out there that can help you measure your task execution and performance, but these two are my favorites for teams that are on a tight budget. (And the cost for Tracealyzer is probably not even noticeable in your company's balance sheet; I know it's not in mine!)

[20] https://percepio.com/docs/OnTimeRTOS32/manual/

# Final Thoughts

Real-time operating systems have found their way into most embedded systems, and it's likely they will increase and continue to dominate embedded systems. How you break your system up into tasks and processes will affect how scalable, reusable, and even host robust your system is. In this chapter, we've only scratched the surface of what it takes to decompose a system into tasks, set our task priorities, and arrive at a functional system task architecture. No design or implementation can be complete without leveraging modern-day tracing tools to measure your assumptions in the real world and use them to feed back into your RMA model and help to verify or tune your application.

---

## ACTION ITEMS

To put this chapter's concepts into action, here are a few activities the reader can perform to start applying RTOS design principles to their application(s):

- Take a few moments to think of a system that would be fun to design. For example, it could be an IoT sensor node, thermostat, weather station, drone, etc. Then, write out a high-level system description and list a few requirements.

- From your system description, decompose the system into tasks using the feature-based decomposition method.

- From your system description, use the outside-in approach to decompose the system into tasks. Then, compare the differences between the two designs.

- Create a data flow diagram for your system.

- Perform an RMA on your system. Use the RMA to identify and set your task priorities. Is your system schedulable?

- Examine your tasks and the supporting objects. Then, break your system up further into processes.

- Review your design and identify areas where it can be optimized.

- Implement a simple RTOS-based application that includes a couple of tasks. The system doesn't need to do anything, just have the tasks execute at different rates and periods.

    - Download SystemView and explore the trace capabilities it provides to analyze your simple RTOS application. What are its strengths and weaknesses?

    - Download Tracealyzer and explore the trace capabilities it provides to analyze your simple RTOS application. What are its strengths and weaknesses?

These are just a few ideas to go a little bit further. Carve out time in your schedule each week to apply these action items. Even minor adjustments over a year can result in dramatic changes!

# CHAPTER 5

# Design Patterns

I've worked on over 100 different embedded projects for the first 20 years of my career. These projects have ranged from low-power sensors used in the defense industry to safety-critical medical devices and flight software to children's toys. The one thing I have noticed between all these projects is that common patterns tend to repeat themselves from one embedded system to the next. Of course, this isn't an instance where a single system is designed, and we just build the same system repeatedly. However, having had the chance to review several dozen products where I had no input, these same patterns found their way into many systems. Therefore, they must contain some basic patterns necessary to build embedded systems.

This chapter will explore a few design patterns for embedded systems that leverage a microcontroller. We will not cover the well-established design patterns for object-oriented programming languages by the "Gang of Four (GoF),"[1] but instead focus on common patterns for building real-time embedded software from a design perspective. We will explore several areas such as single vs. multicore development, publish and subscribe models, RTOS patterns, handling interrupts, and designing for low power.

---

**Tip** Schedule some time to read, or at least browse, *Design Patterns: Elements of Reusable Object-Oriented Software.*

---

## Managing Peripheral Data

One of the primary design philosophies is that data dictates design. When designing embedded software, we must follow the data. In many cases, the data starts in the outside world, interacts with the system through a peripheral device, and is then served

---

[1] https://springframework.guru/gang-of-four-design-patterns/

© Jacob Beningo 2022
J. Beningo, *Embedded Software Design*, https://doi.org/10.1007/978-1-4842-8279-3_5

up to the application. A major design decision that all architects encounter is "How to get the data from the peripheral to the application?". As it turns out, there are several different design mechanisms that we can use, such as

- Polling

- Interrupts

- Direct memory access (DMA)

Each of these design mechanisms, in turn, has several design patterns that can be used to ensure data loss is not encountered. Let's explore these mechanisms now.

## Peripheral Polling

The most straightforward design mechanism to collect data from a peripheral is to simply have the application poll the peripheral periodically to see if any data is available to manage and process. Figure 5-1 demonstrates polling using a sequence diagram. We can see that the data becomes available but sits in the peripheral until the application gets around to requesting the data. There are several advantages and disadvantages to using polling in your design.

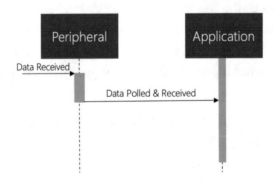

***Figure 5-1.***  *The application polls a peripheral for data on its schedule*

The advantage of polling is that it is simple! There is no need to set up the interrupt controller or interrupt handlers for the peripheral. Typically, a single application thread with very well-known timing is used to check a status bit within the peripheral to see if data is available or if the peripheral needs to be managed.

Unfortunately, there are more disadvantages to using polling than there are advantages. First, polling tends to waste processing cycles. Developers must allocate processor time to go out and check on the peripheral whether there is data there or not. In a resource-constrained or low-power system, these cycles can add up significantly. Second, there can be a lot of jitter and latency in processing the peripheral depending on how the developers implement their code.

For example, if developers decide that they are going to create a while loop that just sits and waits for data, they can get the data very consistently with low latency and jitter, but it comes at the cost of a lot of wasted CPU cycles. Waiting in this manner is polling using blocking. On the other hand, developers can instead not block and use a nonblocking method where another code is executed, but if the data arrives while another code is being executed, then there can be a delay in the application getting to the data adding latency. Furthermore, if the data comes in at nonperiodic rates, it's possible for the latency to vary, causing jitter in the processing time. The jitter may or may not affect other parts of the embedded system or cause system instability depending on the application.

Despite the disadvantages of polling, sometimes polling is just the best solution. If a system doesn't have much going on and it doesn't make sense to add the complexity of interrupts, then why add them? Debugging a system that uses interrupts is often much more complicated. If polling fits, then it might be the right solution; however, if the system needs to minimize response times and latency, or must wake up from low-power states, then interrupts might be a better solution.

# Peripheral Interrupts

Interrupts are a fantastic tool available to designers and developers to overcome many disadvantages polling presents. Interrupts do precisely what their name implies; they interrupt the normal flow of the application to allow an interrupt handler to run code to handle an event that has occurred in the system. For example, an interrupt might fire for a peripheral when data is available, has been received, or even transmitted. Figure 5-2 shows an example sequence diagram for what we can expect from an interrupt design.

The advantages of using an interrupt are severalfold. First, there is no need to waste CPU cycles checking to see if data is ready. Instead, an interrupt fires when there is data available. Next, the latency to get the data is deterministic. It takes the same number of clock cycles when the interrupt fires to enter and return from the interrupt service routine (ISR). The latency for a lower priority interrupt can vary though if a higher priority interrupt is running or interrupts it during execution. Finally, jitter is minimized and only occurs if multiple interrupts are firing simultaneously. In this case, the interrupt with the highest priority gets executed first. The jitter can potentially become worse as well if the interrupt fires when an instruction is executing that can't be interrupted.

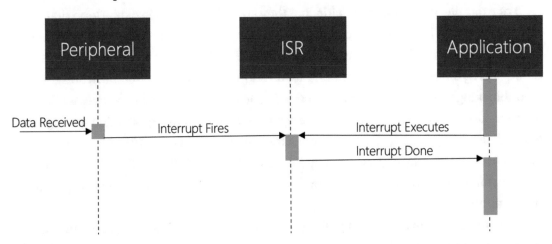

***Figure 5-2.*** *When data is received in the peripheral, it fires an interrupt which stops executing the application, runs the interrupt handler, and then returns to running the application*

Despite interrupts solving many problems associated with polling, there are still some disadvantages to using interrupts. First, interrupts can be complicated to set up. While interrupt usage increases complexity, the benefits usually overrule this disadvantage. Next, designers must be careful not to use interrupts that fire too frequently. For example, trying to use interrupts to debounce a switch can cause an interrupt to fire very frequently, potentially starving the main application and breaking its real-time performance. Finally, when interrupts are used to receive data, developers must carefully manage what they do in the ISR. Every clock cycle spent in the ISR is a clock cycle away from the application. As a result, developers often need to use the ISR to handle the immediate action required and then offload processing and non-urgent activities to the application, causing software design complexity to increase.

# Interrupt Design Patterns

When an interrupt is used in a design, there is a chance that the work performed on the data will take too long to run in the ISR. When we design an ISR, we want the interrupt to

- Run as quickly as possible (to minimize the interruption)

- Avoid memory allocation operations like declaring nonstatic variables, manipulating the stack, or using dynamic memory

- Minimize function calls to avoid clock cycle overhead and issues with nonreentrant functions or functions that may block.

There's also a good chance that the data just received needs to be combined with past or future data to be useful. We can't do all those operations in a timely manner within an interrupt. We are much better served by saving the data and notifying the application that data is ready to be processed. When this happens, we need to reach for design patterns that allow us to get the data quickly, store it, and get back to the main application as soon as possible.

Designers can leverage several such patterns used on bare-metal and RTOS-based systems. A few of the most exciting patterns include

- Linear data store

- Ping-pong buffers

- Circular buffers

- Circular buffer with semaphores

- Circular buffer with event flags

- Message queues

## Linear Data Store Design Pattern

A linear data store is a shared memory location that an interrupt service routine can directly access, typically to write new data to memory. The application code, usually the data reader, can also directly access this memory, as shown in Figure 5-3.

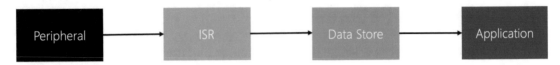

***Figure 5-3.*** *An interrupt (writer) and application code (reader) can directly access the data store*

Now, if you've been designing and developing embedded software for any period, you'll realize that linear data stores can be dangerous! Linear data stores are where we often encounter race conditions because access to the data store needs to be carefully managed so that the ISR and application aren't trying to read and write from the data store simultaneously. In addition, the variables used to share the data stored between the application and the ISR also need to be declared volatile to prevent the compiler from optimizing out important instructions caused by the interruptible nature of the operations.

Data stores often require the designer to build mutual exclusion into the data store. Mutual exclusion is needed because data stores have a critical section where if the application is partway through reading the data when an interrupt fires and changes it, the application can end up with corrupt data. We don't care how the developers implement the mutex at the design level, but we need to make them aware that the mutex exists. I often do this by putting a circular symbol on the data store containing either an "M" for a mutex or a key symbol, as shown in Figure 5-4. Unfortunately, at this time, there are no standards that support official nomenclature for representing a mutex.

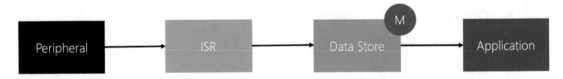

***Figure 5-4.*** *The data store must be protected by a mutex to prevent race conditions. The mutex is shown in the figure by an "M"*

## Ping-Pong Buffer Design Pattern

Ping-pong buffers, also sometimes referred to as double buffers, offer another design solution meant to help alleviate some of the race condition problems encountered with a data store. Instead of having a single data store, we have two identical data stores, as shown in Figure 5-5.

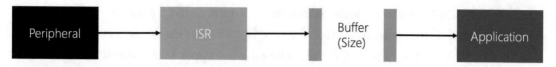

***Figure 5-5.*** *A ping-pong buffer transfers data between the ISR and application*

Now, at first, having two data stores might seem like an opportunity just to double the trouble, but it's a potential race condition saver. A ping-pong buffer is so named because the data buffers are used back and forth in a ping-pong-like manner. For example, at the start of an application, both buffers are marked as write only – the ISR stores data in the first data store when data comes in. When the ISR is done and ready for the application code to read, it marks that data store as ready to read. While the application reads that data, the ISR stores it in the second data store if additional data comes in. The process then repeats.

## Circular Buffer Design Pattern

One of the simplest and most used patterns to get and use data from an interrupt is to leverage a circular buffer. A circular buffer is a data structure that uses a single, fixed-size buffer as if it were connected end to end. Circular buffers are often represented as a ring, as shown in Figure 5-6. Microcontroller memory is not circular but linear. When we build a circular buffer in code, we specify the start and stop addresses, and once the stop address is reached, we loop back to the starting address.

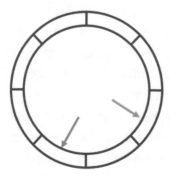

***Figure 5-6.*** *An eight-element circular buffer representation. The red arrow indicates the head where new data is stored. The green arrow represents the tail, where data is read out of the buffer*

The idea with the circular buffer is that the real-time data we receive in the interrupt can be removed from the peripheral and stored in a circular buffer. As a result, the interrupt can run as fast as possible while allowing the application code to process the circular buffer at its discretion. Using a circular buffer helps ensure that data is not lost, the interrupt is fast, and we still process the data reasonably.[2]

The most straightforward design pattern for a circular buffer can be seen in Figure 5-7. In this pattern, we are simply showing how data moves from the peripheral to the application. The data starts in the peripheral, is handled by the ISR, and is stored in a circular buffer. The application can come and retrieve data from the circular buffer when it wants to. Of course, the circular buffer needs to be sized appropriately, so the buffer does not overflow.

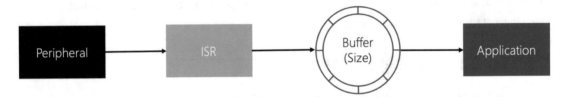

***Figure 5-7.*** *The data flow diagram for moving data from the peripheral memory storage into the application where it is used*

## Circular Buffer with Notification Design Pattern

The circular buffer design pattern is great, but there is one problem with it that we haven't discussed; the application needs to poll the buffer to see if there is new data available. While this is not a world-ending catastrophe, it would be nice to have the application notified that the data buffer should be checked. Two methods can signal the application: a semaphore and an event flag.

A semaphore is a synchronization primitive that is included in real-time operating systems. Semaphores can be used to signal tasks about events that have occurred in the application. We can leverage this tool in our design pattern, as shown in Figure 5-8. The goal is to have the ISR respond to the peripheral as soon as possible, so there is no data loss. The ISR then saves the data to the circular buffer. At this point, the application

---

[2] The definition for reasonable amount of time is obviously open to debate and very application dependent.

doesn't know that there is data to be processed in the circular buffer without polling it. The ISR then signals the application by giving a semaphore before completing execution. When the application code runs, the task that manages the circular buffer can be unblocked by receiving the semaphore. The task then processes the data stored in the circular buffer.

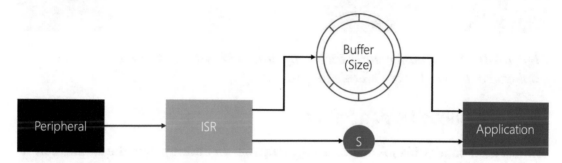

***Figure 5-8.*** *A ring buffer stores incoming data with a semaphore to notify the application that data is available for processing*

Using a semaphore is not the only method to signal the application that data is ready to be processed. Another approach is to replace the semaphore with an event flag. An event flag is an individual bit that is usually part of an event flag group that signals when an event has occurred. Using an event flag is more efficient in most real-time operating systems than using a semaphore. For example, a designer can have 32 event flags in a single RAM location on an Arm Cortex-M processor. In contrast, just a single semaphore with its semaphore control block can easily be a few hundred bytes.

Semaphores are often overused in RTOS applications because developers jump straight into the coding and often don't take a high-level view of the application. The result is a bunch of semaphores scattered throughout the system. I've also found that developers are less comfortable with event flags because they aren't covered or discussed as often in classes or engineering literature.

An example design pattern for using event flags and interrupts can be seen in Figure 5-9. We can represent an event flag version of a circular buffer with a notification design pattern. As you can see, the pattern itself does not change, just the tool we use to implement it. The implementation here results in fewer clock cycles being used and less RAM.

---

**Tip**   Semaphores are often overused in RTOS applications. Consider first using an event flag group, especially for event signals with a binary representation.

---

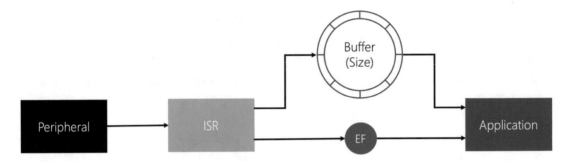

***Figure 5-9.*** *A ring buffer stores incoming data with an event flag to notify the application that data is available for processing*

## Message Queue Design Pattern

The last method we will look at for moving data from an interrupt into the application uses message queues. Message queues are a tool available in real-time operating systems that can take data of a preset set maximum size and queue it up for processing by a task. A message queue can typically store more than a single message and is configurable by the developer.

To leverage the message queue design pattern, the peripheral once again produces data retrieved by an ISR. The ISR then passes the data into a message queue that can be used to signal an application task that data is available for processing. When the application task has the highest priority, the task will run and process the stored data in the queue. The overall pattern can be seen in Figure 5-10.

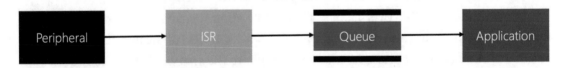

***Figure 5-10.*** *A message queue stores incoming data and passes it into the application for processing*

Using a message queue is like using a linear buffer with a semaphore. While designers may be tempted to jump immediately to using a message queue, it's essential to consider the implications carefully. A message queue typically requires more RAM, ROM, and processing power than the other design patterns we discussed. However, it can also be convenient if there is enough horsepower and the system is not a highly constrained embedded system.

# Direct Memory Access (DMA)

Many microcontrollers today include a direct memory access (DMA) controller that allows individual channels to be set up to move data in the following ways without interaction from the CPU:

- RAM to RAM

- Peripheral to RAM

- Peripheral to peripheral

A typical block diagram for DMA and how it interacts with the CPU, memory, and peripherals can be seen in Figure 5-11.

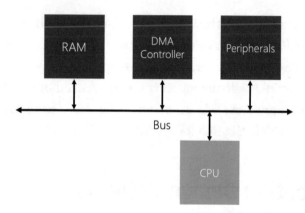

***Figure 5-11.***   *The DMA controller can transfer data between and within the RAM and peripherals without the CPU*

The DMA controller can help to dramatically improve data throughput between the peripheral and the application. In addition, the DMA controller can be leveraged to alleviate the CPU from having to run ISRs to transfer data and minimize wasted compute cycles. For example, one design pattern that is especially common in analog-to-digital converters is to leverage DMA to perform and transfer all the converted analog channels and then copy them to a buffer.

For example, Figure 5-12 shows a design pattern where the DMA controller transfers peripheral data into a circular buffer. After a prespecified number of byte transfers, the DMA controller will trigger an interrupt. The interrupt, in this case, uses a semaphore to signal the application that there is data ready to be processed. Note that we could have used one of the other interrupt patterns, like an event flag, but I think this gives you the idea of how to construct different design patterns.

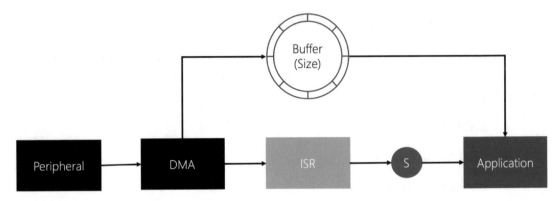

**Figure 5-12.** *An example design pattern that uses the DMA controller to move data from a peripheral into a circular buffer*

Again, the advantage is using the DMA controller to move data without the CPU being involved. The CPU is able to go execute other instructions while the DMA controller is performing transfer operations. Developers do have to be careful that they don't access or try to manipulate the memory locations used by the DMA controller while a transfer is in progress. A DMA interrupt is often used to signal that it is safe to operate on the data. Don't forget that it is possible for the DMA to be interrupted.

# RTOS Application Design Patterns

Design patterns can cover nearly every aspect of the embedded software design process, including the RTOS application. For example, in an RTOS application, designers often need to synchronize the execution of various tasks and data flowing through the application. Two types of synchronization are usually found in RTOS applications, resource synchronization and activity synchronization.

Resource synchronization determines whether the access to a shared resource is safe and, if not, when it will be safe.[3] When multiple tasks are trying to access the same resource, the tasks must be synchronized to ensure integrity. Activity synchronization determines whether the execution has reached a specific state and, if not, how to wait and notify that the state has been reached.[4] Activity synchronization ensures correct execution order among cooperating tasks.

---

[3] *Real-Time Concepts for Embedded Systems* by Qing Li with Caroline Yao, page 231.
[4] *Real-Time Concepts for Embedded Systems* by Qing Li with Caroline Yao, page 233.

Resource synchronization and activity synchronization design patterns come in many shapes and sizes. Let's explore a few common synchronization patterns in more detail.

# Resource Synchronization

Resource synchronization is about ensuring multiple tasks, or tasks and interrupts, that need to access a resource, like a memory location, do so in a coordinated manner that avoids race conditions and memory corruption. For example, let's say we are working on a system that acquires sensor data and uses that data to control a system output. The system might contain two tasks, a sensor task and a control task, in addition to a shared memory location (RAM) that stores the sensor data to be displayed, as shown in Figure 5-13.

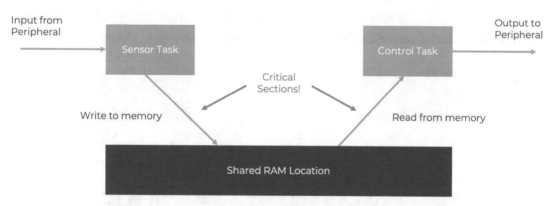

**Figure 5-13.**  *Two tasks looking to access the same memory location require a design pattern to synchronize resource accesses[5]*

In Figure 5-13, you can see that the sensor task acquires data from a device and then writes it to memory. The control task reads the data from memory and then uses it to generate an output. Marked in red, you can see that the write and read operations are critical sections! If the control task is in the process of reading the memory when the sensor task decides to update the memory, we can end up with data corruption and perhaps a wrong value being used to control a motor or other device. The result could be really bad things happening to a user!

---

[5] Modified diagram from *Real-Time Concepts for Embedded Systems* by Qing Li with Caroline Yao, page 232, Figure 15.1.

There are three ways designers can deal with resource synchronization: interrupt locking, preemption locking, and mutex locking.

---

**Tip**    A good architect will minimize the need for resource synchronization. Avoid it if possible, and when necessary, use these techniques (with mutex locking being the preference).

---

## Interrupt Locking

Interrupt locking occurs when a system task disables interrupts to provide resource synchronization between the task and an interrupt. For example, suppose an interrupt is used to gather sensor data that is written to the shared memory location. In that case, the control task can lock (disable) the specific interrupt during the reading process to ensure that the data is not changed during the read operation, as shown in Figure 5-14.

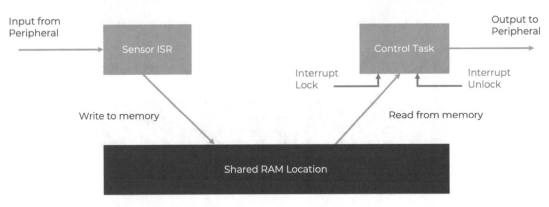

***Figure 5-14.***  *Interrupt locking is used to disable interrupts to prevent a race condition with the sensor task*

Interrupt locking can be helpful but can cause many problems in a real-time embedded system. For example, by disabling the interrupt during the read operation, the interrupt may be missed or, best case, delayed from execution. Delayed execution can add unwanted latency or even jitter into the system.

# Preemption Lock

Preemption lock is a method that can be used to ensure that a task is uninterrupted during the execution of a critical section. Preemption lock temporarily disables the RTOS kernel preemptive scheduler during the critical section, as shown in Figure 5-15.

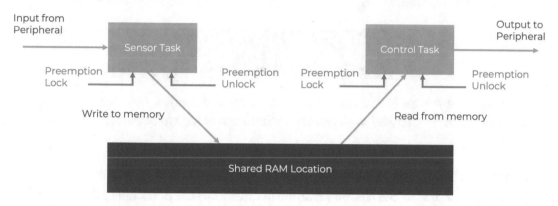

**Figure 5-15.** *A preemption lock protects the critical section by not allowing the RTOS kernel to preempt the executing task*

Preemption lock is a better technique than interrupt locking because critical system interrupts are still allowed to run; however, a higher priority task may be delayed in its execution. Preemption lock can also introduce additional latency and jitter into the system. There is also a possibility that a higher priority task execution could be delayed due to the kernel's preemptive scheduler being disabled.

# Mutex Lock

One of the safest and most recommended methods for protecting a shared resource is to use a mutex lock. A mutex is an RTOS object whose sole purpose is to provide mutual exclusion to shared resources, as shown in Figure 5-16.

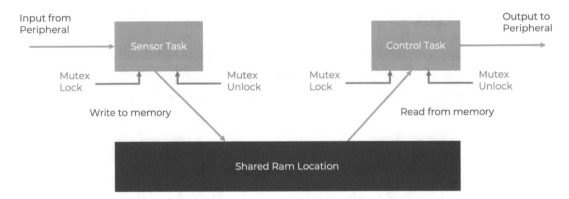

***Figure 5-16.*** *A mutex lock is used to protect a critical section by creating an object whose state can be checked to determine whether it is safe to access the shared resource*

A mutex will not disable interrupts. It will not disable the kernel's preemptive scheduler. It will protect the shared resource in question. One potential problem with a mutex lock protecting a shared resource is that developers need to know it exists! For example, if a developer wrote a third task that would access the sensor data, if they did not know it was a shared resource, they could still just directly access the memory! A mutex is an object that manages a lock state, but it does not physically lock the shared resource.

I highly recommend that if you have a shared resource with which you would like to use a mutex lock, either encapsulate the mutex behind the scenes in the code module or build the mutex into your data structure. Then, when the mutex is built into the data structure when developers use dot notation to access the structure members, the mutex will show up, and they'll realize that the resource is shared and that they need to check to lock the mutex first. (My preference is to encapsulate the module's behavior, but it depends on the use case.)

# Activity Synchronization

Activity synchronization is all about coordinating task execution. For example, let's say we are working on a system that acquires sensor data and uses the sensor values to drive a motor. We most likely would want to signal the motor task that new data is available so that the task doesn't act on old, stale data. Many activity synchronization patterns can be used to coordinate task execution when using a real-time operating system. In this section, we are going to look at just a few.[6]

## Unilateral Rendezvous (Task to Task)

The unilateral rendezvous is the first method we will discuss to synchronize two tasks. The unilateral rendezvous uses a binary semaphore or an event flag to synchronize the tasks. For example, Figure 5-17 shows tasks 1 and 2 are synchronized with a unilateral rendezvous. First, task 2 executes its code to a certain point and then becomes blocked. Task 2 will remain blocked until task 1 reaches the point where it is ready for task 2 to resume executing. Then, task 1 notifies task 2 that it's okay to proceed by giving a semaphore. Task 2 then unblocks, takes the semaphore, and continues to execute.

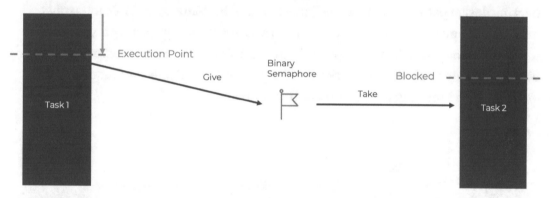

***Figure 5-17.*** *An example of unilateral rendezvous synchronization between tasks using a binary semaphore*

## Unilateral Rendezvous (Interrupt to Task)

The unilateral rendezvous is not only for synchronizing two tasks together. Unilateral rendezvous can also synchronize and coordinate task execution between an interrupt and a task, as shown in Figure 5-18. The difference here is that after the ISR gives the semaphore or the event flag, the ISR will continue to execute until it is complete. In addition, the unilateral rendezvous between two tasks may cause the task that gives the semaphore to be preempted by the other task if the second task has a higher priority.

---

[6] I teach a course entitled Designing and Building RTOS-based Applications that goes further into these patterns.

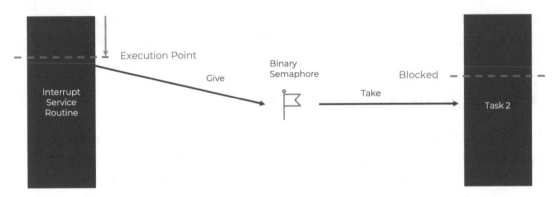

***Figure 5-18.*** *A unilateral rendezvous can be used to synchronize an ISR and task code*

Unilateral rendezvous can also be used with counting semaphores to create a credit tracking design pattern, as shown in Figure 5-19. In this pattern, multiple interrupts might fire before task 2 can process them. In this case, a counting semaphore is used to track how many times the interrupt fired. An example could be an interrupt that saves binary byte data to a buffer. To track how many bytes are available for processing, the interrupt each time increments the counting semaphore.

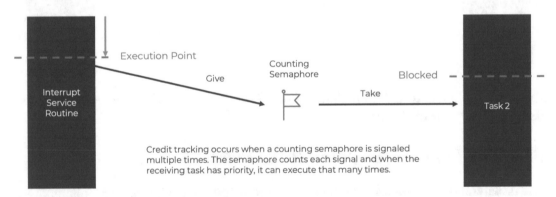

***Figure 5-19.*** *A counting semaphore is used in this unilateral rendezvous to track how many "credits" have been provided by the receiving task to act upon*

## Bilateral Rendezvous

In case you didn't notice, a unilateral rendezvous is a design pattern for synchronizing tasks where one task needs to coordinate with the second task in one direction. On the other hand, a bilateral rendezvous may be necessary if two tasks need to coordinate in both directions between them. An example of bilateral rendezvous can be seen in Figure 5-20.

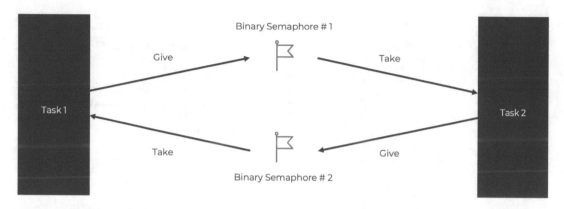

**Figure 5-20.**  *A bilateral rendezvous that uses two binary semaphores to synchronize execution between two tasks*

In Figure 5-20, task 1 reaches a certain point in its execution and then gives a semaphore to task 2. Task 1 might continue to execute to a specific point where it then blocks and waits to receive the second semaphore. The exact details depend on the application's needs and how the tasks are designed. For example, task 2 might wait for the first semaphore and then, upon taking it, runs some code until it reaches its synchronization point and then gives the second binary semaphore to task 1.

## Synchronizing Multiple Tasks

Sometimes, a design will reach a complexity level where multiple tasks need to synchronize and coordinate their execution. The exact design pattern used will depend on the specific application needs. This book will only explore a straightforward design pattern, the broadcast design pattern.

The broadcast design pattern allows multiple tasks to block until a semaphore is given, an event flag occurs, or even a message is placed into a message queue. Figure 5-21 shows an example of a task or an interrupt giving a semaphore broadcast to three other tasks. Each receiving task can consume the semaphore and execute its task code once it is the highest priority task ready to run.

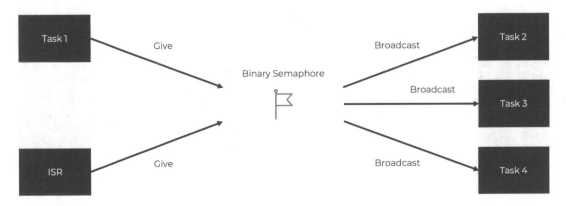

**Figure 5-21.** *A task or an ISR can give a binary semaphore broadcast consumed by multiple tasks*

Broadcasting can be a useful design pattern, but it is essential to realize that it may not be implemented in all real-time operating systems. If you are interested in triggering more than one task off an event, you'll have to check your RTOS documentation to see if it can do so. If broadcasting is not supported, you can instead create multiple semaphores given by the task or the interrupt. Using multiple semaphores isn't as elegant from a design standpoint or efficient from an implementation standpoint, but sometimes that is just how things go in software.

# Publish and Subscribe Models

Any developer working with embedded systems in the IoT industry and connecting to cloud services is probably familiar with the publish/subscribe model. In many cases, an IoT device will power up, connect to the cloud, and then subscribe to message topics it wants to receive. The device may even publish specific topics as well. Interestingly, even an embedded system running FreeRTOS can leverage the publish and subscribe model for its embedded architecture.

The general concept, shown in Figure 5-22, is that a broker is used to receive and send messages below to topics. Publishers send messages to the broker specifying which topic the message belongs to. The broker then routes those messages to subscribers who request that they receive messages from the specific topics. It's possible that there would be no subscribers to a topic, or that one subscriber subscribes to multiple topics, and so forth. The publish and subscribe pattern is excellent for abstracting the system architecture. The publisher doesn't know who listens to its messages and doesn't care. The subscriber only cares about messages that it is subscribed to. The result is a scalable system.

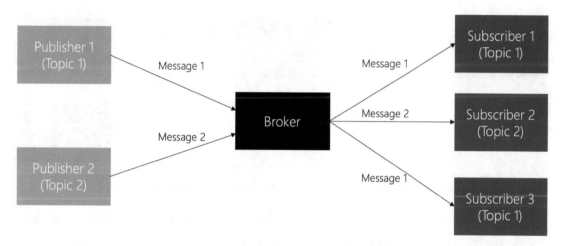

***Figure 5-22.*** *An example publish/subscribe system where two publishers publish messages to the broker, which then routes the messages to the subscribers of the message topics*

Let's, for a moment, think about an example. Let's say we have a system that will collect data from several sensors. That sensor data is going to be stored in nonvolatile memory. In addition, that data needs to be transmitted as part of a telemetry beacon for the system. On top of that, the data needs to be used to maintain the system's orientation through use in a control loop.

We could architect this part of the system in several ways, but one exciting way would be to use the publish and subscribe design pattern. For example, we could add a message broker to our system, define the sensor acquisition task as a publisher, and then have several subscribers that receive data messages and act on that data, as shown in Figure 5-23.

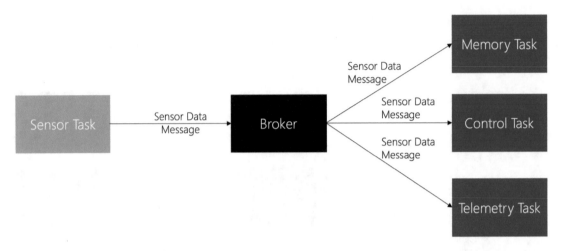

***Figure 5-23.*** *A sensor task delivers sensor data to subscribers through the broker*

What's interesting about this approach is that the dependencies are broken between the sensor tasks and any tasks that need to use the data. The system is also entirely scalable. If we suddenly need to use the sensor data for fault handling, we don't have to go back and rearchitect our system or add another dependency. Instead, we create a fault task that subscribes to the sensor data. That task will get the data it needs to operate and detect faults.

While the publish and subscribe model can be interesting, it obviously will result in a larger memory footprint for the application. However, that doesn't mean that parts of the system can't use this design pattern.

# Low-Power Application Design Patterns

Designing a device today for low power has become a familiar and essential design parameter to consider. With the IoT, there are a lot of battery-operated devices that are out there. Designers want to maximize how long the battery lasts to minimize the inconvenience of changing batteries and having battery waste. There are also a lot of IoT devices that are continuously powered that should also be optimized for energy consumption. For example, a connected home could have upward of 20 connected light switches! Should all those light switches be continuously drawing 3 Watts? Probably not. That's a big waste of energy and nonrenewable energy sources.

Regarding low-power design patterns, the primary pattern is to keep the device turned off as much as possible. The software architecture needs to be event driven. Those events could be a button press, a period that has elapsed, or another trigger. In between those events, when no practical work is going to be done, the microcontroller should be placed into an appropriate low-power state, and any non-essential electronics should be turned off. A simple state diagram for the design pattern can be seen in Figure 5-24.

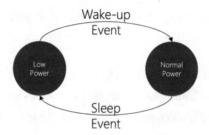

**Figure 5-24.**  *The system lives in the low-power state unless a wake-up event occurs. The system returns to a low-power state after practical work is complete*

The preceding design pattern is, well, obvious. The devil is in the details of how you stay in the low-power state, wake up fast enough, and then get back to sleep. Like many things in embedded systems, the specifics are device and application specific. However, there are at least a few tips and suggestions that I can give to help you minimize the energy that your devices consume.

First, if your system uses an RTOS, one of the biggest struggles you'll encounter is keeping the RTOS kernel asleep. Most RTOS application implementations I've designed and the ones I've seen set the RTOS timer tick to one millisecond. This is too fast when you want to go into a low-power state. The kernel will wake the system every millisecond when the timer tick expires! I'd be surprised if the microcontroller and system had even settled into a low-power state by that time. Designers need to use an RTOS with a tickless mode built-in or one where they can scale their system tick to keep the microcontroller asleep for longer periods.

An example is FreeRTOS. FreeRTOS does have a tickless mode that can be enabled. The developer can add their custom code to manage the low-power state. When the system goes to sleep, the system tick is automatically scaled, so fewer system ticks occur! Figure 5-25 shows some measurements from an application I was optimizing for energy. On channel 00, the system tick is running once every millisecond. On channel 01, task 1, there is a blip of activity and then a hundred milliseconds of no activity. The system was not going to sleep because the system tick was keeping it awake, drawing nearly 32 milliamps.

***Figure 5-25.*** *An unoptimized system whose RTOS system tick prevents the system from going to sleep*

In this system, I enabled the FreeRTOS tickless mode and set up the low-power state I wanted the system to move into. I configured the tickless mode such that I could get a system tick once every 50 milliseconds, as shown in Figure 5-26. The result was that the system could move into a low-power state and stay there for much longer. The power consumption also dropped from 32 milliamps down to only 11 milliamps!

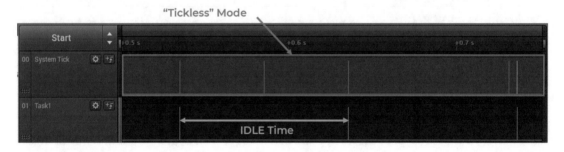

***Figure 5-26.*** *Enabling a tickless mode removed the constant system wake-up caused by the system tick*

Another exciting design pattern to leverage is an event-driven architecture. Many developers don't realize it, but there is no requirement that an RTOS or an embedded system has a constantly ticking timer! You don't need a system tick! If an event causes everything that happens in the system, then there is no reason even to use the RTOS time base. The time base can be disabled. You can use interrupts to trigger tasks that synchronize with each other and eventually put the system to sleep.

# Leveraging Multicore Microcontrollers

Over the last couple of years, as the IoT has taken off and machine learning at the edge has established itself, an interest in multicore microcontrollers has blossomed. Microcontrollers have always been the real-time hot rods of microprocessing. They fill a niche where hard deadlines and consistently meeting deadlines are required. However, with the newer needs and demands being placed on them, many manufacturers realized that either you move up to a general-purpose processor and lose the real-time performance, or you start to put multiple microcontrollers on a single die. Multiple processors in a single package can allow a low-power processing to acquire data and then offload it for processing on a much more powerful processor.

While multicore microcontrollers might seem new and trendy, there have been microcontrollers going back at least 20 years that have had secondary cores. In many cases, though, these secondary cores were threaded cores, where the central processor could kick off some code to be executed on the second core, and the second core would plug away at it, uninterrupted until it completed its task. A few examples are the Freescale S12X and the Texas Instruments C2000 parts. In any event, as the processing needs of the modern system have grown, so has the need for processing power. So in the microcontroller space, the current trend is to add more cores rather than making them faster and more general to compute.

Today, in 2022, there are many multicore microcontroller offerings, such as the Cypress PSoC 6, the STMicroelectronics STM32H7, and the Raspberry Pi Pico, to name a few. A multicore microcontroller is interesting because it provides multiple execution environments that can be hardware isolated where code can run in parallel. I often joke, although prophetically, that one day microcontrollers may look something like Figure 5-27, where they have six, eight, or more cores all crunching together in parallel.

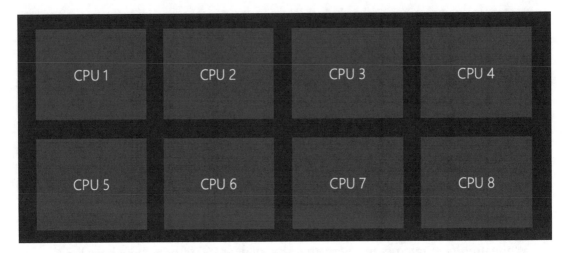

***Figure 5-27.*** *A future microcontroller is envisioned to have eight separate cores*

Today's architectures are not quite this radical yet. A typical multicore microcontroller has two cores. However, two types of architectures are common. First, there is homogenous multiprocessing. In these architectures, each processing core uses the same processor architecture. For example, the ESP32 has Dual Xtensa 32-bit LX6 cores, as shown in Figure 5-28. One core is dedicated to Wi-Fi/Bluetooth, while the other is dedicated to the user application.

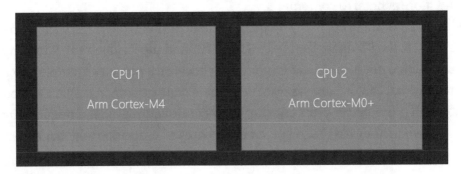

***Figure 5-28.*** *Symmetric multicore processing has two cores of the same processor architecture*

The alternative architecture is to use heterogeneous multiprocessing. In these architectures, each processing core has a different underlying architecture. For example, the Cypress PSoC 64 has an Arm Cortex-M4 for user applications and an Arm Cortex-M0+ to act as a security processor (see Figure 5-29). The two cores also don't have to run at the same clock speed. For example, in the previous example, the Arm Cortex-M4 runs at 150 MHz, while the Arm Cortex-M0+ runs at 100 MHz.

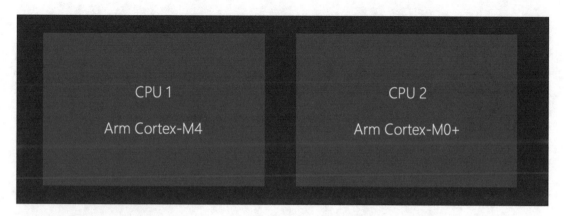

***Figure 5-29.*** *Heterogeneous multiprocessors use multiple-core architectures*

With multiple processing cores, designers can leverage various use cases and design patterns to get the most effective behavior out of their embedded system. Let's explore what a few of them are.

# AI and Real-Time Control

Artificial intelligence and real-time control are the first use case that I see being pushed relatively frequently. Running a machine learning inference on a microcontroller can undoubtedly be done, but it is compute cycle intensive. The idea is that one core is used to run the machine learning inference, while the other does the real-time control, as shown in Figure 5-30. For example, in the STM32H7 family, the multicore parts have an Arm Cortex-M7 and an Arm Cortex-M4. The M7 runs the machine learning inference, while the M4 does the standard real-time stuff like motor control, sensor acquisition, and communication. The M4 core can feed the M7 with the data it needs to run the AI algorithm, and the M7 can feed the M4 the result.

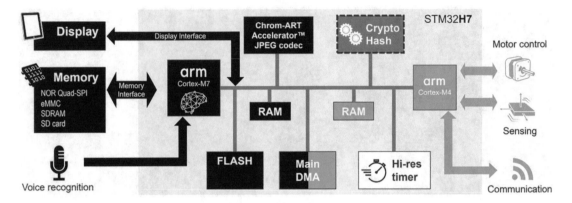

***Figure 5-30.*** *A multicore microcontroller use case where one core is used for running an AI inference, while the other core manages real-time application control (Source: STM32H7 MCUs for rich and complex applications, Slide 16)*

## Real-Time Control

Another use case often employed in multicore microcontroller systems is to have each core manage real-time control capabilities, as shown in Figure 5-31. For example, I recently worked on a ventilator system that used multiple cores on an STM32H7. The M7 core was used to drive an LCD and manage touch inputs. The M4 core was used to run various algorithms necessary to control pumps and valves. In addition, the M4 interfaced with a third processor that performed all the sensor data acquisition and filtering. Splitting different domain functions into different cores and processors dramatically simplified the software architecture design. It also helped with assigning responsibility to different team members and allowed for parallel development among team members.

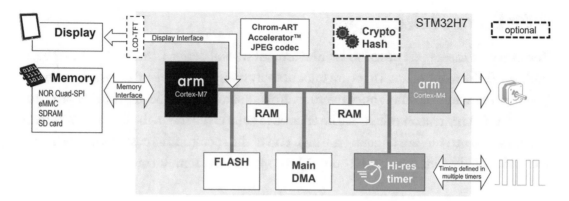

**Figure 5-31.** *A multicore microcontroller use case where one core runs cycle-intensive real-time displays and memory, while the other manages real-time application control (Source: STM32H7 MCUs for rich and complex applications, Slide 29)*

# Security Solutions

Another popular use case for multiple cores is managing security solutions with an application. For example, developers can use the hardware isolation built into the multiple cores to use one as a security processor where the security operations and the Root-of-Trust live. In contrast, the other core is the normal application space (see Figure 5-32). Data can be shared between the cores using shared memory, but the cores only interact through interprocessor communication (IPC) requests.

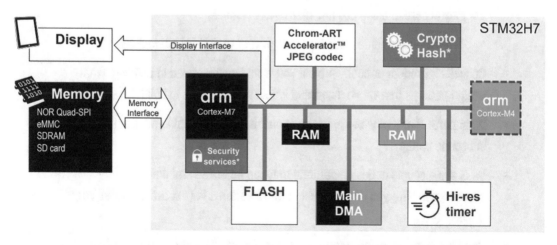

**Figure 5-32.** *A multicore microcontroller use case where one core is used as a security processor, while the other core manages real-time application control (Source: STM32H7 MCUs for rich and complex applications, Slide 28)*

# A Plethora of Use Cases

The three use cases that I described earlier are just several design patterns that emerge for multicore microcontrollers. There are undoubtedly many other potential design patterns. There will certainly be others once microcontrollers start to move beyond having just two cores. Multiple microcontrollers will be an area to keep an eye on in the future. While many applications currently are targeting the high end of the spectrum, as costs come down, we will undoubtedly start seeing 8-bit and 16-bit applications for multicore parts.

# Final Thoughts

Design patterns are commonly occurring patterns in software that designers and developers can leverage to accelerate their design work. There often isn't a need to design a system from scratch. Instead, the architecture can be built from the ground up using fundamental patterns. Not only does this accelerate design, but it also can help with code portability.

---

## ACTION ITEMS

To put this chapter's concepts into action, here are a few activities the reader can perform to get more familiar with design patterns:

- Examine how your most recent application accesses peripherals. For example, are they accessed using polling, interrupts, or DMA?

  - What improvements could be made to your peripheral interactions?

- Consider a product or hobby project you would be interested in building. What design patterns that we've discussed would be necessary to build the system?

- What parts, if any, of your application could benefit from using publish and subscribe patterns?

- What are a couple of design patterns that can be used to minimize energy consumption? Are you currently using these patterns in your architecture? Why or why not?

- What are some of the advantages of using a multicore microcontroller? Are there disadvantages? If so, what are they?

---

# PART II

# Agile, DevOps, and Processes

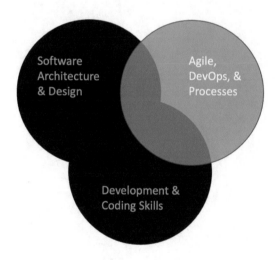

# CHAPTER 6

# Software Quality, Metrics, and Processes

I suspect, if I were to ask you this question, you would respond that you write quality software. Now, you may have some reservations about some of your code, but I don't think any of you will say that you purposely write poor code or that you are a terrible programmer. On the other hand, I would expect that those of you who say you don't write quality code probably have some reason for it. Management pushes you too hard; you're inexperienced and still wet behind the ears, but you're doing the best you can.

What if I asked you to define what quality software is? Could you define it? I bet many of you would respond the way Associate Justice Potter Stewart wrote in the 1964 Supreme Court case (Jacobellis v. Ohio) about hard-core pornography, "I know it when I see it."[1] Yet, if you were to give the same code base to several dozen engineers, I bet they wouldn't be able to agree on its quality level. Without a clear definition, quality is in the eye of the beholder!

In this chapter, we will look at how we can define software quality, how quality can be more of a spectrum than a destination, and what embedded software developers can do to hit the quality levels they desire more consistently. We will find that to hit our quality targets, we need to define metrics that we can consistently track, and put in place processes that catch bugs and defects early, before they become costly and time-consuming.

---

[1] www.thefire.org/first-amendment-library/decision/jacobellis-v-ohio/

© Jacob Beningo 2022
J. Beningo, *Embedded Software Design*, https://doi.org/10.1007/978-1-4842-8279-3_6

# Defining Software Quality

Let's explore the definition of software quality bestowed upon us by that great sage Wikipedia:

> *Software Quality refers to two related but distinct notions:*
>
> 1. *Software **functional quality** reflects how well it complies with or conforms to a given design based on functional requirements or specifications.[2] That attribute can also be described as the fitness for the purpose of a piece of software or how it compares to competitors in the marketplace as a worthwhile product.[3] Finally, it is the degree to which the correct software was produced.*
>
> 2. *Software **structural quality** refers to how it meets non-functional requirements that support the delivery of the functional requirements, such as robustness or maintainability. It has much more to do with the degree to which the software works as needed.[4]*

From this definition, it's obvious why developers have a hard time defining quality; there are two distinct but related components to it. Many of us probably look at the first notion focusing on the "fitness of purpose." If it works and does what it is supposed to, it's high quality (although, to many, the need to occasionally restart a device still seems acceptable, but I digress).

When I started working in the embedded systems industry, my employers loved the software quality I produced. I was that young, hard-working engineer who consistently delivered a working product, even if it meant working 15-hour days, seven days a week. Demos, tests, and fielded units met all the functional requirements. Behind the scenes, though, structurally, the software was, well, not what I would have wanted it to be. In fact, the words big ball of spaghetti and big ball of mud come to mind. (Thankfully, I quickly matured to deliver on both sides of the quality equation.)

---

[2] Board (IREB), International Requirements Engineering. "Learning from history: The case of Software Requirements Engineering – Requirements Engineering Magazine." *Learning from history: The case of Software Requirements Engineering – Requirements Engineering Magazine.* Retrieved 2021-02-25.

[3] Pressman, Roger S. (2005). *Software Engineering: A Practitioner's Approach* (Sixth International ed.). McGraw-Hill Education. p. 388. ISBN 0071267824.

[4] https://en.wikipedia.org/wiki/Software_quality

In many situations, the customer, our boss, and so forth view the functional behavior of our software as the measure of quality. However, I would argue that the objective measure of software quality is structural quality. Anyone can make a product that functions, but that doesn't mean it is not buggy and will work under all the desired conditions. It is not uncommon for systems to work perfect under controlled lab conditions only to explode in a companies face when it is deployed to customers.

Genuine quality is exhibited in the structural code because it directly impacts the product's robustness, scalability, and maintainability. Many products today evolve, with new features added and new product SKUs branching off the original. To have long-term success, we must not just meet the functional software requirements but also the structural quality requirements.

---

**Note**    The disconnect between management and the software team is often that management thinks of functional quality, while developers think of structural quality.

---

Suppose the defining notion of software quality is in the structural component. In that case, developers and teams can't add quality to the product at the end of the development cycle! Quality must be built into the entire design, architecture, and implementation. Developers who want to achieve quality must live the software quality life, always focusing on building quality at every stage.

# Structural Software Quality

When examining the structural quality of a code base, there are two main areas that we want to explore: architectural quality and code quality. Architectural quality examines the big picture of how the code is organized and structured. Code quality examines the fine details of the code, such as whether best practices are being followed, the code is testable, and that complexity has been minimized.

# Architectural Quality

In the first part of this book, we explored how to design and architect embedded software. We saw that software quality requirements often encompass goals like portability, scalability, reusability, etc. Measuring the quality of these requirements can be difficult because they require us to track how successfully we can change the code base over time. Developers can make those measurements, but it could take a year of data collection before we know if we have a good software architecture or not. By that time, the business could be in a lot of trouble.

Instead of relying upon long-term measurements to determine if you have reached your architectural quality goals, you can rely on software architecture analysis techniques. Software architecture analyses are just techniques that are used to evaluate software architectures. Many tools are available to assist developers in the process, such as Understand,[5] Structure101,[6] etc. (these tools are further discussed in Chapter 15). The analysis aims to detect any noncompliances with the architecture design, such as tight coupling, low cohesion, inconsistent dependencies, and so forth.

As a team builds their software, it's not uncommon for the architecture to evolve or for ad hoc changes to break the goals of good architecture. Often, tight coupling or dependency cycles enter the code with the developer being completely unaware. The cause is often a last-minute feature request or a change in direction from management. If we aren't looking, we won't notice these subtle issues creeping into our architecture, and before we know it, the architecture will be full of tightly coupled cycles that make scaling the code a nightmare. The good news is that we can monitor our architecture and discover these issues before they become a major issue.

I recently had a project that I was working on that used FreeRTOS Kernel v10.3.1 that had a telemetry task that collected system data and then transmitted it to a server. Using Understand, I plotted a dependency graph of the telemetry task to verify I didn't have any unwanted coupling or architectural violations, as seen in Figure 6-1. I was surprised to discover a cyclical dependency within the FreeRTOS kernel, as seen by the red line between FreeRTOS_tasks.c and trcKernelPort.c in Figure 6-1.

---

[5] www.scitools.com/

[6] https://structure101.com/

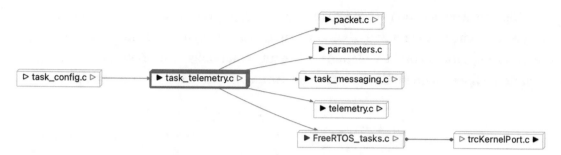

**Figure 6-1.** *The task_telemetry.c call graph has a cyclical dependency highlighted by the red line between FreeRTOS_tasks.c and trcKernelPort.c*

Since the violation occurred within a third-party library and is contained to it, I normally would ignore it. However, I was curious as to what was causing the cycle. The red line violation tells us, as architects, that the developer has code in trcKernelPort.c that is calling code in FreeRTOS_tasks.c and vice versa! Hence, a cycle has been created between these code modules. As architects, we would want to drill a bit deeper to determine what is causing this violation in the architecture. You can see in Figure 6-2 that the cyclical calls are caused by two calls within trcKernelPort.c.

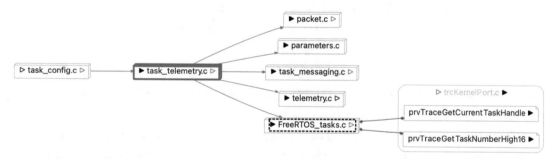

**Figure 6-2.** *Expanding trcKernelPort.c reveals that two functions are involved in the cyclical call tree*

We want to avoid cyclical calls like this because it tightly couples our software, decreases scalability, and starts to create a giant ball of mud. If this cycle had occurred in higher-level application code, there is the potential that other code would use these modules, which in turn are used by others. The result is a big ball of mud from an architectural standpoint.

Digging even deeper, we can expand FreeRTOS_tasks.c to see what functions are involved, as shown in Figure 6-3. Unfortunately, the tool that I am using at that point removes the coloring of the violations, so we must manually trace them. However, examining the expanded diagram, you can see clearly why the two violations occur.

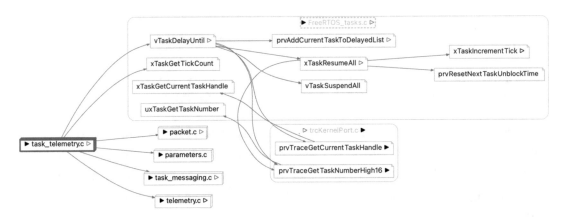

***Figure 6-3.*** *Expanding the modules to trace the cyclical violation fully reveals the functions involved: xTaskGetCurrentTaskHandle and uxTaskGetTaskNumber*

First, vTaskDelayUntil (FreeRTOS_tasks.c) makes a call to prvTraceGetCurrentTaskHandle (trcKernelPort.c) which then calls xTaskGetCurrentTaskHandle (FreeRTOS_tasks.c).

Second, vTaskDelayUntil (FreeRTOS_tasks.c) makes a call to xTaskResumeAll (FreeRTOS_tasks.c) which calls prvTraceGetTaskNumberHigh16 (trcKernelPort.c) which then calls uxTaskGetTaskNumber (FreeRTOS_tasks.c).

These two modules have cyclical dependencies, which we want to avoid in our architectures! Now in FreeRTOS, we are less concerned with this small cycle because it doesn't affect our use of FreeRTOS because it is abstracted away behind the interfaces. Since the violation occurs in the third-party library and is contained to it, there is little risk of the violation impeding our application's scalability or causing other architectural issues.

---

**Note**    If you use FreeRTOS, the cyclical dependency we are looking at does not affect anything in our application and is no cause for any genuine concern.

---

If you discover a cyclical dependency in your code, it usually means that the code structure needs to be refactored to better group functionality and features. For example, looking back at the FreeRTOS example, xTaskGetCurrentTaskHandle and uxTaskGetTaskNumber should not be in FreeRTOS_tasks.c but, instead, be relocated to trcKernelPort.c. Moving these functions to trcKernelPort.c would break this cycle and improve the code structure and architecture.

---

**Caution**    We are examining FreeRTOS in a very narrow scope! Based on the larger picture of the kernel, these changes might not make sense. However, the technique for finding issues like this is sound.

---

Architectural quality is key to improving a product's quality. The architecture is the foundation from which everything is built upon in the code base. If the architecture is faulty or evolves in a haphazard manner, the result will be lower quality, more time-consuming updates, and greater technical debt for the system. The architecture alone won't guarantee that the software product is high quality. The code itself also has to be written so that it meets the quality level you are attempting to achieve.

# Code Quality

Software architecture quality is essential and helps to ensure that the technical layering and domains maintain high cohesion and low coupling. However, software architecture alone won't guarantee that the software will be of high quality. How the code is constructed can have just as much of an impact, if not more, on the software's quality.

When it comes to code quality, there's quite a bit that developers can do to ensure the result is what they desire. For example, low-hanging fruit that is obvious but I often see overlooked is making sure that your code compiles without warnings! Languages like C set low standards for compiler checking, so if the compiler is warning you about something, you should pay attention and take it seriously.

When I look at defining code quality, there are several areas that I like to look at, such as

- The code adheres to industry best practices and standards for the language.

- The code's function complexity is minimized and meets defined code metrics.

- The code compiles without warnings and passes static code analysis.

- The code unit test cases have 100% branch coverage.

- The code has gone through the code review processes.

Let's look at each of these areas and how they help to contribute to improving the quality of embedded software.

## Coding Standards

Coding standards and conventions are a set of guidelines for a specific programming language that recommend programming style, practices, and methods for each aspect of a program written in that language.[7] Developers should not think of standards as rules handed down by God that every programmer must follow to avoid programmer hell, but instead as best practices to improve quality and make the code more maintainable. Sometimes, the rules make sense for what you are doing; other times, they don't and should be cast aside. The idea behind the standards is that if developers follow them, they will avoid common mistakes and issues that are hell to debug!

The C programming language, which is still the most common language used in embedded systems development, has several standards developers can use to improve their code quality, such as GNU,[8] MISRA C,[9] and Cert C.[10] Typically, each standard has its focus. For example, MISRA C is a standard that focuses on writing safe, portable, and reliable software. On the other hand, Cert C focuses on writing secure software. In many cases, developers may decide to follow more than one standard if the standards do not conflict.

In many cases, a static code analysis tool can check adherence to a standard. For example, I'll often use a tool like Understand to perform MISRA C checking on the code I am developing. For MISRA, a static code analysis can be used to verify many directives. However, some require a code review and must be performed manually. In many cases, static analysis can even be used to confirm that the code meets the style guide the team

---

[7] https://en.wikipedia.org/wiki/Coding_conventions
[8] www.gnu.org/prep/standards/html_node/Writing-C.html
[9] www.misra.org.uk/
[10] https://wiki.sei.cmu.edu/confluence/display/c

uses. Attempting to manually check that a code base is meeting a standard is foolish. An automated static analysis tool is the only way to do this efficiently and catch everything. Be warned though; the output from a static analysis tool will vary by vendor. It is not uncommon for teams focused on high quality to use more than one static analysis tool.

---

**Best Practice**    Use an analysis tool to verify that your software meets your following standards.

---

Standards help developers ensure that their code follows industry best practices. While we often think that we will catch everything ourselves, if you take a minute to browse Gimpel Software's old "The Bug of the Month"[11] archive, you'll see how easily bugs sneak through. Did you know, if you examine the C standard(s) appendix, you'll find around 200 undefined implementation-dependent behaviors? If you aren't careful, you could have code that behaviors one way with compiler A, and then the same code behaves totally different using compiler B. That's some scary (insert your preferred profanity statement)!

Standards can also help to ensure that no matter how many developers are working on the project, the code is consistent and looks like just a single developer wrote it. There are a few things more difficult than trying to review code that is poorly organized and commented and varies wildly in its look and feel. When the code looks the same no matter who wrote it, it becomes easier to perform code reviews and focus on the intent of the code rather than the style of the code.

The goal of a standard is to provide developers with best practices that will help them to avoid problems that would have otherwise bogged them down for hours or weeks of painful debugging. It's critical that developers take the time to understand the standards that are pertinent to their industry and put in place the processes to avoid costly mistakes.

---

[11] www.gimpel.com/archive/bugs.htm

# Software Metrics

A software metric is a measurement that developers can perform on their code that can be used to gain insights into the behavior and quality of their software. For example, a standard metric that many teams track is lines of code (LOC). The team can use LOC for many purposes, such as

- Determining the rate at which the team generates code

- Calculating the cost to develop software per LOC

- Examining how the code base changes over time (lines added, modified, removed, etc.)

- Identifying where a module may be growing too large and refactoring may be necessary

Developers do need to be careful with their metrics, though. For example, when we talk about LOC, are we talking about logical lines, comment lines, or both? It's essential to be very clear in the definition of the metric so that everyone understands exactly what the metric means and what is being measured. Before using a metric, developers also need to determine why the metric is important and what the changes in that metric mean over time. Tracking the trend in metrics can be as critical taking the measurement. A trend will tell you if the code is heading in the right direction or slowly started to fall over a cliff! Also, don't put too much stock in any one metric. Collectively examining the trends will provide far more insights into the software quality state.

Developers can track several quality metrics that will help them understand if potential issues are hiding in their software. For example, some standard metrics that teams track include

- McCabe Cyclomatic Complexity

- Code churn (lines of code modified)

- CPU utilization

- Assertion density

- Test case code coverage

- Bug rates

- RTOS task periodicity (minimum, average, and max)

There are certainly plenty of metrics that developers can track. The key is to identify metrics that provide your team insights into the characteristics that will benefit you the most. For example, don't select metrics just to collect them! If LOC doesn't tell you anything constructive or exciting, don't track it! No matter what metrics you choose, make sure you take the time to track the trends and the changes in the metrics over time.

---

**Tip**    If you want to get some ideas on metrics you can track, install a program like Metrix++, Understand, etc., and see the giant list of metrics they can track. Then select what makes sense for you.

---

An important metric that is often looked by teams involve tracking bugs. Bug tracking can show how successful a team is at creating quality code. Capers Jones has written about this extensively in several of his books and papers. Jones states that "defect potential and defect removal efficiency are two of the most important metrics" for teams to track. It's also quite interesting to note that bugs tend to cluster in code. For example:

- Barry Boehm found that 80% of the bugs are in 20% of the modules.[12]

- IBM found 57% of their defects in 7% of their modules.[13]

- The NSA found 95% of their bugs in 2.5% of their code.[14]

The trends will often tell you more about what is happening than just a single metric snapshot. For example, if you monitor code churn and LOC, you may find that over time, code churn starts to rise while the LOC added every sprint slowly declines. Such a case could indicate issues with the software architecture or code quality issues if, most of the time, developers are changing existing code vs. adding new code. The trend can be your greatest insight into your code.

One of my favorite metrics to track, which tells us quite a bit about the risk our code presents, is McCabe's Cyclomatic Complexity. So let's take a closer look at this metric and see what it tells us about our code and how we can utilize it to produce higher-quality software.

---

[12] Barry Boehm, *Software Engineering Economics.*

[13] *Facts and Fallacies of Software Engineering*, Robert Glass et al.

[14] Thomas Drake, Measuring Software Quality: A Case Study.

# McCabe Cyclomatic Complexity

Several years ago, I discovered a brain training application for my iPad that I absolutely loved. The application would test various aspects of the brain, such as speed, memory, agility, etc. I was convinced that my memory was a steel trap; anything entering it stayed there, and I could recall it with little to no effort. Unfortunately, from playing the games in the application, I discovered that my memory is not as good as I thought and that it could only hold 10–12 items at any given time. As hard as I might try, remembering 13 or 14 things was not always reasonable.

The memory game is a clear demonstration that despite believing we are superhuman and capable of remembering everything, we are, in fact, quite human. What does this have to do with embedded software and quality? The memory game shows that a human can easily handle and track 8–12 items at a time on average.[15] Or in software terms, developers can remember and manage approximately 8–12 linearly independent paths (branches) at any given time. Writing functions with more than ten branches means that a developer may suddenly struggle to remember everything happening in that function. This is a critical limit, and it tells developers that to minimize risk for bugs to creep into their software, the function should be limited to ten or fewer independent paths.

McCabe Cyclomatic Complexity (Cyclomatic Complexity) is a measurement that can be performed on software that measures the number of linearly independent paths within a function. Cyclomatic Complexity is a valuable measure to determine several factors about the functions in an application, which include

- The *minimum* number of test cases required to test the function

- The risk associated with modifying the function

- The likelihood that the function contains undiscovered bugs

The value assigned to the Cyclomatic Complexity represents the complexity of the function. The higher the value, the higher the risk (along with a larger number of test cases to test the function and paths through the function).

Table 6-1 provides a general rule of thumb for mapping the McCabe Cyclomatic Complexity to the functions' associated risks. Notice that the risk steadily rises from minimal risk to untestable when the Cyclomatic Complexity reaches 51! Any value over

---

[15] Miller, G. (1956). Found the "magical" number is 7 +/– 2 things in short-term memory.

50 is considered to be untestable! There is a simple explanation for why the risk increases with more complexity; humans can't track all that complexity in their minds at once, hence the higher the likelihood of a bug in the function.

***Table 6-1.*** *McCabe Cyclomatic Complexity value vs. the risk of bugs*

| Cyclomatic Complexity | Risk Evaluation |
| --- | --- |
| 1–10 | A simple function without much risk |
| 11–20 | A more complex function with moderate risk |
| 21–50 | A complex function of high risk |
| 51 and greater | An untestable function of very high risk |

**Best Practice**    The Cyclomatic Complexity for a function should be kept to ten or less to minimize complexity and the risk associated with the function.

As the Cyclomatic Complexity rises, the risk of injecting a new defect (bug) into the function while fixing an existing one or adding functionality increases. For example, look at Table 6-2. You can see that there is about a 1 in 20 chance of injecting a new bug into a function if its complexity is between one and ten. However, that risk is four times higher for a function with a complexity between eleven and twenty!

Lower Cyclomatic Complexity for a function has two effects:

1) The less complex the code is, the less likely the chance of a bug in the code.

2) The less complex the code is, the lower the risk of injecting a bug when modifying the function.

Wise developers looking to produce quality software while minimizing the amount of time they spend debugging will look to Cyclomatic Complexity measurements to help guide them. Managing complexity will decrease the number of bugs that are put into the software in the first place! The least expensive and least time-consuming bugs to fix are the ones that are prevented in the first place.

**Table 6-2.**  *The risk of bug injection as the Cyclomatic Complexity rises*[16,17]

| Cyclomatic Complexity | Risk of Bug Injection |
| --- | --- |
| 1–10 | 5% |
| 11–20 | 20% |
| 21–50 | 40% |
| 51 and greater | 60% |

What happens when a developer discovers that they have a high Cyclomatic Complexity for one of their functions? Let's look at the options. First, a developer should determine the function's actual risk. For example, if the Cyclomatic Complexity value is 11–14, it's higher than desired, but it may not be worth resolving if the function is readable and well documented. That range is "up to the developer" to use their best judgment.

Next, for values higher than 14, developers have a function that is trying to do too much! The clear path forward is to refactor the code. Refactoring the code, in this case, means going through the function and breaking it up into multiple functions and simplifying it. The odds are you'll find several different blocks of logic in the function that are doing related activities but don't need to be grouped into a single function. You will discover that some code can be broken up into smaller "helper" functions that make the code more readable and maintainable. If you are worried about the overhead from extra function calls, you can inline the function or tell the compiler that those functions

---

[16] McCabe, Thomas Jr. Software Quality Metrics to Identify Risk. Presentation to the Department of Homeland Security Software Assurance Working Group, 2008.

[17] (www.mccabe.com/ppt/SoftwareQualityMetricsToIdentifyRisk.ppt#36) and Laird, Linda and Brennan, M. Carol. Software Measurement and Estimation: A Practical Approach. Los Alamitos, CA: IEEE Computer Society, 2006.

should be compiled for speed. With most modern microcontrollers and toolchains today, the concern for inlining a function is generally not a concern because the compiler will inline where it makes sense to optimize execution and microcontrollers are very efficient.

# Code Analysis (Static vs. Dynamic)

Developers can use two types of code analysis to improve the quality of their software: static and dynamic analyses. Static code analysis is an analysis technique performed on the code without executing the code. Dynamic analysis, on the other hand, is an analysis technique performed on the code while the code is running.

Static analysis is often performed on the code to identify semantic and language issues that the compiler would not otherwise detect. For example, a compiler is not going to detect that you are going to overflow the stack, while a static analysis tool might be able to do so. Static analysis can also verify that code meets various industry standards and coding style guides. For example, it's not uncommon for a team to use a static code analyzer to confirm that their code meets the MISRA-C or MISRA-C++ standards. Static analysis tools can vary in price from free to tens of thousands of dollars. Typically, the more expensive the tool is, the more robust the analysis.

Dynamic code analysis is performed on the code to understand the runtime characteristics of the system. For example, a developer might use a tool like Percepio Tracealyzer[18] to collect and analyze the runtime behavior of their RTOS application. Developers can check the periodicity of their tasks, the delay time between when the task is ready to run and when it runs, CPU utilization, memory consumption, and a bunch of other exciting and valuable metrics.

In some cases, static and dynamic analyzers can give you two different pictures of the same metric. For example, if you are interested in knowing how to set your stack size, a static analysis tool can calculate how deep your stack goes during analysis. On the other hand, developers can also use a dynamic analysis tool to monitor their stack usage during runtime while running all their worst test cases. The two results should be very similar and can then help developers properly set their stack size.

---

[18] https://percepio.com/

# Achieving 100% Branch Coverage

Another great metric, although one that can be very misleading, is branch coverage. Branch coverage tells a developer whether their test cases cause every branch in their code to be executed or not. Branch coverage can be misleading because a developer sees the 100% coverage and assumes that there are no bugs! This is not the case. It means the tests cover all the branches, but the code may or may not meet the system requirements, perform the function the customer wants, or cover boundary conditions related to variable values.

The nice thing about branch coverage is that it requires developers to write their unit tests for their modules! When you have unit tests and use a tool like gcov, the tools will tell you if you haven't covered all your branches. For example, Figure 6-4 shows the results of running unit tests on a code base where you can see that not every branch was executed.

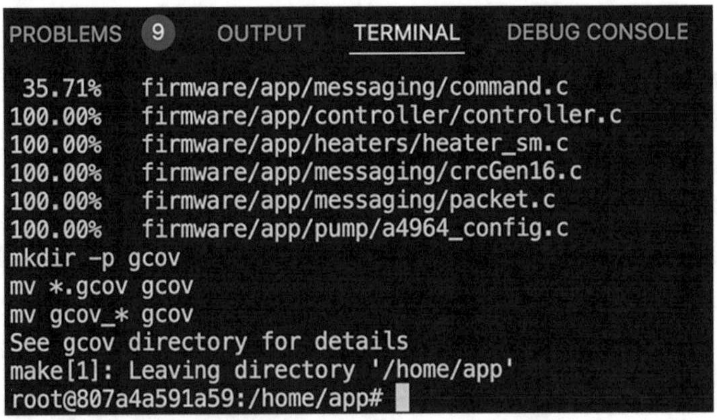

*Figure 6-4. gcov reports the coverage that unit tests reach in the code*

Notice from Figure 6-4 that most of the modules have 100% coverage, but there is one module named command.c that only has 35.71% coverage. The nice thing about this report is that I now know that command.c is only tested 35.71%. There very well could be all kinds of bugs hiding in the command.c module, but I wouldn't know because all the branches were not covered in my tests! The gcov tool will produce an analysis file with the extension gcov. We can go into the file and search ####, which will be prepended to the lines of code that were not executed during our tests. For example, Figure 6-5 shows that within command.c, there is a function named Command_Clear that was not executed! It looks like a series of function stubs that have not yet been implemented!

```
268        -:    264:*******************************************************************************/
269   #####:    265:static void Command_Clear(uint8_t const * const Data)
270        -:    266:{
271        -:    267:    // This is just to use the passed parameters to remove static analysis errors
272        -:    268:    // that we don't use the passed variable.
273        -:    269:    (void)Data;
274   #####:    270:}
```

*Figure 6-5.* gcov identifies an untested function and line of code using ####

Writing test cases that get 100% branch coverage won't remove all the bugs in your code, but it will go a long way in minimizing the bugs, making it easier to detect bugs as they are injected, and improving the quality of the code. Don't let it lure you into a false sense of security, though! 100% code coverage is not 100% bug free!

During one of my advisory/mentoring engagements with a customer, I was working with an intelligent junior engineer who disagreed with me on this topic. "If the code has 100% coverage, then the code is 100% bug free," he argued. I politely listened to his arguments and then assigned him to write a new code module with unit tests and a simulation run to show that the module met requirements. In our next meeting, he conceded that 100% test coverage does not mean 100% bug free.

## Code Reviews

When one thinks about getting all the bugs out of an embedded system, you might picture a team of engineers frantically working at their benches trying to debug the system. Unfortunately, debugging is one of the least efficient methods for getting the bugs out of a system! Single-line stepping is the worst, despite being the de facto debugging technique many embedded software developers choose. A better alternative is to perform code reviews!

A code review is a manual inspection of a code base that is designed to verify

- There are no logical or semantic errors in the software.

- The software meets the system requirements.

- The software conforms to approved style guidelines.

Code reviews can come in many forms. First, they can be large, formalized, process-heavy beasts where there are four to six people in a dimly lit room pouring over code for hours on end. These code reviews are inefficient and a waste of time, as we will soon see. Next, code reviews can be lightweight reviews where developers perform short, targeted code reviews daily or when code is available to review. These code reviews are preferable as they find the most defects in the shortest time. Finally, code reviews fall throughout the spectrum between these two. The worst code reviews are the ones that are never performed!

Several characteristics make code reviews the most efficient mechanism to remove bugs, but only if the right processes are followed. For example, SmartBear performed a study where they found examining more than 400 LOC at a time resulted in finding fewer defects! Look at Figure 6-6 from their research. The defect density, the number of defects found per thousand lines of code (KLOC), was dramatically higher when less code was reviewed. Conversely, the more code reviewed in a session resulted in a lower defect density.

Taking a moment to think through this result does make a lot of sense. When developers start to review code, their minds are probably engaged and ready to go. At first, they spot a lot of defects, but as time goes on in the review and they look at more and more code, the mind probably starts to glaze over the code, and defects start to get missed! I do a fair number of code reviews, and I know that after looking in detail at several hundred lines of code, my mind starts to get tired and needs a break!

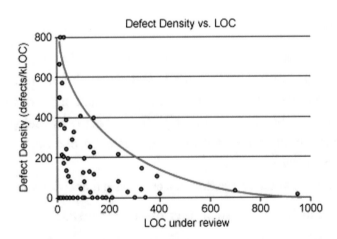

**Figure 6-6.**  *Defect density per KLOC based on the number of lines reviewed in a session*[19]

---

[19] https://smartbear.com/learn/code-review/best-practices-for-peer-code-review/

A successful code review requires developers to follow several best practices to maximize the benefit. Code review best practices include

- Review no more than 400 LOC at a time.

- Review code at a rate of less than 500 LOC per hour, although the optimal pace is 150 LOC per hour.

- Perform code reviews daily.

- Perform code reviews when code is ready to be merged.

- Avoid extensive, formalized code reviews; instead, perform smaller team or individual code reviews.

- Keep a checklist of code review items to watch for.

# Case Study – Capstone Propulsion Controller

Now that we've examined a few aspects of software quality, let's look at an example. I recently worked on a project that I found to be quite interesting. The project involved architecting and implementing the software for a propulsion controller used on NASA's Capstone mission. Capstone is a microwave oven–sized CubeSat weighing just 55 pounds that will serve as the first spacecraft to test a unique, elliptical lunar orbit as part of the Cislunar Autonomous Positioning System Technology Operations and Navigation Experiment (CAPSTONE)[20] mission. I worked with my customer, Stellar Exploration,[21] to help them design, build, and deploy the embedded software for their propulsion controller in addition to advising and mentoring their team.

From the 30,000-foot view, a propulsion system consists of a controller that receives commands and transmits telemetry back to the flight computer. The controller oversees running a pump that moves fuel through the fuel lines to valves that can be commanded to an open or closed position. (It's much more complicated than this, but this level of detail serves our purpose.) If the valve is closed, the fuel can't flow to the combustion chamber, and no thrust is produced. However, if the valve is opened, fuel can flow to interact with the catalyst and produce thrust. A simplified system diagram can be seen in Figure 6-7.

---

[20] www.nasa.gov/directorates/spacetech/small_spacecraft/capstone
[21] www.stellar-exploration.com/

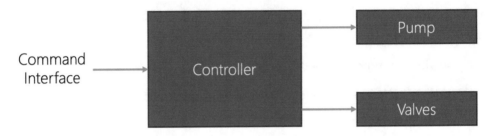

***Figure 6-7.***  *A simplified block diagram of a propulsion system*

Even though the controller seems simple, it consists of ~113,000 LOC. That's a lot of code for a little embedded controller! If we were to take a closer look, we'd discover why you can't just blindly accept a metric like LOC! Figure 6-8 shows the breakdown of the code that was measured using Understand. Notice that there is only 24,603 executable LOC, a reasonable number for the complexity of the control system.

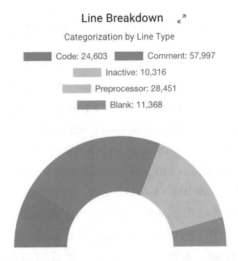

***Figure 6-8.***  *Capstone propulsion controller code breakdown for the core firmware*

The LOC breakdown can help us gain a little insight into the code without looking at a single LOC! For example, we can see that the code base is most likely relatively well documented. We can see that there are ~60,000 LOC that are comments.

One metric I like to track on projects that some people would argue about is the comment-to-code ratio. Many argue that you don't need code comments because the code should be self-documenting. I agree, but I've also found that LOC does not explain how something should work, why it was chosen to be done that way, and so forth. For this reason, I prefer to see a comment-to-code ratio greater than 1.0. For this code base, you can see that the ratio is ~2.36 comments per LOC. We don't know the quality of the comments, but at least there are comments!

---

**Best Practice**    Comment code with your design decisions, intended behavior, and so forth. Your future self and colleagues will be thankful later!

---

Someone being picky might look at Figure 6-8 and ask why there are ~10,000 lines of inactive code. The answer is quite simple. A real-time operating system was used in the project! An RTOS is often very configurable, and that configuration usually uses conditional preprocessor statements to enable and disable functionality. Another contributor to the inactive lines could be any middleware or vendor-supplied drivers that rely on configuration. Configuration tools often generate more code than is needed because they are trying to cover every possible case for how the code might be used. Just because there are inactive lines doesn't mean there is a quality issue. It may just point to a very configurable and reusable code base!

One metric we've been looking at in this chapter is Cyclomatic Complexity. So how does the propulsion controller code base look from a Cyclomatic Complexity standpoint? Figure 6-9 shows a heat map of the maximum Cyclomatic Complexity by file. The heat map is set so that the minimum is 0 and the maximum is 10. Files in green have functions well below the maximum. As functions in a file reach the maximum, they begin to turn a shade of red. Finally, when the file's function complexity exceeds the maximum, they turn red. At a quick glance, we can see what files may have functions with a complexity issue.

---

**Note**    If you can't read the filenames in Figure 6-9, that is okay! I just want you to get a feel for the technique vs. seeing the details of the code base.

---

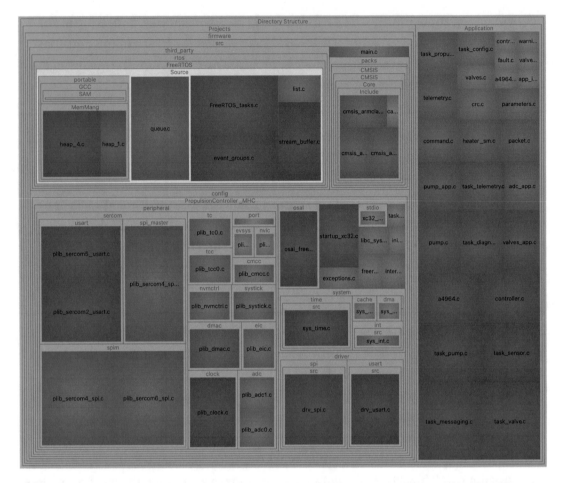

**Figure 6-9.** *An example Cyclomatic Complexity heat map to identify complex functions*

A glance at the heat map may draw a lot of attention to the red areas of the map! The attention drawn is a good thing because it immediately puts our sights on the modules that contain functions that may require additional attention! For example, five files are a deep red! Don't forget; this does not mean that the quality of these files or the functions they contain is poor; it just means that the risk of a defect in these functions may be slightly higher, and the risk of injecting a defect when changing these functions is elevated.

We can investigate the details of the functions that are catching our attention in several ways. One method is to generate another heat map, but, this time, generate the heat map on a function basis rather than a file basis. The new heat map would look something like Figure 6-10.

172

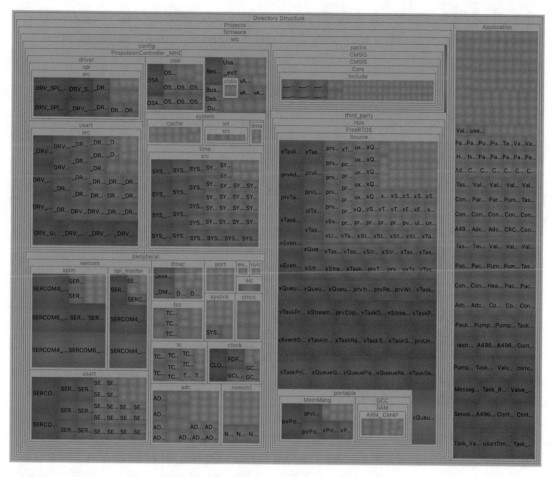

**Figure 6-10.** *A Cyclomatic Complexity heat map displayed by function*

We can see from the new heat map which functions are causing the high complexity coloring. Two areas are causing the high complexity readings. First, several serial interrupts have a Cyclomatic Complexity of 17. These serial interrupt handlers are microcontroller vendor–supplied handlers. The interrupts are generated automatically by the toolchain that was used. The second area where the complexity is higher than ten is in the RTOS! The RTOS has high complexity in the functions related to queues and task priority settings.

I think it is essential at this point to bring the reader's attention to something important. The heat maps appear to show a lot of red, which catches a human's eye. However, out of the 725 functions that make up the software, less than 15 functions have a complexity more significant than 10! Examine the breakdown shown in Figure 6-11. You'll see that there are 13 in total with not a single function with a complexity more significant than 17.

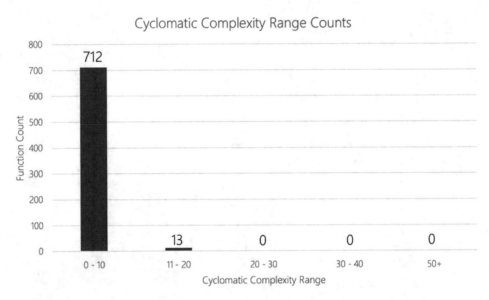

**Figure 6-11.** *The number of functions per Cyclomatic Complexity range for the propulsion controller software*

The controller software was written to minimize complexity and followed at least the quality techniques we've discussed so far in this chapter. Without context, it can be easy to look at the complexity and say, "Yea, so it should be that way!". I've been blessed with the opportunity to analyze a lot of embedded software in my career. An average representation of what I typically see can be seen in Table 6-3. This table averages projects in the same size range for a total number of functions.

**Table 6-3.** *An average project representation for several projects with their average function Cyclomatic Complexity ranges*

| Project | Function Total | 1 - 10 | 11 - 20 | 21 - 50 | 51+ |
|---------|----------------|--------|---------|---------|-----|
| A | 530 | 494 | 26 | 8 | 2 |
| B | 1058 | 1016 | 31 | 7 | 4 |
| C | 3684 | 3378 | 205 | 72 | 29 |
| D | 4190 | 4086 | 61 | 32 | 11 |
| E | 4973 | 4884 | 53 | 28 | 8 |

When I first put this table together, I was expecting to see the worst results in the largest projects; however, I would have been wrong. The projects that I've analyzed to date seem to suggest that mid-sized projects struggle the most with function complexity. All it takes though is one poor project to skew the results. In any event, notice on average even the smallest projects can end up with functions that are untestable and may hold significant risk for bugs.

High complexity in third-party components brings a development team to an interesting decision that must be made: Do you live with the complexity, or do you rewrite those functions? First, I wouldn't want to modify an RTOS kernel that I had not written. To me, that carries a much higher risk than just living with a few functions with higher complexity. After all, if they have been tested, who cares if the risk of injecting a bug is high because we will not modify it! If we are concerned with the functions in questions, we can also avoid using them the best we can.

The serial interrupt handlers also present themselves as an interesting conundrum. The complexity in these handlers is higher because the drivers are designed to be configurable and handle many cases. They are generic handlers. The decision is not much different. Do you go back and rewrite the drivers and handlers, or do you test them to death? The final decision often comes down to these factors:

- What is the cost of each?

- Is there time and bandwidth available?

- Can testing reduce the risk?

In this case, after examining the complex functions, it would be more efficient and less risky to perform additional testing and, when the time was available on future programs, to go back and improve the serial drivers and handlers. After all, they performed their function and purpose to the requirements! Why rewrite them and add additional risk just to appease a metric?

---

**Best Practice**   Never add cost, complexity, or risk to a project just to meet some arbitrary metric.

---

So far, I've just walked you through the highest level of analysis performed regularly as the controller software was developed. I realize now that I would probably need several chapters to give all the details. However, I hope it has helped to provide you with some ideas about how to think about your processes, metrics, and where your defects may be hiding. The following are a few additional bullet points that give you some idea of what else was done when developing the controller software:

- Consistent architectural analysis and evolution tracking.

- Unit tests were developed to test the building blocks with Cyclomatic Complexity being used to ensure there were enough tests.

- Regression testing using automated methodologies.

- Algorithm simulations to verify expected behavior with comparisons to the implemented algorithms.

- Runtime performance measurements using Percepio Tracealyzer to monitor task deadlines, CPU utilization, latency, and other characteristics.

- Metrics for architecture coupling and cohesion.

- System-level testing (lots and lots and lots of it).

- Test code coverage monitoring.

- Code reviews.

A lot of sweet and effort goes into building a quality software system. Even with all the effort, it doesn't mean that it is bug or defect free. Situations can come up that are unexpected or expected and reproducible like a single event upset or a cosmic ray corrupting memory. With the right understanding of what quality means, and backing it up with processes, metrics, and testing, you can decrease the risk that an adverse event will occur. Systems can be complex, and unfortunately there are never guarantees.

(As of this writing, Capstone is successfully on its path to orbit the Moon and has successfully completed several burns as scheduled. If all goes according to plan, it will enter its final orbit in November 2022.)

# Final Thoughts

Defining software quality for you and your team is critical. We've explored several definitions for quality and explored the metrics and processes that are often involved in moving toward higher-quality development. Unfortunately, we've just been able to scratch the surface; however, the details we've discussed in this chapter can help put you and your team on a firm footing to begin developing quality embedded software. In the rest of this book, we will continue to explore the modern processes and techniques that will help you be successful.

---

## ACTION ITEMS

To put this chapter's concepts into action, here are a few activities the reader can perform to start improving the quality of their embedded software:

- How do you define software quality? First, take some time to define what software quality is to you and your team and what level is acceptable for the products you produce.

- What characteristics tell you that you have a quality architecture? Take some time this week to analyze your software architecture. What areas of improvement do you find?

- Improve your metrics tracking:

  - Make a list of code metrics that should be tracked.

  - Note which ones you currently track and which ones you don't.

  - Add another note for the period at which those metrics should be reviewed.

  - What changes can you make to your processes to make sure you track the metrics and the trends they produce? (Do you need a checklist, a check-in process, etc.?)

- How complex is your code? Install a free Cyclomatic Complexity checker like pmccabe and analyze your code base. Are there functions with a complexity greater than ten that you need to investigate for potential risk?

- What coding standards do you use? Are there any new standards you should start to follow? What can you do to improve standard automatic checking in your development processes?

- Code reviews are the most effective method to remove defects from software. How often are you performing your code reviews? Are you following best practices? What changes can you make to your code review process to eliminate more defects faster?

- What processes are you missing that could help improve your software quality? Are the processes missing or is it the discipline to follow the processes? What steps are you going to take today and in the future to improve your quality?

# CHAPTER 7

# Embedded DevOps

If you've been paying any attention to the software industry over the last several years, you know that DevOps has been a big industry buzzword. DevOps is defined as a software engineering methodology that aims to integrate the work of software development and software operations teams by facilitating a culture of collaboration and shared responsibility.[1] In addition, DevOps seeks to increase the efficiency, speed, and security of the software development and delivery process. In other words, DevOps promises the same dream developers have been chasing for more than half a century: better software, faster. DevOps encompasses a lot of agile philosophy and rolls it up into a few actionable methodologies, although if you read the highlights, it'll give you that warm and fuzzy feeling with few obvious actionable items.

DevOps processes for several years have primarily been implemented in mobile and cloud applications that are "pure applications" with few to any hardware dependencies. Software giants like Google, Facebook, and Amazon have used these philosophies for several years[2] successfully. DevOps success by the tech giants has caused interest in it to filter down to smaller businesses.

In general, embedded software teams have often overlooked DevOps, choosing to use more traditional development approaches due to the hardware dependencies. However, the interest has been growing, and it's not uncommon to hear teams talk about "doing DevOps." From an embedded perspective, delivering embedded software faster is too tempting to ignore. Many teams are beginning to investigate how they can leverage DevOps to improve their development, quality assurance, and deployment processes.

---

[1] https://about.gitlab.com/topics/devops/
[2] https://bit.ly/3aZJkXW

J. Beningo, *Embedded Software Design*, https://doi.org/10.1007/978-1-4842-8279-3_7

This chapter will look at how embedded software teams can modernize their development and delivery processes. We will explore Embedded DevOps, applying modern DevOps processes to embedded software. In doing so, you'll learn what the Embedded DevOps processes entail and how you might be able to leverage them to develop and deploy a higher-quality software product.

# A DevOps Overview

DevOps is all about improving the efficiency, speed, and quality of software that is delivered to the customer. To successfully accomplish such a feat requires the involvement of three key teams: developers, operations, and quality assurance.

Developers are the team that architects, designs, and constructs the software. Developers are responsible for creating the software and creating the first level of tests such as unit tests, integration tests, etc. I suspect that most of you reading this book fall into this category.

Operations is the team that is responsible for managing the effectiveness and efficiency of the business. Where DevOps is concerned, these are the team members that are managing the developers, the QA team, schedules, and even the flow of customer feedback to the appropriate teams. Operations often also encompasses IT, the members who are handling the cloud infrastructure, framework managements, and so forth.

Quality assurance (QA) is the team that is responsible for making sure that the software product meets the quality needs of the customer. They are often involved in testing the system, checking requirements, and identifying bugs. In some instances, they may even be involved in security testing. The QA team feeds information back to the developers.

Traditionally, each team works in their own silo, separate from the other teams. Unfortunately, when having separate teams it's not uncommon for competition to get involved and an "us vs. them" mentality to arise. Mentalities like this are counterproductive and decrease effectiveness and efficiency. The power of DevOps comes from the intersection, the cooperation, and the communication between these three teams. DevOps becomes the glue that binds these teams together in a collaborative and effective manner as shown in Figure 7-1.

***Figure 7-1.*** *The three teams involved in successfully delivering a software product collaborate and communicate through DevOps processes*

When teams discuss DevOps, they often immediately jump to conversations about continuous integration and continuous deployment (CI/CD). While CI/CD is an essential tool in DevOps, it does not define what DevOps is on its own, even though it encompasses many DevOps principles and provides one path for collaboration between the various teams. Four principles guide all DevOps processes:

1. Focus on providing incremental value to the users or customers in small and frequent iterations.

2. Improve collaboration and communication between development and operations teams. (These improvements can also be within a single team and raise awareness about the software product itself.)

3. Automate as much of the software development life cycle as possible.

4. Continuously improve the software product.

Do these principles sound familiar? If you are at all familiar with agile methods, then it should! The four primary values associated with Agile from the Manifesto for Agile Software Development[3] include

1. **Individuals and interactions** over processes and tools

2. **Working software** over comprehensive documentation

---

[3] https://agilemanifesto.org/

3.  **Customer collaboration** over contract negotiation

4.  **Responding to change** over following a plan

DevOps is the new buzzword that simply covers the modern-day practices of Agile. I'd go one step further to say that it hyperfocuses on specific values and principles from the agile movement particularly where "automation" and "continuous" are involved. In case you are not familiar with agile principles, I've included them in Appendix B. Note that principles 1, 3, 4, 8, 9, and 12 sound exactly like the principles that DevOps has adopted, just in a different written form.

# DevOps Principles in Action

DevOps principles are often summarized using a visualization that shows the interaction between the company and the customer as shown in Figure 7-2. The relationships contain two key points. First, a delivery pipeline is used to deploy software with new features to the customer. Next, a feedback pipeline is used to provide customer feedback to the company which can then in turn be used to create actions that feed the delivery pipeline. In theory, you can end up with an infinite loop of new software and feedback.

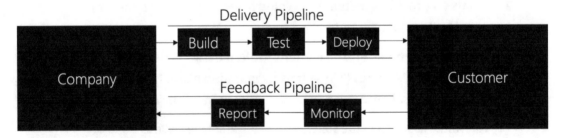

***Figure 7-2.*** *A high-level overview of the DevOps process*

The delivery pipeline is where most developers focus their attention. The delivery pipeline provides a mechanism for developers to build, test, and deploy their software. The delivery pipeline is also where the QA team can build in their automated testing, perform code reviews, and approve the software before it is delivered to the customer. When people discuss DevOps, in my experience, most of the focus isn't on customer relations or "facilitating a culture of collaboration and shared responsibility" but on continuous integration and continuous deployment (CI/CD) processes that are used to automate the delivery process. We'll talk more about what this is and entails later in the chapter.

There are at least five parts to every successful delivery pipeline, which include

- Source control management through a tool like Git

- A CI/CD framework for managing pipelines and jobs

- Build automation tools such as Docker containers, compilers, metric analyzers, etc.

- Code testing and analysis frameworks

- Review, approval, and deployment tools

There are certainly several ways to architect a delivery pipeline through CI/CD tools, but the five points I've mentioned tend to be the most common. We'll discuss this later in the chapter as well in the CI/CD section.

So far, I've found that when embedded teams focus on DevOps processes, they jump in and highly focus on the delivery pipeline. However, the feedback pipeline can be just as crucial to a team. The feedback pipeline helps the development team understand how their devices are operating in the field and whether they are meeting the customers' expectations or not. The feedback from the devices and the customers can drive future development efforts, providing a feedback loop that quickly helps the product reach its full value proposition.

While the user feedback from the feedback pipeline can sometimes be automated, it is often a manual process to pull feedback from the user. However, in many cases, the engineering team may be interested in how the device performs if there are any crashes, bugs, memory issues, etc. Getting technical engineering feedback can be automated in the deployed code! Developers can use tools like Percepio DevAlert[4] or the tools from MemFault[5] to monitor and generate reports about how the device is performing and if it is encountering any bugs.

---

**Note**    When using device monitoring tools, you must account for your system's extra memory and processor overhead. In general, this is usually less than 1–2%. Well worth the cost!

---

[4] https://percepio.com/devalert/
[5] https://memfault.com/

Now that we have a general, fundamental understanding of what DevOps is supposed to be, let's take a look at what the DevOps delivery pipeline process might look like.

# Embedded DevOps Delivery Pipeline

It's critical that as part of Embedded DevOps, we define processes that we can use to achieve the goals and principles that DevOps offer. After all, these principles should provide us with actionable behaviors that can improve the condition of our customers and our team. Before we dive into details about tools and CI/CD, we should have a process in place that shows how we will move our software from the developer to QA and through operations to the customer. Figure 7-3 shows a process that we are going to be discussing in detail and for now consider our Embedded DevOps process.

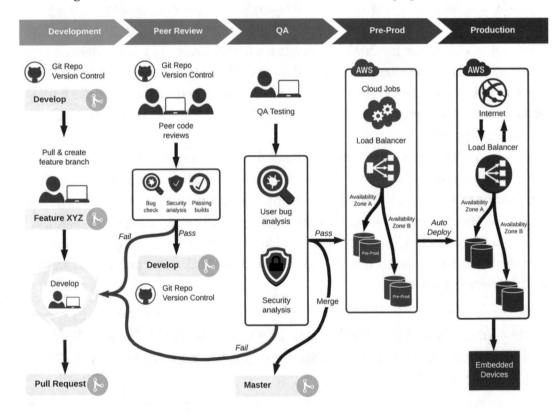

**Figure 7-3.** *An example Embedded DevOps delivery pipeline process[6]*

---

[6] The process was adapted from an existing Lucidchart template from lucidchart.com.

The Embedded DevOps process is broken up into five steps which include

- Development

- Peer review

- QA

- Preproduction

- Production

Each step assists the business in coordinating collaboration between the developers, QA, and operations. The process can also be done in quick intervals that result in reaching production several times per day, or at a slower pace if it makes more sense for the business. Let's discuss this process in a bit more detail.

First, development is where we are creating new features and fixing product bugs. The development team leverages a Git repo that manages the software version. Typically, we save the master branch for code that is nearing an official release. Official versions are tagged. We use the Develop branch as our primary integration branch. In development, we create a new feature branch off the Develop branch. Once a feature is completed by the developer, they create a pull request to kick off the peer review process.

The peer review process occurs on the feature branch prior to merging the feature into Develop, the integration branch. Peer reviews should cover several things to ensure that the feature is ready to merge into the Develop branch including

- Verifying the feature meets requirements

- Performing a code review

- Testing and code analysis

- A security analysis (assuming security is important)

- Verifying the feature builds without warnings and errors

Once the team is convinced that the feature is ready and meets the requirements for use and quality, a merge request can be created to merge the new feature into the Develop branch. If any problems are discovered, the development team is notified. The development team will then take the feedback and adjust. Eventually, a pull request is created, and the process repeats until successful.

Once a new feature is ready, a new version might be immediately released, or a release may not take place for an indefinite period of time while additional features are developed. When a new version is going to be released, the QA team will get involved to perform their testing. Typically, the QA team is testing at a system level to ensure that all the desired features work as expected and that the system meets the customer requirements. The QA team will search for bugs and defects in addition to potentially performing additional security testing.

If the QA team discovers any bugs or problems, the discovery is provided to the development team. The development team can then schedule fixes for the bugs and the process resumes with a pull request. If the software passes the QA testing, then the new features will be merged into the master branch, tagged and versioned. The new software then heads to preproduction where the new version is prepared to be pushed out to device fleet.

Keep in mind that the preproduction and production processes will vary dramatically depending on your product. In our process, I've assumed that we are working with an Internet-connected device. If our devices instead were stand-alone devices, we would not have a cloud service provider and jobs that push out the new firmware. Instead, we might have a process that emails updated versions to the customer or just posts it to a website for customers to come and find.

DevOps folks get really excited about automating everything, but just because people get excited about the buzz doesn't mean that it makes business sense for you to jump on board as well. It is perfectly acceptable, and often less complex and costly, to delegate downloading the latest software to the end customer. It's not as sexy, but in all honesty we are going for functional.

The big piece of the puzzle that we are currently missing is how to implement this process that we have been discussing. From a high-level standpoint, this looks great, but what are the mechanisms, tools, and frameworks used to build out the delivery pipeline in Embedded DevOps? To answer this question, we need to discuss continuous integration and continuous deployment (CI/CD).

# CI/CD for Embedded Systems

CI/CD is a process of automating building, testing, analyzing, and deploying software to end users. It's the tool that is used by developers, QA, and operations to improve collaboration between the various teams. CI/CD can provide development teams with a variety of value, such as

- Improving software quality

- Decreasing the time spent debugging

- Decreasing project costs

- Enhancing the ability to meet deadlines

- Simplifying the software deployment process

While CI/CD offers many benefits to embedded developers and teams, like any endeavor, there is an up-front cost in time and budget before the full benefits of CI/CD are achieved. In addition, teams will go through the learning curve to understand what tools and processes work best for them, and there will be growing pains. In my experience, while the extra up-front effort can seem painful or appear to slow things down, the effort is well worth it and has a great return on investment.

Implementing a CI/CD solution requires developers to use a CI/CD framework of one sort or another. There are several tools out there that can be used to facilitate the process. The tools that I most commonly see used are Jenkins and GitLab. No matter which CI/CD tool you pick, there is still a commonality to what is necessary to get them up and running. For example, nearly every CI/CD tool has the following components:

- Git repo integration

- Docker image(s) to host building, analyzing, testing, etc.

- Pipeline/job manager to facilitate building, analyzing, testing, etc.

- A job runner

A high-level overview of the components and their interactions can be seen in Figure 7-4. Let's walk through the CI/CD system to understand better how CI/CD works. First, the team has a Git repository that contains the product's source code. Next, developers branch the repositories' source code to develop a feature or fix a bug locally on their development machine. In the local environment, the developer may use tools like Docker for a containerized build environment along with testing and debugging tools. A container is a standard unit of software that packages up code and all its dependencies, so the application runs quickly and reliably from one computing environment to another.[7] Containers make it very easy to pass development environments around and ensure that each team member is working with the same libraries and toolchains. Once the feature is ready, the developer will commit changes to their branch in the repository.

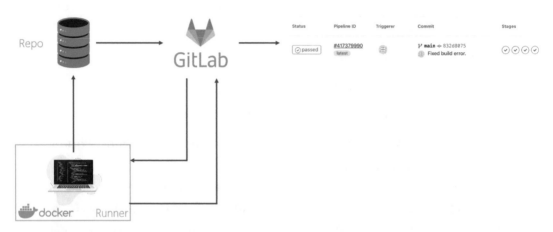

**Figure 7-4.** *An overview of the components necessary in a CI/CD system for embedded systems (Source: Beningo Embedded Group)*

The Git repository is linked to a CI/CD framework like Jenkins or GitLab. There are quite a few CI/CD frameworks available. I like Jenkins because it is the most widely used open source tool. GitLab is a commercial tool but offers a lot of great integrations and is free for small teams up to a certain number of "server minutes" per month. I've found that commercial tools are often easier to get up and running, so if you are new to CI/CD, I recommend using something like GitLab to get your feet wet.

The CI/CD framework will have a defined pipeline containing a series of automated steps performed whenever code is committed. Usually, teams will configure their CI/CD framework to run when any code on any branch is committed. The process is not just run on the master branch. Allowing the CI/CD process to run on any branch helps developers get feedback on their commits and features before merging them into an established and functional branch.

---

[7] `www.docker.com/resources/what-container/`

**Best Practice**   In 2010, Vincent Driessen wrote a blog post entitled "A successful Git branching model." The model has become very popular and is worth checking out if you need an excellent Git process.

The job that runs in the pipeline first will ultimately depend upon how the pipeline was designed and configured. However, building the application image is usually the first job. Building the application image requires having all the tools necessary to build the application successfully. In general, CI/CD tools don't have the capacity to support building. This is where Docker comes in!

Docker is an open source platform for building, deploying, and managing containerized applications.[8] Docker allows a development team to install all their development tools and code into a controlled container that acts like a mini virtual machine. Docker manages resources efficiently, is lighter weight than virtual machines, and can be used to scale applications and/or build CI/CD pipelines.

The CI/CD tool can use a Docker image with the build toolchain already set up to verify that the code can build successfully. To build successfully, the CI/CD tool often needs a runner set up that can dispatch commands to build and run the Docker image. The runner is itself a containerized application that can communicate with the CI/CD framework. The runner is responsible for building the CI/CD pipeline images and running the various jobs associated with it. For small teams, the runner may exist on a single developer's laptop, but it's highly recommended that the runner be installed on a server either on-premises or in the cloud. This removes the dependency of the developer's laptop being on when someone needs to commit code.

When the runner has completed the pipeline stage, it will report the results to the CI/CD tool. The CI/CD tool may dispatch further commands with multiple pipeline stages. Once completed, the CI/CD tool will report the results of running the pipeline. If everything goes well, then the status will be set to pass. If something did not complete successfully, it would be set to fail. If the pipeline failed, developers could go in and review which job in the pipeline did not complete successfully. They are then responsible for fixing it.

---

[8] www.ibm.com/ae-en/cloud/learn/docker

# Designing a CI/CD Pipeline

Designing a CI/CD pipeline often requires careful coordination between the developers, QA, and operations. A CI/CD pipeline is just a collection of executed jobs. However, these jobs specify essential aspects of the software application build, analysis, and deployment process. Each team is often responsible for managing a piece of the CI/CD pipeline. For example, developers manage the build jobs, unit tests, and basic analysis. QA manages the detailed system-level testing and analysis. Operations manages the server, frameworks, tools, and the deployment process.

The end goal is to verify and test the software in an automated fashion before releasing the software into the wild. The jobs executed in an Embedded DevOps process are completely product and company specific. Teams looking to get started with CI/CD often get overwhelmed because they look at the end goal, which can often look quite complicated. The best way to get started with CI/CD pipelines is to start simple and build additional capabilities into your process only as the previous capabilities have been mastered.

A simple, getting started pipeline will often perform the following jobs:

- Build the image(s)

- Statically analyze the source code

- Perform unit tests

- Deploy the application

When implemented in a tool like GitLab, the pipeline dashboard will look like Figure 7-5. You can see that the pipeline is broken up into three main pieces: a build stage, a test stage, and a deploy stage. I recommend not concerning yourself with a deployment job during a first deployment. Instead, start simple and then build additional capabilities into your pipeline.

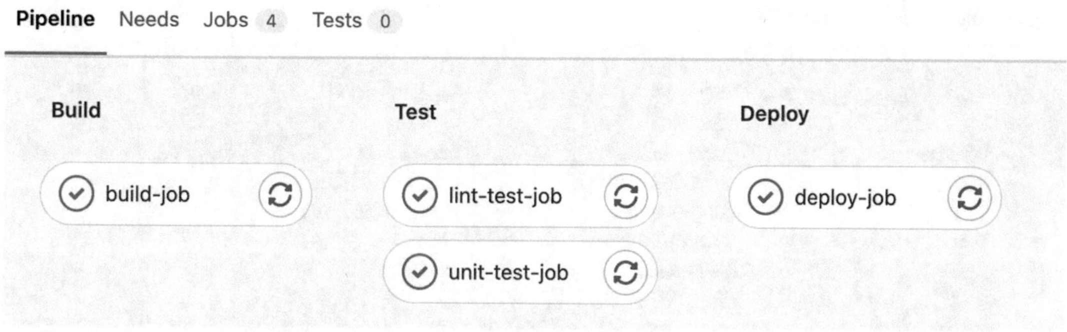

*Figure 7-5.* A CI/CD pipeline for first-time DevOps developers will contain four jobs for building, testing, and deploying the application

In general, before you start to implement a pipeline for your product, you will want to design what your pipeline looks like. While the four-job getting started pipeline is excellent, you'll most likely want to design what your final pipeline would look like before getting started. The goal would be to implement one job at a time as needed until the entire pipeline is built.

For embedded software developers, I often recommend a four-stage pipeline that includes

- Build

- Analyze

- Test

- Deploy

Each stage covers an essential aspect of the software development life cycle and can be seen in Figure 7-6.

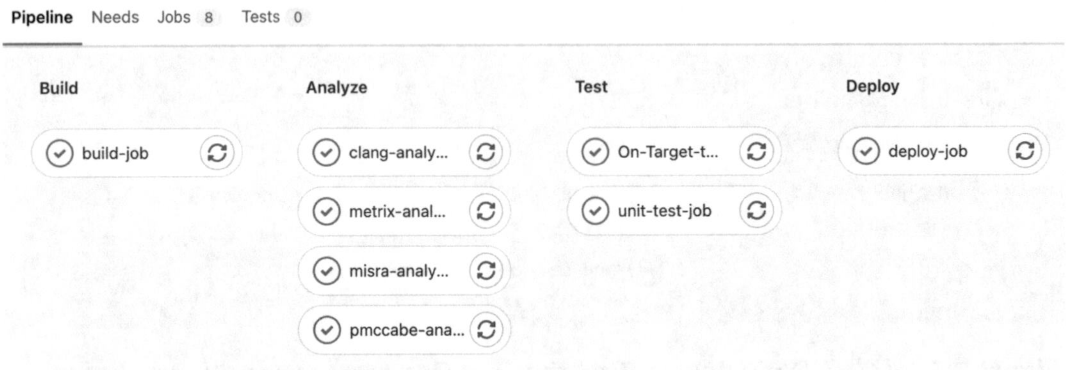

Pipeline   Needs   Jobs 8   Tests 0

| Build | Analyze | Test | Deploy |
|---|---|---|---|
| ✓ build-job ⟳ | ✓ clang-analy... ⟳ | ✓ On-Target-t... ⟳ | ✓ deploy-job ⟳ |
| | ✓ metrix-anal... ⟳ | ✓ unit-test-job ⟳ | |
| | ✓ misra-analy... ⟳ | | |
| | ✓ pmccabe-ana... ⟳ | | |

*Figure 7-6.* *A more advanced embedded software pipeline adds code analysis and on-target testing*

The first stage builds the application image(s). I like to have the first stage build the software because if the code won't build, there is no need to run the other jobs. Developers first need to commit code that fully compiles without errors and warnings. A highly recommended industry best practice is to only commit code when it builds successfully with zero warnings. If there are warnings, then it means there is something in the code that the compiler is making an assumption about and the resulting code may not behave as we expect it to. I can't tell you how often I see code compile with warnings and developers just ignore it! A successful build is zero errors *and* zero warnings. When this is done successfully, then we can move forward to analyzing the code base. If the build completes with errors, the pipeline will halt.

---

**Tip**    Add the compiler option -Werror to turn all warnings into errors. You'll be less tempted to commit code with a warning.

---

The second stage analyzes the application code. It's common knowledge that whatever gets measured gets managed. The analysis phase is the perfect place to perform static code analysis through tools like clang, gather metrics through tools like Metrix++, and verify that the code meets the desired coding standards. The CI/CD pipeline can be configured to generate the results in a report that developers can then use to verify the code meets their quality expectations.

The third stage is the testing stage. There are at least two types of testing that we want to automate. First, we want to rerun our unit tests and ensure they all pass. Running unit tests will often involve running a tool like CppUTest along with gcov to check our test coverage. The second type of testing that we want to run is on-target testing. On-target testing can involve running unit and integration tests on the target, or it could be a system-level test where the latest build is deployed to the target and external tools are used to put the product through its paces.

Finally, the last stage deploys the embedded software to the customers. The deploy stage can be fully automated so that when the testing completes successfully, it is automagically pushed to the customer or such that it requires manual approval. How the final deployment is performed will dramatically vary based on the team, product, and comfort level with automated deployment.

Usually, the CI/CD pipeline is executed on code branches without the ability to deploy the software. When a branch successfully passes the tests, the code will go through a final inspection. If everything goes well, then a request will be made to merge the feature into the mainline. When the merge occurs, the CI/CD process will run again except this time the deploy job can be enabled to allow the new version to be pushed to the customer.

CI/CD provides a lot of value to embedded software teams and holds a lot of potential for the future. At first, the process can seem complicated, especially since embedded systems often require custom hardware. However, once the process is set up, it can help identify issues early, which saves time and money in the long run. In addition, the process can help keep developers accountable for the code they write, especially the level of quality that they write. A good CI/CD process starts with designing what capabilities your pipeline needs and then incrementally adding those capabilities to the pipeline.

# Creating Your First CI/CD Pipeline

Talking about all these topics is great, but learning the high-level theory and what the potential benefits can be is one thing. Actually creating a pipeline successfully is something totally different. I've put together a simple tutorial in Appendix C that you can use to gain some hands-on experience creating a CI/CD pipeline for an STM32 microcontroller using GitLab. The tutorial covers a few basics to get you started. You may find though that you need additional assistance to design and build a CI/CD system for your team. If you run into issues, feel free to reach out to me and we can discuss how best to get you and your team up and running.

# Is DevOps Right for You?

If you ask anyone who has bought into the DevOps buzz, the answer will be that you can't create a quality software product without DevOps. It's a resounding YES! Like most things in life though, the real answer may be more in the gray.

Embedded DevOps, at least parts of the processes and principles, can benefit a lot of embedded teams both big and small. However, you do need to perform an analysis on the return on investment and the risk. Investing in DevOps does not mean you are going to achieve all the benefits. In fact, I've heard stories through the grape vine of teams that tried to implement DevOps only to fail miserably.

I have had success internal to Beningo Embedded Group and with several of my clients implementing Embedded DevOps. With some clients, we only implemented processes within the development team to improve their quality and visibility into the code for the rest of the team. In others, we integrated with the QA team to carefully coordinate testing. In others, it made sense to implement the whole set of processes and tools.

In some situations, it can make little sense to continuously deploy. Do you really want to manage complex firmware updates for entire fleets of devices? Does your customer who just bought your microwave product really want new features added daily? Honestly, that's a little bit terrifying from a user's perspective.

Is DevOps right for you? It's a loaded question and only one that you or a professional (like me) can really answer. In my experience, everyone can benefit from adopting some of the principles and processes, but only a few companies benefit from going all in.

# Final Thoughts

Embedded DevOps has become a major buzzword in software circles. There are critical processes and principles included in Embedded DevOps that every team can benefit from. The trick is to not get caught up in the buzz. Selectively and carefully develop your Embedded DevOps processes. If you do it right, you can do it piecemeal and manage your investment and your return on investment to maximize benefits to both you and your customers. We've only scratched the surface of what can be done, but we will look to build on these concepts as we explore additional processes that are crucial to the success of every embedded software team.

## ACTION ITEMS

To put this chapter's concepts into action, here are a few activities the reader can perform to start improving their Embedded DevOps:

- What can you do to improve how your developers, QA, and operations teams interact with each other?

- How might you modify the Embedded DevOps process that we examined in Figure 7-3? What would it take to implement a similar process in your own team?

- Schedule time to walk through Appendix C step by step if you haven't already done so. How easy or difficult is it to get a build pipeline up and running? What challenges did you encounter that you weren't expecting?

- What CI/CD tool is right for you? Jenkins? GitLab? Something else? Investigate the capabilities of the various CI/CD tools and select which tool best fits you and your team.

- What is your dream CI/CD pipeline? (Yes, who thought you'd ever be asked that question!) Then, spend 30 minutes designing the stages and the jobs in your dream pipeline.

  - What changes/improvements do you need to make to your DevOps processes over the next three months to move toward it? Then, map out a plan and start executing it.

- Common tools that are used for communication, like Slack, can be integrated into the CI/CD process. Take some time to research how to connect your CI/CD pipeline so that the team can be notified about successful and failed commits. This can help keep you as a developer accountable.

- What can you do to further develop your feedback pipeline? While we focused mostly on the delivery pipeline, what could you do to get good feedback about your customers' experience?

# CHAPTER 8

# Testing, Verification, and Test-Driven Development

My experience from working with several dozen companies worldwide over the last 20 years has been that embedded software teams are not great at testing. Testing embedded software tends to be sporadic and ad hoc. Developers tend to spot-check the code they are writing and, once checked, assume that code is frozen and will never have a defect in it again. While I wish this were the case, the truth is that it is pretty standard for new code to break tested and verified code. In addition, many modern embedded systems have reached a complexity level where manually testing that code works is inefficient and nearly impossible without a team of engineers working around the clock.

The modern embedded software team should look to test automation to improve their software quality. Test automation integrated into a CI/CD pipeline allows developers to elevate the quality of their software, decrease development time, and test their software! There are many levels of testing and several modern processes that can be adopted, like Test-Driven Development (TDD). Unfortunately, like the promises of so many processes, the gap between achieving these results in practice vs. theory can be considerable and, in some cases, elusive.

This chapter will explore testing processes with a primary focus on TDD. First, we will see where it makes sense to use TDD and where developers should avoid it like the plague (or, perhaps, COVID?). Like with any other tool or process, we want to use it where it provides the most benefit. Once we understand TDD, I would recommend you dig deeper into Appendix C and test-drive TDD with the CI/CD example.

© Jacob Beningo 2022
J. Beningo, *Embedded Software Design*, https://doi.org/10.1007/978-1-4842-8279-3_8

# Embedded Software Testing Types

Testing should not be performed for the sake of testing. Every test and test type should have an objective to accomplish by creating the test. For example, the primary purpose of most embedded software testing is to verify that the code we have written does what it is supposed to under the conditions we expect the system to operate under. We might write tests designed to verify code behavior through an interface or write test cases to discover defects. We could even write tests to determine how well the system performs.

If you were to go and search the Internet for different types of testing that can be performed on software, you would discover that your list, at a minimum, contains several dozen test types. It is unrealistic for developers to attempt to perform every possible test on their code and system. The second goal of testing is to reduce risk. What risk? The danger is that a user will get hurt using the product. The risk is that the system will ruin the company's reputation. The risk is that the objectives of the system have not been met. We don't write tests or select test types to make management feel better, although perhaps that is the third goal. We trade off the amount and types of testing we perform to minimize the risks associated with our products.

There are several different ways to think about the types of testing that need to be performed. The first is to look at all the possible tests and break them into functional and nonfunctional testing groups. Functional tests are a type of testing that seeks to establish whether each application feature works per the software requirements.[1] Nonfunctional tests are tests designed to test aspects of the software that are not related to functional aspects of the code, such as[2]

- Performance

- Usability

- Reliability

- Scalability

---

[1] https://bit.ly/3zPKv6r

[2] www.guru99.com/non-functional-testing.html

Figure 8-1 shows some typical types of tests that one might encounter.

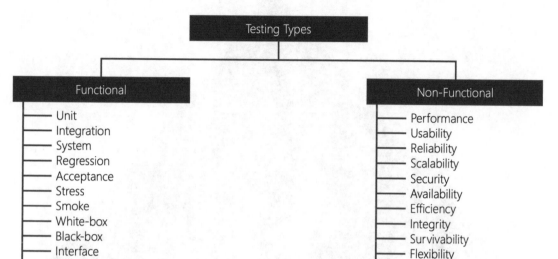

***Figure 8-1.*** *The types of testing are broken down into functional and nonfunctional tests*

I'm confident that if you take the time to investigate and perform each type of testing on your product, by the end, you'll need your own sanity testing. As you can see, the potential types of tests that can be performed are huge, and I didn't even list them all in Figure 8-1. I got tired and bored and decided it was time to move on! While each of these tests has its place, it isn't required that every test type be performed on every system.

# Testing Your Way to Success

With so many different types of testing available, it's more than enough to make any developer and team's head spin. However, all types of tests aren't required to design and build an embedded system successfully. When we start to look at the minimum testing types that are needed to ship a product successfully, you'll discover that there are, in fact, six tests that every team uses which can be seen in Figure 8-2. Let's examine each of these test types and how developers can use them.

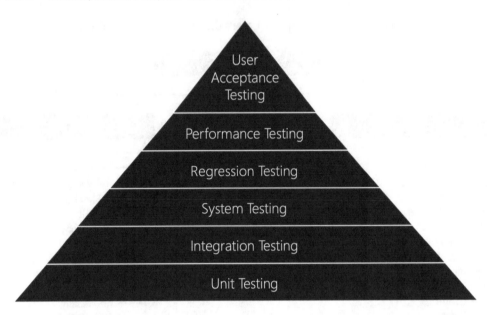

***Figure 8-2.*** *The core testing methodologies that every embedded developer should leverage*

The first test developers should use, which forms the foundation for testing your way to success, is unit testing. Unit tests are designed to verify that individual units of source code, often individual functions, or objects, are fit for use in the application.[3] Unit tests are foundational because they test the smallest possible units in an application. Given that a (well written) function implements one behavior, a unit test ensures that the function does indeed correctly implement that behavior. You can't build a bridge with shoddy bricks, and you can't build a system with unstable functions. We test every function to ensure we are building our application on a solid foundation.

Integration tests are just a single level above unit tests. Once we have proven that the individual pieces of code work, we can start to integrate the pieces to build the application; undoubtedly, as components interact with each other, there will be defects and unexpected behaviors. Integration testing is conducted to evaluate the compliance of a system or component with specified functional requirements for the application.[4] Integration testing shows that our building blocks are all working together as expected.

---

[3] https://en.wikipedia.org/wiki/Unit_testing
[4] https://en.wikipedia.org/wiki/Integration_testing

As an embedded system is developed, the end goal is to create specific features that meet the provided system requirements. To ensure those requirements are met, we create system-level tests. System testing is conducted on a complete integrated system to evaluate its compliance with specified requirements.[5] In many cases, you may find that system-level tests are not written in C/C++. Instead, system testing often requires building out hardware systems that can automatically interact with the system and measure how it behaves. Today, many of these applications are written in Python.

So far, the first three tests we've discussed are everything you need to ensure that your code is checked from its lowest levels to the system level. However, we still need several types of tests to ensure that our system meets our clients' needs and that new features are breaking existing code.

Regression tests are automated tests that rerun the functional and nonfunctional tests to ensure that previously developed and tested software still performs after the software has been modified.[6] I can't tell you how many times early in my career I would manually test a feature, see that it worked, and move on to the next feature, only to discover later I had broken a feature I had already tested. Unfortunately, in those days, I was not using automated tests, so it was often weeks or months before I discovered something was broken. I'm sure you can imagine how long it then took to debug. Too long! Regression tests should be performed often and include your unit, integration, and system-level tests.

An essential and often-overlooked test that every developer should perform on their software is performance testing. Embedded developers work on a microcontroller (or low-end general-purpose processor) in resource-constrained environments. Understanding how many clock cycles or milliseconds the code takes to run is important. You may find that you only use about 10% of your processor early in your development cycle. However, as new features are added, the utilization often goes up. Before you know it, unexpected latencies can appear in the system, degrading performance. I highly recommend that developers take the time to measure their system performance before every merge. It'll tell you if you made a big blunder and help you understand how much bandwidth the processor has left.

---

[5] https://en.wikipedia.org/wiki/System_testing
[6] https://en.wikipedia.org/wiki/Regression_testing

The last minimum test that developers should make sure they use is user acceptance tests. A user acceptance test is where the team releases their product to their intended audience. User acceptance tests can shed a lot of light on whether the users like the product and how they use it and often discover defects with the product. I always joke that there are two times when a well-tested system is bound to fail: when giving a demo to your boss and releasing the product to your customer. In these situations, Murphy's law runs rampant. If you want to have a quality system, you must also let your users beat the thing up. They will use the system and exploit flaws in your product that you would never have imagined (because you designed and wrote it!).

# What Makes a Great Test?

Edsger W. Dijkstra stated that "Testing shows the presence, not the absence of bugs." No matter how much we might try, we can never use testing to show that there are no bugs in our software. There is a limit to what our tests can tell us. Testing helps us to show that under specific circumstances the software behaves as we expect it to. If those conditions change, then all bets are off. Before we understand what makes a great test, we need to understand that tests have their limits and don't guarantee the absence of bugs or defects.

With that said, tests can in fact help us to show that a software system under specific conditions does meet the system requirements and behaves as we expect it to. Part of the problem is identifying what exactly to test. For any specific embedded system, we can define a nearly infinite number of tests. With an infinite number of tests, how do we decide what should be tested? After all, if we can create an infinite number of test cases, we'll undoubtedly bankrupt our company by never being ready to ship a product.

Every industry has their own requirements for how much and what testing should be performed. For example, the requirements for a consumer electronics product like an electronic light switch will have far less requirements than a Boeing 737 that includes safety-critical flight systems. An initial decision on what needs to be tested can be gleaned from industry standards and even federal laws. However, there are also some general rules of thumb we can follow as well.

First, we should be fully testing our interfaces with unit tests. Unit tests can be written so that they run on hardware and exercise peripherals and drivers. Unit tests can also be run on host machines that test hardware-independent software. When we test our interfaces, we should be asking ourselves several questions such as

- What are the inputs to the interface?

- What are the outputs from the interface?

- Are there errors that are reported or can be tracked?

- What ranges can be inputs and outputs be within?

When I define an interface, for every function I document three to five key features of the function's interface:

- Inputs and their expected ranges

- Outputs and their expected ranges

- Preconditions – What should happen before calling the function

- Postconditions – What should happen after calling the function

- Range of function return values (typically error codes)

As you can see, these five documented features tell any developer exactly how the function can be used and what to expect from it. With this information, I can also design test cases that verify all those inputs, outputs, and postconditions.

Next, testing should also go well beyond our own code and apply to any libraries or third-party software that we use. It's not uncommon for this code to not be well tested. Library interfaces should be carefully examined and then test cases created to test the subset of features that will be used by the application. Again, this is helping to ensure that we are building our application upon a good, quality, foundation. Unless the library is coming a commercial source that has a test harness already available, develop tests for your library code, don't just assume that they work.

There are several characteristics that make up a great test. First, a great test should be limited in scope and test only one thing at a time. If a test fails, we shouldn't have to go and figure out what part of a test failed! Tests are specific and have just a singular purpose. Next, tests need to communicate their intent directly and clearly. Each test should clearly belong to a test group and have clear name associated with it. For example, if I have a communication packet decoding module, I might have a test group called PacketDecoding with a test that verifies the CRC algorithm produces the correct results:

```
TEST(PacketDecoding, crc_correct)
```

Next, tests should also be self-documenting. Since we are writing more code to test the code we already wrote, the last thing we need to do is write more documentation. A good test will have a few comments associated with it, but read very well. For example, if I am examining a test case, I should find that I can just read the code and understand what the test is doing. When I write a test, I need to choose clear variable names and be verbose in how I name things. Where things look complicated, I should refactor the test and add a few comments.

Finally, great tests should also be automated. Manual testing is a time-consuming and expensive process. At first, manual testing may seem like a good idea, but over time, as more and more tests are performed, it becomes an activity that has a diminishing return. Teams are better off putting time in up front to develop an automated test harness that can then report the results as to whether the tests have passed or failed.

# Regression Tests to the Rescue

A major problem that often crops up with software development is that new features and code added to a project may break some existing feature. For example, I was working on a project once where we had some code that used a hardware-based CRC generator that used a direct memory access (DMA) controller to transfer data into the CRC calculator. It was a cool little device. A colleague made some changes to a seemingly unrelated area of code, but one that used an adjacent DMA channel. The CRC code was accidentally broken. Thankfully, I had several unit tests in place that, when executed as part of the regression tests, revealed that the CRC tests were no longer passing. We were then able to dig in and identify what changed recently and found the bug quickly.

If I hadn't created my unit tests and automated them, it might have been weeks or months before we discovered that the CRC code was no longer working. At that point, I wouldn't have known where the bug was and very well could have spent weeks trying to track down the issue. Using regression testing, the issue appeared nearly immediately which made fixing the problem far faster and cheaper.

Regression testing can sound scary to set up, but if you have a good test set up, then regression testing should require no effort on the development team's part. When you set up testing, make sure that you use a test harness that allows you to build test groups that can be executed at any time. Those test groups can then be executed as part of an automated build process or executed prior to a developer committing their code to a Git repository. The important thing is that the tests are executed regularly so that any code changes that inject a bug are discovered quickly.

# How to Qualify Testing

An interesting problem that many teams encounter when it comes to testing their embedded software is figuring out how to qualify their testing. How do you know that the number of tests you've created is enough? Let's explore this question.

First, at the function level, you may recall from Chapter 6 that we discussed McCabe's Cyclomatic Complexity. If you think back to that chapter, you'll remember that McCabe's Cyclomatic Complexity tells us the minimum number of test cases required to cover all the branches in a function. Let's look at an example.

Let's say that I have a system that has an onboard heater. The user can request the heater to be on or off. If the request is made to turn the heater on, then a heater timeout timer is reset and calculates the time that the timer should be turned off. The code has two parts: first, a section to manage if a new request has come in to turn the heater on or off and, second, a section to manage if the heater state should be HEATER_ON or HEATER_OFF. Listing 8-1 demonstrates what this function code might look like.

***Listing 8-1.*** An example heater state machine function that controls the state of a heater

```
HeaterState_t HeaterSm_Run(uint32_t const SystemTimeNow)
{
    // Manage state transition behavior
    if (HeaterState != HeaterStateRequested || UpdateTimer == true)
    {
        if (HeaterStateRequested == HEATER_ON)
        {
            HeaterState = HEATER_ON;
            EndTime = SystemTimeNow + HeaterOnTime;
            UpdateTimer = false;
        }
```

```
        else
        {
            EndTime = 0;
        }
    }
    else
    {
        // Do Nothing
    }

    // Manage HeaterState
    if(SystemTimeNow >= EndTime)
    {
        HeaterState = HEATER_OFF;
    }

    return HeaterState;
}
```

How many test cases do you think it will take to test Listing 8-1? I'll give you a hint; the Cyclomatic Complexity measurements for HeaterSm_Run can be seen in Listing 8-2. From the output, you can see that the Cyclomatic Complexity is five, and there are nine statements in the function. Listing 8-2 is the output from pmccabe, a Cyclomatic Complexity metric analyzer for HeaterSm_Run.

The minimum number of test cases required to test HeaterSm_Run is five. The test cases that we need are

1.  Two test cases for the first if/else pair

2.  Two test cases for the if/else pair of the HeaterStateRequested conditional

3.  One test case to decide if the heater should be turned off

***Listing 8-2.*** The Cyclomatic Complexity results for HeaterSm_Run using the pmccabe application

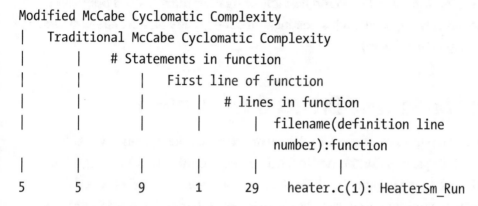

```
Modified McCabe Cyclomatic Complexity
|   Traditional McCabe Cyclomatic Complexity
|       |    # Statements in function
|       |        First line of function
|       |        |    # lines in function
|       |        |    |    filename(definition line
|       |        |    |                number):function
|       |        |    |    |           |
5       5        9    1    29      heater.c(1): HeaterSm_Run
```

Cyclomatic Complexity tells us the minimum number of test cases to cover the branches! If you look carefully at the HeaterSm_Run code, you may realize that there are several other tests that we should run such as

- A test case for the "HeaterState != HeaterStateRequested" side of the first if statement

- A test case for the "|| UpdateTimer == true" side of the first if statement

- At least one or maybe two test cases for what happens if "SystemTimeNow + CommandedOnTime" rolls over

At this point, we have five test cases just to cover the branches and then at least another three test cases to cover specific conditions that can occur in the function. The result is that we need at least eight test cases! That's just for a simple function that manages a state machine that can be in the HEATER_ON or HEATER_OFF state.

From the example, I believe you can also see now why if you have tests that get you 100% test coverage, you might still have bugs in your code. A code coverage tool will analyze whether each branch of code was executed. The number of branches is five, not eight! So, if the first five test cases to cover the branches are covered by the tests, then everything can look like you have fully tested your function. In reality, there are at least another three test cases which could have a bug hiding in them.

Now that we have a better understanding of testing, what makes a good test, and how we can make sure we are qualifying our tests, let's examine an agile process known as Test-Driven Development. I've found that following the principles and premises of TDD can dramatically improve not just testing, but code quality and the speed at which developers can deliver software.

# Introduction to Test-Driven Development

My first introduction to Test-Driven Development for embedded systems was at the Boston Embedded Systems Conference in 2013. James Grenning was teaching a four-hour course based on his popular book *Test-Driven Development for Embedded C*. (By the way, I highly recommend that every embedded developer and manager read this book.) My first impression was that TDD was a super cool technique that held a lot of promise, but I was highly skeptical about whether it could deliver the benefits that it was promising. TDD was so different from how I was developing software at the time that it felt awkward, and I shelved it for nearly half a decade. Today though, I can't imagine writing embedded software any other way.

Test-Driven Development is a technique for building software incrementally that allows the test cases to drive the production code development.[7] TDD is all about writing small test cases that are automated, fail first, and force developers to write the minimum amount of code required to make the test case pass. TDD is very different from traditional embedded software development because we typically bang out a whole bunch of code and then figure out how to test it (similar to what we did in the last section). The problem with this approach is that there are usually huge gaping holes in our test suites, and we are typically designing our tests in such a way that they may or may not catch a problem.

Test-Driven Development uses a simple, repeating process known as a microcycle to guide developers in writing their code. The TDD microcycle was first introduced in Kent Beck's *Test-Driven Development*[8] book but also appeared in James Grenning's book on page 7. A high-level overview of the TDD microcycle can be seen in Figure 8-3.

---

[7] *Test-Driven Development for Embedded C*, James Grenning, 2011, page 4.

[8] www.amazon.com/Test-Driven-Development-Kent-Beck/dp/0321146530

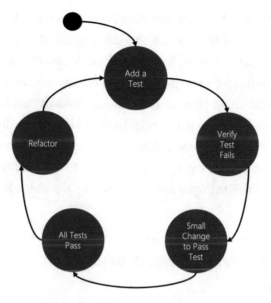

**Figure 8-3.** *The TDD microcycle*

The TDD microcycle involves five simple, repeating steps:[9]

1.  Add a small test.

2.  Run all the tests and see the new one fail. (The test might not even compile.)

3.  Make the small change(s) needed to pass the test.

4.  Run all the tests and see the new one pass.

5.  Refactor to remove duplication and improve the expressiveness of the tests.

As you can see from this simple, repeating process, developers build their software one test case at a time. They can verify that their test case fails first, which means if something happens to the resulting production code, the test case has been verified that it can detect the problem. Developers are continuously running all their tests, so if new code added to the production code breaks something, you know immediately and can fix it. Catching a defect the moment you introduce it can dramatically decrease how much time you spend debugging.

---

[9] These steps are taken nearly verbatim from *Test-Driven Development for Embedded C,* page 7.

Another exciting aspect of TDD is that developers can test a lot of code on a host environment rather than the intended embedded target. If the application is designed properly with hardware abstractions, the application functions can be executed on the host. Fake inputs and outputs can be generated in the test harness to more fully test the code than trying to duplicate the same behavior in hardware. Leveraging the host environment allows developers to work quickly, avoid long compile and deployment cycles, and continue development even if the hardware is not yet available! Now, developers do have to be careful with potential issues like endianness and other incompatibility issues with the target device. However, development can be dramatically accelerated.

---

**Definition**   Refactoring is the process of restructuring code while not changing its original functionality.[10]

---

TDD can now present additional problems, such as dealing with hardware and operating system dependencies. If you've followed the design philosophies discussed in this book, you'll discover that you can avoid quite a few of these issues. Where you can't, you can use tools to get around these issues, such as using

- Abstraction layers

- Test doubles

- Mocks

---

**Definition**   A mock allows a test case to simulate a specific scenario or sequence of events[11] that minimizes external dependencies.

---

At this point, I will warn you; I recommend being very careful about how far down the mock rabbit hole you go. TDD can dramatically improve the testability of your code, decrease debug times, and so forth, but don't get so pulled into the idea that you spend

---

[10] www.techtarget.com/searchapparchitecture/definition/refactoring
[11] *Test-Driven Development for Embedded C*, James W. Grenning, page 177.

lots of time and money trying to "do TDD perfectly." On the other hand, if you get to an area where it will be a lot of time and effort to write your tests, it might just make sense to write your test cases so that they run on your target device.

---

**Tip**    A test harness can run on a microcontroller. When hardware dependencies become too much to run on a host, run them on your target!

---

I have found that using TDD improves the software I write. I spend less time debugging. I also have found that it's fun! There is an almost addictive property to it. When you make a test case fail, write your production code, and then see your test case pass, I'm pretty sure the brain gives us a little dopamine hit. We are then encouraged to continue writing the code using a methodology that benefits our team, company, and end users.

# Setting Up a Unit Test Harness for TDD

To get started with Test-Driven Development, embedded developers need a test harness that works in the language of their choice. Typically, embedded developers are going to be using C or C++. Several test harnesses are available to developers that range from free to commercial. A few examples of test harnesses that I've worked with include

- Unity

- CppUTest

- Google Test

- Ceedling

- Parasoft C/C++ test

In general, my go-to test harness is CppUTest. Undoubtedly, I am biased, though. Everything I've learned and know about TDD and test harnesses I've learned from my interactions with James Grenning. I mention this only so that you know to try several of your own and see which harness best fits your purposes.

CppUTest is a C/C++-based testing framework used for unit testing and test driving code. In general, CppUTest has been used for testing C and C++ applications. The framework provides developers with a test harness that can execute test cases. CppUTest also offers a set of assertions that can be used to test assumptions such as

- CHECK_FALSE

- CHECK_EQUAL

- CHECK_COMPARE

- BYTES_EQUAL

- FAIL

- Etc.

If the result is incorrect, the test case is marked as having failed the test.

CppUTest provides a free, open source framework for embedded developers to build unit tests to prove out application code. With a bit of extra work, developers can even run the tests on target if they so desire. In general, I use CppUTest to test my application code that exists above the hardware abstraction layer. CppUTest allows developers to build test groups that are executed as part of the test harness. CppUTest comes with a set of assertions that can be used to verify that a test case has completed successfully or not which is then reported to the developer. CppUTest also includes hooks into gcov which can provide the status of branch coverage as well.

## Installing CppUTest

Several different installation methods can be used to set up CppUTest that can be found on the CppUTest website.[12] The first is to install it prepackaged on Linux or macOS. (If you want to install on Windows, you'll need to use Cygwin, use a container, or a similar tool.) Alternatively, a developer can clone the CppUTest git repository.[13] Cloning the repository doesn't provide any real added benefit unless you are setting up the test harness within a Docker container as part of a CI/CD pipeline. In this case, it is just a simpler way to automatically install CppUTest as part of the Docker image.

---

[12] https://cpputest.github.io/

[13] https://github.com/cpputest/cpputest.github.io

I recommend a different approach if you want to get started quickly and experiment a bit. James Grenning has put together a CppUTest starter project[14] with everything a developer needs. The starter project includes a Dockerfile that can be loaded and a simple command to install and configure the environment. If you want to follow along, clone the CppUTest starter project to a suitable location on your computer. Once you've done that, you can follow James' instructions in the README.md or follow along with the rest of this section.

Before getting too far, it's essential to make sure that you install Docker on your machine. Mac users can find instructions here.[15] Windows users can find the instructions here.[16] For Linux users, as always, nothing is easy. The installation process varies by Linux flavor, so you'll have to search a bit to find the method that works for you.

Once Docker is installed and running, a developer can use their terminal application to navigate to the root directory of the CppUTest starter project directory and then run the command:

```
docker-compose run cpputest make all
```

The first time you run the preceding command, it will take several minutes for it to run. After that, the command will download Docker images, clone and install CppUTest, and build the starter project. At this point, you would see something like Figure 8-4 in your terminal. As you can see in the figure, there was a test case failure in tests/MyFirstTest.cpp on line 23 along with an ERROR: 2 message. This means that CppUTest and James' starter project is installed and working correctly.

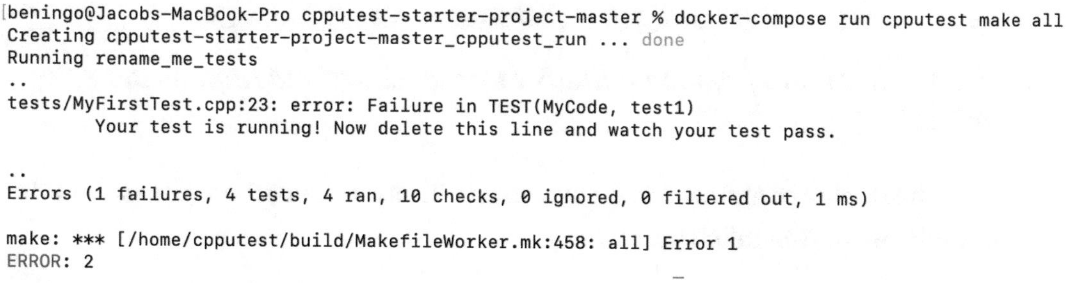

```
[beningo@Jacobs-MacBook-Pro cpputest-starter-project-master % docker-compose run cpputest make all
Creating cpputest-starter-project-master_cpputest_run ... done
Running rename_me_tests
..
tests/MyFirstTest.cpp:23: error: Failure in TEST(MyCode, test1)
        Your test is running! Now delete this line and watch your test pass.

..
Errors (1 failures, 4 tests, 4 ran, 10 checks, 0 ignored, 0 filtered out, 1 ms)

make: *** [/home/cpputest/build/MakefileWorker.mk:458: all] Error 1
ERROR: 2
```

***Figure 8-4.*** *The output from successfully building the CppUTest Docker image*

---

[14] https://github.com/jwgrenning/cpputest-starter-project
[15] https://docs.docker.com/desktop/mac/install/
[16] https://docs.docker.com/desktop/windows/install/

# Leveraging the Docker Container

The docker-compose run command causes Docker to load the CppUTest container and then make all. Once the command has been executed, it will leave the Docker container. In the previous figure, that is why we get the ERROR: 2. It's returning the error code for the exit status of the Docker container.

It isn't necessary to constantly use the "docker-compose run CppUTest make all" command. A developer can also enter the Docker container and stay there by using the following command:

```
docker-compose run --rm --entry point /bin/bash cpputest
```

By doing this, a developer can simply use the command "make" or "make all." This advantage is that it streamlines the process a bit and removes the ERROR message returned when exiting the Docker image from the original command. So, for example, if I run the Docker command and make, the output from the test harness now looks like what is shown in Figure 8-5.

```
[root@e0384cff4bf3:/home/src# make
compiling MyFirstTest.cpp
Linking rename_me_tests
Running rename_me_tests
..
tests/MyFirstTest.cpp:23: error: Failure in TEST(MyCode, test1)
        Your test is running! Now delete this line and watch your test pass.

..
Errors (1 failures, 4 tests, 4 ran, 10 checks, 0 ignored, 0 filtered out, 0 ms)

make: *** [/home/cpputest/build/MakefileWorker.mk:458: all] Error 1
```

***Figure 8-5.***  *The output from mounting the Docker image and running the test harness*

To exit the Docker container, I need to type exit. This is because I prefer to stay in the Docker container to streamline the process.

# Test Driving CppUTest

Now that we have set up the CppUTest starter project, it's easy to go in and start using the test harness. We should remove the initial failing test case before we add any tests of our own. This test case is in /tests/MyFirstTest.cpp. The file can be opened using your favorite text editor. You'll notice from the previous figure that the test failure occurs at line 23. The line contains the following:

```
FAIL("Your test is running! Now delete this line and watch your test pass.");
```

FAIL is an assertion that is built into CppUTest. So, the first thing to try is commenting out the line and then running the "make" or "make all" command. If you do that, you will see that the test harness now successfully runs without any failed test cases, as shown in Figure 8-6.

```
[root@001496277db5:/home/src# make
compiling MyFirstTest.cpp
Linking rename_me_tests
Running rename_me_tests
....
OK (4 tests, 4 ran, 9 checks, 0 ignored, 0 filtered out, 0 ms)
```

***Figure 8-6.*** *A successfully installed CppUTest harness that runs without failing test cases*

Now you can start building out your unit test cases using the assertions found in the CppUTest manual. The developer may decide to remove MyFirstTest.cpp and add their testing modules or start implementing their test cases. It's entirely up to what your end purpose is.

I do have a few tips and recommendations for you to get started with CppUTest before you dive in and start writing your own code. First, I would create separate test modules for each module in my code. For example, if I have a module named heater, I would also have a module named heaterTests. All my heater tests would be located in that one module. I will often start by copying the MyFirstTest.cpp module and then renaming it.

Next, when creating a new module, you'll notice that each test module has a TEST_ GROUP. These test groups allow you to name the tests associated with the group and also define common code that should be executed before and after each test in the group. The setup and teardown functions can be very useful in refactoring and simplifying tests so that they are readable.

You will also find that when you define a test, you must specify what group the test is part of and provide a unique name for your test. For example, Listing 8-3 shows an example test for testing the initial state of the heater state machine. The test is part of the HeaterGroup and the test is InitialState. We use the CHECK_EQUAL macro to verify that the return state from HeaterSm_StateGet is equal to HEATER_OFF. We then run the state machine with a time index of zero and check the state again. We can use this test case as an example to build as many test cases as we need.

**Listing 8-3.** An example test that verifies the initial state of the heater

```
TEST(HeaterGroup, InitialState)
{
  CHECK_EQUAL(HeaterSm_StateGet(), HEATER_OFF);
  HeaterSm_Run(0);
  CHECK_EQUAL(HeaterSm_StateGet(), HEATER_OFF);
}
```

As a first place to start, I would recommend writing a module that can manage a heater. I would list the requirements for the module as follows:

- The module shall have an interface to set the desired state of the heater: HEATER_ON or HEATER_OFF.

- The module shall have an interface to run the heater state machine that takes the current system time as a parameter.

- If the system is in the HEATER_ON state, it should not be on for more than HeaterOnTime.

- If the heater is on, and the heater is requested to be on again, then the HeaterOnTime will reset and start to count over.

Go ahead and give it a shot. I've provided my solutions to the code and my test cases in Appendix D.

---

**Beware**    Always make a test case fail first! Once it fails, write only enough code to pass the test. Then "rinse and repeat" until you have produced the entire module. (The tests drive what code you write, not the other way around.)

---

# Final Thoughts

Testing is a critical process in developing modern embedded software. As we have seen in this chapter, there isn't a single test scheme that developers need to run. Instead, there are several different types of tests at various levels of the software stack that developers need to develop tests for. At the lowest levels, unit tests are used to verify individual functions and modules. Unit tests are most effectively written when TDD is leveraged. TDD allows developers to write the test, verify it fails, then write the production code that passes the test. While TDD can appear tedious, it is a very effective and efficient way to develop embedded software.

Once we've developed our various levels of testing, we can integrate those tests to run automatically as part of a CI/CD pipeline. Connecting the tests to tools like GitLab is nearly trivial. Once integrated, developers have automated regression tests that easily run with each check-in, double-checking and verifying that new code added to the system doesn't break any existing tests.

## ACTION ITEMS

To put this chapter's concepts into action, here are a few activities the reader can perform to start improving their Embedded DevOps:

- Make a list of the risks associated with the types of products you design and build. What types of testing can you perform to minimize those risks?

- What testing methodologies do you currently use? Are there additional methodologies that need to be implemented?

- What are some of the benefits of using Test-Driven Development? What are some of the issues? Make a pros and cons list and try to decide if TDD could be the right technique for you to use.

- Carefully examine Appendix D. Were you able to write a module and test cases that are similar to the solutions?

    - How many test cases were required?

    - Run a Cyclomatic Complexity scan on your code.

    - How does the complexity value compare to the number of test cases you wrote? Do you need more?

- If you have not already done so, read through Appendix C. There is an example on how to connect a test harness to a CI/CD pipeline.

  - Write your own cpputest.mk file.

  - Test-drive TDD by writing your own custom application module.

# CHAPTER 9

# Application Modeling, Simulation, and Deployment

Embedded software developers traditionally do most of their work on development boards and product hardware. Unfortunately, there are many problems with always being coupled to hardware. For example, product hardware is often unavailable early in the development cycle, forcing developers to create "Franken boards" that cobble together enough hardware to move the project forward slowly. There are also often inefficiencies in working on the hardware, such as longer debug times. Modern developers can gain an incredible advantage by leveraging application modeling and simulation.

This chapter will explore the fundamentals of modeling and simulating embedded software. We will look at the advantages and the disadvantages. By the end of this chapter, you should have everything you need to explore further how you can leverage modeling and simulation in your development cycle and processes.

## The Role of Modeling and Simulation

Chapter 1 shows that modern embedded software development comprises many parts. In fact, throughout this book, we have looked at the makeup of modern embedded software to be represented by the diagram shown in Figure 9-1. When we write embedded software, we use configurators, models, and handwritten code to create our application. Then, that application can be run in three scenarios: in our test harnesses, a simulator or emulator, and finally on our target hardware.

© Jacob Beningo 2022
J. Beningo, *Embedded Software Design*, https://doi.org/10.1007/978-1-4842-8279-3_9

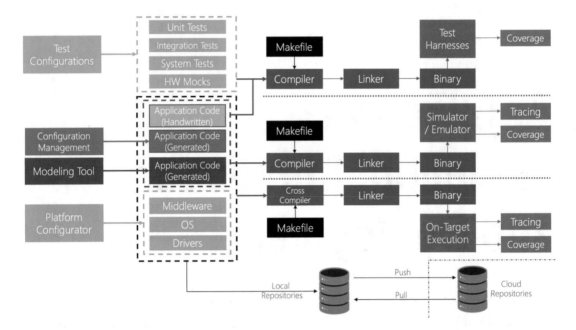

***Figure 9-1.*** *Modern embedded software applies a flexible approach that mixes modeling and configuration tools with hand-coded development. As a result, the code is easily tested and can be compiled for on-target or simulated execution*

Traditionally, embedded software developers skip the modeling and simulation and go directly to the hardware. The tradition dates to the early days of embedded software when the software was tightly coupled to the hardware, and the use of objects, abstractions, encapsulation, and other modern software techniques wasn't used. As a result, developers had no choice but to get the hardware running first and then work on the application. Unfortunately, going this route often leads to lower-quality application code.

Each piece in the modern embedded software development diagram helps lead teams to create firmware faster with a higher level of quality. However, two pieces are overlooked far too often today: the use of modeling to prove out software and running the software in a simulator or an emulator. Both techniques provide several benefits that are overlooked by teams.

First, teams can become more efficient. Modeling and simulating the application code allows the team and customers to run the application code at the highest logic levels. Getting sign-off early on the high-level behavior can help to minimize scope creep and better focus developers on the features that need to be developed.

Next, modeling and simulation often lead to early defect discovery. The cheapest point in a development cycle to fix bugs is as soon as they happen or, better yet, before they ever happen! For example, suppose a requirement specification defect can be found early. In that case, it can save considerable money and help optimize the delivery schedule by removing painful debug time.

Finally, at least for our discussions, the last benefit is that modeled code can be autogenerated. Hand coding by humans is an error-prone endeavor. I like to believe that I'm a pretty good software developer, yet, if it were not for all the processes that I put in place, there would be a lot of defects and a lot of time spent debugging. Using a tool to model and generate the software makes writing and maintaining that code less error-prone.

There are too many benefits to leveraging modeling and simulation for teams to ignore and skip over them. So let's start our exploration of modeling, simulating, and deploying by looking at software modeling.

# Embedded Software Modeling

The greatest return on investment in modeling embedded software is in the application's high-level, hardware abstracted pieces. The application code is the area of the code base that contains the business logic for the device. The application shouldn't have low-level dependencies or be tightly coupled to anything that requires the hardware. When we design software this way, we can model and even simulate the application and test our customers' requirements before we get too far down the development path.

Modeling is a method of expressing the software design that uses an abstract, picture-based language. The most used language to model software systems is the Unified Markup Language (UML). UML has been standardized for decades, providing standard models that can express nearly any software design.[1] For example, expressing a software system using state machines or state diagrams is very common. UML provides the standardized visual components that one would expect to see represented in these types of diagrams.

Embedded software applications can be modeled in several different ways. First, designers can use a stand-alone UML modeling tool. A stand-alone tool will allow the designer to create their models. The more advanced tools will even generate code based

---

[1] The discussion of UML is beyond the scope of this book, but you can learn more about it at `www.uml.org`.

on that model. The second method designers can use is UML modeling tools, including the capability to simulate the model. Finally, advanced tools often allow a designer to model, simulate, and generate code. Let's look at how some models can look with different toolchains.

## Software Modeling with Stand-Alone UML Tools

The first modeling technique, and the one that I see used the most, is to use a stand-alone tool to model the software system in UML. UML provides a visual framework for developers to design various aspects of the software system. For example, if we were designing a system with two states, a SYSTEM_DISABLED and a SYSTEM_ENABLED state, we could use a state machine diagram to represent our system, as shown in Figure 9-2.

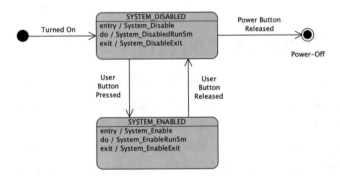

***Figure 9-2.***  *A simple state machine model demonstrates how the system transitions from the SYSTEM_ENABLED state to the SYSTEM_ENABLED state*

The state diagram in Figure 9-2 is drawn using Visual Paradigm,[2] a low-cost modeling tool. The model shows how we transition initially to the SYSTEM_DISABLED state through the system being powered on (starting from the solid black circle). When the system is first powered on, the SYSTEM_DISABLED state's entry function, System_ Disable, is run once to disable the system. After the entry function is executed, the state machine will run System_DisabledRunSm during each state machine loop. It will continue to do so until the user button is clicked.

---

[2]www.visual-paradigm.com/

When the user button is clicked, the SYSTEM_DISABLED exit function will run, SystemDisableExit, to run any state cleanup code before transitioning to the SYSTEM_ ENABLED state. Once there, the System_Enable entry function runs once, and then the System_EnableRunSm function runs while the system remains in SYSTEM_ ENABLED. The transition from the SYSTEM_ENABLED state to the SYSTEM_DISABLED state follows a similar process.

As you can see, the state machine diagram creates a simple visualization that is very powerful in conveying what the system should be doing. In fact, what we have done here is modeled a state machine architecture diagram. Therefore, with some guidance, every engineer and even customers or management should be able to read and verify that the system model represents the requirements for the system.

# Software Modeling with Code Generation

With a software system model, we can leverage the model to generate the system software if the correct tools are used. For example, in our previous state machine example, we could generate the C/C++ code for the state machine. That state machine code could then be integrated into our code base, compiled, and run on our target. If changes were required, we could then update our model, generate the state machine code again, and compile our code to get the latest update. Using a tool to generate our code is often lower risk and has a lower chance for a defect to be in the code. Unfortunately, I've not had much success with the code generation features of Visual Paradigm. However, I have had more success with IAR Visual State software.

Visual State isn't a complete UML modeling tool. It focuses explicitly on state machines. If we wanted to create a class diagram of our software, Visual State would not be the tool to use. However, if we want to create, model, simulate, verify, and generate code for a state machine, Visual State will work just fine. Let's use an example to see how we can perform some of this functionality.

When I work on space systems, nearly every system has a heater to manage the spacecraft's temperature. A heater state machine can be simple with just three states: HEATER_OFF, HEATER_ON, and HEATER_IDLE. Modeling the heater controller using a UML diagram would look like Figure 9-3. The figure shows the three states and the events that cause the state transitions. For example, if the heater controller is in the HEATER_OFF state, the On_Command event will cause a transition of the controller to the HEATER_ON state. The next transition depends on the event and the controller's state.

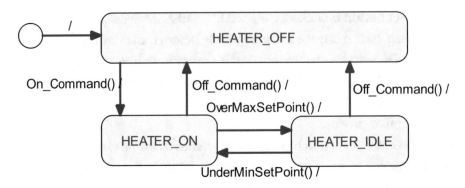

***Figure 9-3.*** *A heater controller state machine modeled in the IAR Visual State tool*

I think the reader can see how powerful a model and code generation can be. However, using a model, though it is not all rainbows and sunshine, there are some issues that need to be thought through up front. One of the issues designers will often encounter with software modeling tools that generate code is that they need to find a way to connect their existing code outside the modeled code with the generated model. For example, if a state machine is designed that uses interrupts or waits for a message in an RTOS queue, some "glue logic" needs to be provided.

Figure 9-4 shows a simplified example of how the developer code converts hardware- or software-generated inputs into events that the state machine code can use. First, the developer inputs events, makes calls to the state machine API, and passes any data to the state machine. The state machine then acts on the events and generates action. Next, the developer needs to add code to take action and convert it to the desired output that the embedded system can use.

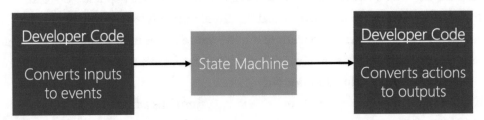

***Figure 9-4.*** *Glue logic in user code converts inputs and outputs to/from the state machine*

---

**Beware**    Generated code can be convenient, but defects can exist in the generator tool. So proceed with caution and test everything!

---

# Software Modeling with Matlab

A second modeling technique designers can take advantage of is leveraging a tool like Matlab. Matlab is a powerful, multidisciplinary tool that provides designers and developers with many toolkits to model state machines, algorithms, and systems and even train machine learning models. Matlab is extremely powerful and can be used by multidisciplinary teams to model far more than just software; however, their Stateflow toolbox is an excellent tool for modeling state machines.

Figure 9-5 shows a simple state machine diagram created in state flow to represent the various heating states of an oven. A designer can quickly visually show three possible states: HEATER_OFF, HEATER_ON, and HEATER_IDLE. The transition into and out of each state is specified using the [] notation. For example, transitioning from the heating state to the idling state can occur if the temperature is [too hot] or from idling to heating if the temperature is [too cold]. Those definitions are not specific for an engineering application and would undoubtedly have some logic associated with them, but I think the reader can see the point.

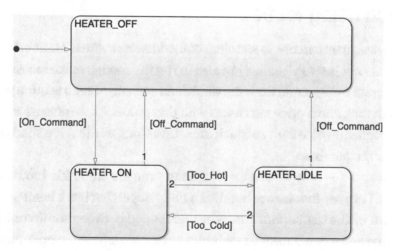

***Figure 9-5.*** *An example Matlab state machine visualization that shows the various states for a heater and causes for the state transitions*

Visualizing what the system and the software are supposed to do is critical to ensuring that the right software is developed. For example, the diagram in Figure 9-5 can be presented to management and the customer and easily verified that it is the required state behavior for the system. If it is wrong, the designer can make a few quick adjustments to the diagram and get the sign-off. Then, the developers can either hand-code based on the diagram, or if they have the Embedded Coder toolbox, they can generate the state machine code themselves.

# Embedded Software Simulation

Creating a software system model is a great way to understand the system that will be built. The ability to run that model in a simulation can provide insights into the design that otherwise may not have been possible without running production software on the device. In addition, simulation can provide insights to fix problems quickly and efficiently before they are allowed to balloon out of control.

There are several ways embedded developers can simulate their embedded software applications. The number of options is nearly dizzying! Let's explore a few options to get you started. Just keep in mind that there are many options out there.

## Simulation Using Matlab

One tool that designers can use to simulate embedded software is Matlab. Matlab allows designers not just to visualize but also to run the model, make measurements, and even generate C/C++ code from the model. As a result, Matlab is quite powerful as a development tool, and a price tag comes with that power. As we discuss throughout this book, though, the price isn't so much an issue as long as there is a solid return on investment for the purchase.

The best way to see how we can use Matlab to simulate embedded software is to look at an example. Let's say that we are building a new widget that has a heating element associated with it. The temperature must be maintained between some configurable minimum value and a configurable maximum value. The current temperature will determine whether the heater should be enabled or disabled. If the heater is enabled, then the temperature will rise. If it is disabled, then the temperature will fall.

We can view our system as having several parts. First, a state machine named Heater Controller manages whether the heater is on or off. The Heater Controller will input several temperature values: minimum, maximum, and current. The Heater Controller also can be enabled or disabled. Second, the system also has a Temperature state machine that determines whether the temperature rises or falls based on the Heater Controller's HeaterState. Figure 9-6 demonstrates what the Simulink model might be like.

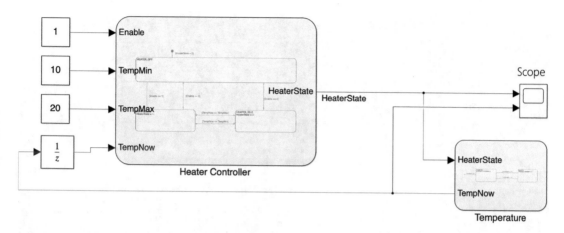

***Figure 9-6.*** *Heater Controller state machine with supporting Simulink functionality to simulate the state machines' behavior fully*

There are two things I'd like to point out in Figure 9-6. First, notice the input to the Heater Controller TempNow port. I've placed a unit delay, represented by the 1/z block, to allow the simulation to run. Since Heater Controller depends on TempNow and Temperature relies on HeaterState, we have a logic loop that creates a "chicken and egg" issue. Placing the unit delay removes this issue and allows the model to simulate successfully. Second, on the upper right, I've added a scope to the model so that we can see the output of the temperature and the HeaterState throughout a 25-second simulation.

At this point, we've just defined the inputs and outputs of the two state machines needed for our simulation. We also need to define these state machines. Figure 9-7 shows the implementation for the Heater Controller state machine. Notice that it comprises just three states: HEATER_OFF, HEATER_ON, and HEATER_IDLE. We enter the state machine in the HEATER_OFF state and initialize HeaterState to 0. When the Enable input port signal is 1, which we've hard-coded using a constant, the state machine will transition from HEATER_OFF to HEATER_ON. In HEATER_ON, the HeaterState is changed to 1, representing that the heating element should be enabled.

Our controller is designed to maintain the temperature within a specific range. If TempNow exceeds or is equal to TempMax, then we transition to the HEATER_IDLE state and set HeaterState to 0. There is a transition back to HEATER_ON if the temperature falls below or equal to TempMin. That is it for the Heater Controller design!

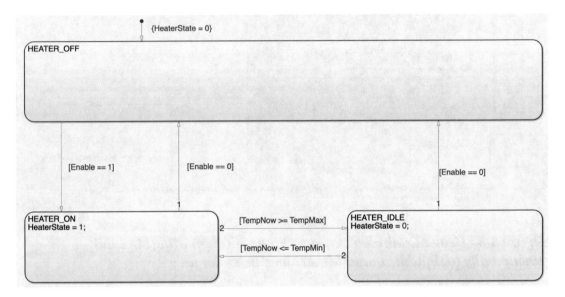

***Figure 9-7.*** *The Heater Controller state machine design*

Next, we must define how the temperature will behave in the Temperature state machine. The temperature will have two possible states; it falls when the heater is off or rises when the heater is on. We will set the initial temperature to 2 degrees when the system starts. After that, the temperature increase or decrease will be in 2-degree increments. Figure 9-8 shows how the temperature state machine is implemented.

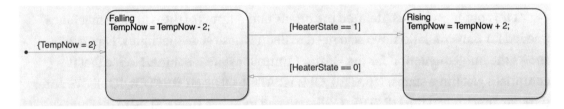

***Figure 9-8.*** *The Temperature feedback state machine design*

We have all the pieces implemented in the model to represent how the system behaves and simulate it! Pressing the green run button in Simulink compiles the model and runs it. Let's check in and see if the system behaves as we want it to. Let's first check out the Sequence Viewer, a tool within Matlab. The Sequence Viewer allows us to see that state transitions in our state machines during the simulation. Figure 9-9 provides an example of the system we have just been designing.

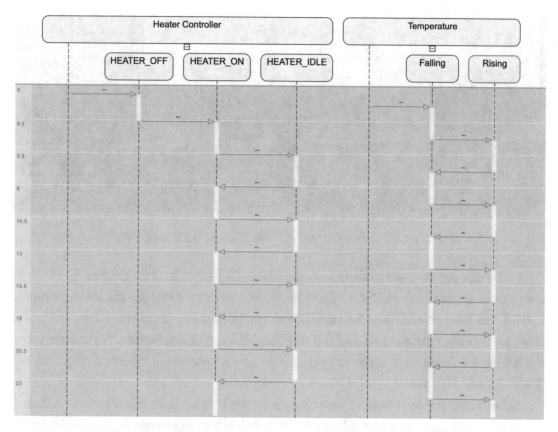

***Figure 9-9.*** *The sequence diagram representing the state transitions for the Heater Controller and Temperature simulation*

Notice that in the simulation, we enter the HEATER_OFF state and then transition to the HEATER_ON state. For the rest of the simulation, the Heater Controller bounces between the HEATER_ON and the HEATER_OFF states. The Temperature state machine in lockstep bounces between the temperature falling and the temperature rising. The state transitions behave as we expect, but does the temperature? Figure 9-10 shows the output on our scope for the HeaterState and TempNow.

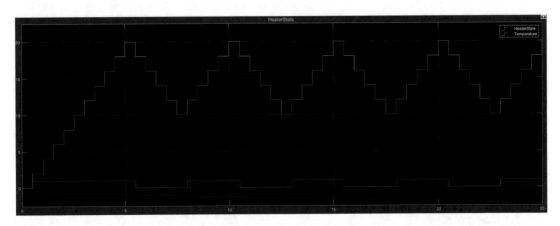

***Figure 9-10.*** *The scope measurements for the HeaterState and Temperature*

The temperature starts to rise from zero when the HeaterState is enabled. The temperature rises to the 20-degree TempMax value, where we can see the HeaterState turnoff. The temperature then falls until it reaches 10 degrees, TempMin. The HeaterState then returns to an enabled state, and the temperature rises. We can see for the simulation duration that the temperature bounces between our desired values of TempMin and TempMax!

Before we move on, I want to point out that there is more than one way to do things and that models can be as sophisticated as one would like. For example, the model we just looked at is completely state machine, Stateflow, based with hard-coded parameters. I asked a colleague of mine, James McClearen, to design the same model within any input from me. His Matlab model for the Heater Controller can be seen in Figure 9-11.

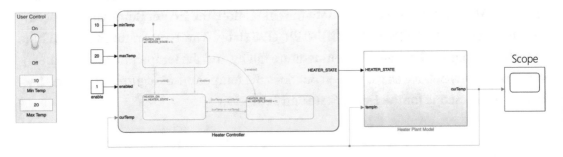

***Figure 9-11.*** *Another version of the heater model*

There are a few subtle differences between his model and mine. First, instead of modeling the temperature response as a state machine, he just created it as a standard Simulink block. Next, he added a user control box to manage the enable/disable state and set the minimum and maximum temperatures. Finally, he didn't need the unit delay block because his Heater Plant Model included a different transfer function on the output as shown in Figure 9-12.

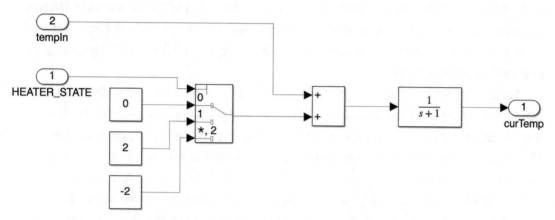

***Figure 9-12.*** *The Heater Plant Model in Simulink which uses "discrete" logic blocks rather than a state machine to simulate the temperature response*

Which model is right? Both! Each model produces the same results. The first model was developed by an embedded software consultant with basic to intermediate Matlab skills focusing on demonstrating state behavior. The second model was developed by a mechanical engineering consultant, focused on modeling it the way a Matlab expert would.

We've just successfully simulated how we want our Heater Controller embedded software application to behave, and we've done it without any hardware in the loop! At this point, we have several options available on how we proceed. First, we can continue adjusting the simulation to test different conditions and use cases to ensure it behaves as needed under all conditions. Next, we can use our visualization to hand-code the state machine in C/C++. Alternatively, we can use the Embedded Coder toolbox, another tool within Matlab, to generate the C/C++ code for the Heater Controller and then deploy it to an embedded target. How you proceed is entirely up to your own needs and comfort level.

---

**Note**   We don't need to generate code for the Temperature state machine. TempNow would be generated by a thermocouple fed into the Heater Controller API.

---

# Software Modeling in Python

Using a visualization and simulation tool like Matlab is not the only way to model embedded software. Designers can also model the behavior of their systems by using a software language that is not an embedded language and runs on a host computer. For example, a designer might decide that they can rapidly model the system by writing the application code in Python!

I know that writing code in Python leaves a bad taste in many embedded software developers' mouths, but Python can be a game-changing tool for developers if they are willing to embrace it. For example, I had a customer who we were designing an electronic controller with. The controller had several states, telemetry outputs, and commands that it would respond to under various conditions. To ensure that we understood the controller's inputs, outputs, and general behaviors, we started by spending several weeks creating a model of how the controller should behave in Python.

The Python controller was written to simulate the electronic controller's behavior. We provided a host computer running the Python model that used a USB to serial adapter to communicate with the other subsystems like it was the actual controller. The customer verified that the model behaved as needed from an input, output, and command standpoint. It didn't have the full real-time performance or energy constraints the end controller would have; however, we were able to get the customer to sign off. Once we had the sign-off, we started to design and build the real-time embedded controller that would provide those same modeled inputs, outputs, and behaviors.

At first glance, modeling the system in another language in code that we essentially just threw away may seem like a waste. However, it was far more effective to build a rapid prototype of the system and shake out all the interfaces to the controller so that the customer could sign off on what we would build. If that had not been done, we undoubtedly would have started to design and build something that would not have met the customers' requirements. We then would have had to incrementally adjust and probably would have ended with a rat's nest of code as the customer changed their minds and evolved the product. Instead, we got early sign-off with fixed requirements we could work from. The customer received a representative model they could then use to develop and test their systems against, which then takes the pressure off the need to deliver the actual product as quickly as possible. With the pressure off, the chances of rushing and creating defects dramatically decrease.

---

**Note**   There isn't any such static project where requirements don't change. Modeling the system up front can help minimize changes and fully define the core product features.

---

Using Python is just one example. There are certainly plenty of additional methods that can be used to simulate an embedded system.

# Additional Thoughts on Simulation

There are several other ways that developers simulate their embedded software applications without the need for the underlying embedded hardware. I don't want to go into too much detail on all these methods. I could easily put together another full book on the topic, but I at least would like the reader to know that they exist.

One tool that I have found to be helpful in creating simulations is to use wxWidgets.[3] wxWidgets is a cross-platform GUI library that provides language bindings for C++, Python, and several other languages. wxWidgets is interesting because with the C++ compiler, it's possible to compile an RTOS. For example, FreeRTOS has a Windows port that can be compiled into wxWidgets. Once this is done, a developer can create various visualizations to simulate and debug embedded software. Dave Nadler gave a great talk about this topic at the 2021 Embedded Online Conference[4] entitled "How to Get the Bugs Out of your Embedded Product."[5]

Another tool to consider is using QEMU.[6] QEMU is a popular generic and open source machine emulator and virtualizer. Over the last several years, I've seen several ports and developers pushing to emulate Arm Cortex-M processors using QEMU. For example, if you visit `www.qemu.org/docs/master/system/arm/stm32.html`, you'll find an Arm system emulator for the STM32. The version I'm referring to supports the netduino, netduinoplus2, and stm32vldiscovery boards.

---

[3] `www.wxwidgets.org/`

[4] `https://embeddedonlineconference.com/`

[5] `https://embeddedonlineconference.com/session/How_to_Get_the_Bugs_Out_of_your_Embedded_Product`

[6] `www.qemu.org/`

Leveraging QEMU can provide developers with emulations of low-level hardware. For embedded developers who like to have that low-level support, QEMU can be a great way to get started with application simulation without removing the low-level hardware from the mix. The STM32 version I mentioned earlier has support for ADC, EXTI interrupt, USART, SPI, SYSCFG, and the timer controller. However, there are many devices missing such as CAN, CRC, DMA, Ethernet, I2C, and several others. Developers can leverage the support that does exist to get further down the project path than they would have otherwise.

---

**Note**    There are many QEMU emulator targets. Do a search to see if you can find one for your device (and a better STM32 version than I found).

---

There are certainly other tools out there for simulation, but I think the core idea is that the tool doesn't matter. The idea, the concept, and the need within teams are that we don't need to be as hardware dependent as we often make the software out to be. Even without a development board, we can use a host environment to model, simulate, debug, and prove our systems long before the hardware ever arrives. Today, there aren't enough teams doing this, and it's a technique that could dramatically transform many development teams for the better.

# Deploying Software

The final output option for our software that isn't a test harness or a simulator is to deploy the embedded software to the hardware. I suspect that deploying software to the hardware is the technique that most readers are familiar with. Typically, all that is needed is to click the debug button in an IDE, magic happens, and the software is now up and running on the target hardware. For our purposes in this book, we want to think a little bit beyond the standard compile and debug cycle we are all too familiar with.

In a modern deployment process, there are several mechanisms that we would use to deploy the software such as

- The IDE run/debug feature

- A stand-alone flash tool used for manufacturing

- A CI/CD pipeline job to run automatic hardware-in-loop (HIL) tests

- A CI/CD pipeline job to deploy the software to devices in the field through firmware-over-the-air (FOTA) updates or similar mechanism

Since you are undoubtedly familiar with the IDE run/debug feature, let's examine a few of the other options for deploying software.

# Stand-Alone Flash Tools for Manufacturing

Once developers get away from their IDE to deploy software, the software deployment process begins to get far more complex. When manufacturing an embedded product, a big concern that comes up is how to program units on the assembly line at the manufacturing facility. Having a developer with their source and a flashing tool doesn't make sense and is in fact dangerous! When we begin to look at how to deploy code at the manufacturing facility, we are looking to deploy compiled binaries of our application.

The biggest concern with deploying software at the manufacturing facility is the theft of intellectual property. The theft can come in several different ways such as

- Access to a company's source code

- Reverse-engineering compiled binaries

- Shadow manufacturing[7]

It's interesting to note that with today's tools and technology, someone can pull your compiled code off a system and have readable source code in less than ten minutes! A big push recently has been for companies to follow secure device manufacturing processes.

What is necessary for manufacturing is for companies to purchase secure flash programmers. A secure flash programmer can hold a secure copy of the compiled binary and be configured to only program a prespecified number of devices. A secure programmer solves all the concerns listed earlier in a single shot.

For example, the secure flash programmers will not allow a contract manufacturer to have access to the binaries. The binaries are stored on the flash programmer, but not accessible. In addition, the fact that the secure flash programmer will only allow a specified number of devices to be programmed means that the devices programmed can be easily audited. A company doesn't have to worry about 1000 units being produced that they don't know about. The secure flash programmer tracks each time it programs a device, creating a nice audit trail.

---

[7] Shadow manufacturing is when the contract manufacturer builds your product during the day and by night is building your product for themselves.

The secure manufacturing process is quite complex, and beyond the book's scope, but I at least wanted to mention it since many of you will encounter the need to deploy your firmware and embedded software to products on the assembly line.

## CI/CD Pipeline Jobs for HIL Testing and FOTA

A very common need is to deploy embedded software within a CI/CD pipeline as part of Embedded DevOps processes. The CI/CD pipeline will typically have two job types that need to be performed. First, software will need to be deployed to the target that is part of a HIL setup. The second is that the software will need to be deployed to the target that is in the field. Let's explore what is required in both situations.

Figure 9-13 shows an example architecture that one might use to deploy software to a target device. As you can see, we have the target device connected to a flash programming tool. That programming tool requires a programming script to tell it what to program on the target device. The program script can be as simple as a recipe that is put together as part of a makefile that knows what commands to execute to program the target.

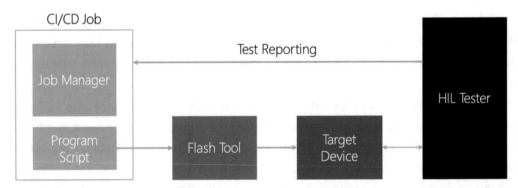

***Figure 9-13.*** *Example architecture for deploying to a target device in a CI/CD HIL test job*

For example, using a typical setup using a SEGGER J-Link Ultra+, I would install the J-Link tools on a Docker image that is part of my CI/CD framework. My Git repository would contain a folder that has the J-Link command script included. The command script is just the commands necessary to program a device such as commands to reset, erase, program, and verify the device. The script is necessary because there is no high-level IDE driving the programming process.

With the command file in place, a simple recipe can then be used within the main makefile to call the J-Link utilities and run the command script with the desired application to program. If you've lived in a Windows IDE environment, the process can seem complicated at first. Rest assured, a little bit of digging and it's relatively simple. In fact, you'll find an example for how to add a deployment job to a CI/CD pipeline using a SEGGER J-Link in Appendix C. Once it's set up for the first time, there is usually very few changes that need to be made to maintain it.

Adding production deployment to a CI/CD pipeline can be complicated. It often involves careful collaboration with the operations team. There are several different approaches I've used in the past to perform production deployments. First, we've used our CI/CD pipeline in conjunction with the tools in Amazon Web Services. As part of our deploy job, we can create AWS jobs for pushing firmware to our deployed devices. I've found this approach has worked well. There is a bit of a learning curve, so I recommend starting with small batches of deployments and then eventually moving to larger ones once you have the operational experience.

The second method that I've seen used but have not yet personally used is to leverage a tool like Pelion.[8] Pelion provides companies with the mechanisms to manage fleets of IoT devices in addition to managing their device life cycles. Teams can integrate their deployment pipelines into the Pelion IoT management platform and then go from there. Like I mention, I've seen the solution but I have not used it myself, so make sure if you do use a platform like this that you perform your due diligence.

# Final Thoughts

The need to model, simulate, and deploy software systems are not techniques reserved for big businesses with multimillion-dollar budgets. Instead, modeling and simulation are tools that every development team can use to become more efficient and deliver software that meets requirements at a higher-quality level. At first, modeling and simulation can appear intimidating, but steady progress toward implementing them in your development cycles can completely transform how you design and build your systems.

Using a CI/CD pipeline to deploy software to hardware devices locally and in the field can dramatically change how companies push firmware to their customers. I can't tell you how many times per day I get a notification that Docker has a new software

---

[8] https://pelion.com/

update. Docker is continuously deploying new features and capabilities as they are ready. Could you imagine doing that with your customers or team? Providing a base core functionality and then continuously deploying new features as they become available? It might just help take the heat off, delivering the entire product if it can be updated and enhanced over time.

There is a lot more that can be done with simulation and the CI/CD pipeline. There are improvements from an organizational standpoint, along with additional capabilities that can be added. For example, the deployToTarget we created only works on a local embedded device. A fun next step would be integrating this into an over-the-air update system that can be used to push production code to thousands of devices. Granted, you might want to start by deploying to just a half dozen boards scattered throughout the office.

I think that you will find that modeling, simulation, and automated deployment are great additions to the current processes that you are using. As I've mentioned, you'll find that these enable you to resolve countless problems well before the hardware is required. When the hardware is needed, you now have a simple process to deploy your application to the hardware. With some modification, hardware-in-loop testing can even be performed, adding a new level of feedback to the application development.

## ACTION ITEMS

To put this chapter's concepts into action, here are a few activities the reader can perform to get more familiar with modeling, simulation, and deployment:

- Examine Figure 9-1 again. What areas of this diagram are you missing? What will you do over the coming quarters to add in the missing pieces?

- Research and identify several tools that will allow you to model your software. Pick a few to download trials for. Model a simple system like the Heater Controller. Share the model with several colleagues.

  - How quickly do they understand the design intent?

  - How quickly and accurately can they hand-code the model?

  - If the model changes, what issues are encountered trying to update the code by hand?

  - What tool best fits your design needs?

- Use a tool like Matlab to design and simulate a simple system.

    - How did the tool affect your understanding of the system?

    - How did the tool affect your efficiency, bug quality, and evolution of the software product?

- Create a simple model of a product using Python on a host computer. Can you see the benefits of having a functionally equivalent system in a host environment? What are the benefits? Any cons?

- Schedule time to work through the deployment example in Appendix C. You'll find that you can easily create a CI/CD pipeline that can use a J-Link to program a board automatically.

# CHAPTER 10

# Jump-Starting Software Development to Minimize Defects

How much of your time or team's time is spent debugging?

When I ask this question at conferences, I find that the average time an embedded software developer spends debugging their software is around 40%! Yes, you read that correctly! That means that, on average, developers spend 4.5 months fighting with their systems, trying to remove bugs and make them work the way they're supposed to! If I had to spend that much time debugging, I would find a different career because I loathe debugging! It's stressful, time-consuming, and did I mention stressful?

I often look at debugging as performing failure work. That's right. I didn't do it right the first time, and now I must go back and do it again, just like in school. Obviously, a lot of teams can do a lot better. So far in Part 2 of this book, we've been exploring what can be done to prevent bugs from getting into your embedded software in the first place. For example, having design reviews, performing code reviews, modeling, and simulating your software. No matter how good your team gets at preventing bugs, sometimes bugs will still get into your software. When that happens, we want to make sure that our processes allow us to discover them quickly to minimize delays and costs.

In this chapter, we will look at what developers and development teams can do to jump-start their embedded software development to minimize defects. The goal will be to provide you with some process ideas that you can use from the start of a project that should not just help you find bugs faster but prevent them in the first place!

© Jacob Beningo 2022
J. Beningo, *Embedded Software Design*, https://doi.org/10.1007/978-1-4842-8279-3_10

# A Hard Look at Bugs, Errors, and Defects

When you carefully consider the typical bug rates of a team and the time developers spend debugging their software, it seems a bit outlandish. Jack Ganssle once said, "No other industry on the planet accepts error rates this high!". He's right! Any other industry would fire a worker with failure rates of 40%! In the software industry, we record our failures as bugs and then celebrate how hard we worked to remove the bugs from the list!

To build out a successful process that minimizes "bugs" in software, we must first take a hard look at the actual definitions of bugs, defects, and errors. The very terminology we use when developing embedded software can set our perspective and potentially devastate our ability to deliver successfully. The most used term to describe an issue with software is "there is a bug," or the code is "buggy." The term bug implies that some external entity is fighting against us and preventing us from being successful. When things go wrong, responsibility is displaced from the engineer, who should be in control of the situation, to a phantom force, and the engineer is simply going along for the ride.

Developers, teams, and companies involved in software development and supporting toolchains need to start shifting their perspectives and mindsets away from bugs and instead creating terminology that places the responsibility where it belongs, on themselves! Now I know that this sounds uncomfortable. In fact, I wouldn't be surprised if you are a bit squeamish now and shifting around in your chair. But, at the end of the day, what we currently call bugs are errors or defects in the software, so why not just call them what they truly are?

A defect is a software attribute or feature that fails to meet one of the desired software specifications.[1] An error is a mistake or the state of being wrong.[2] An error could also be considered a discrepancy in the execution of a program.[3]

These are some excellent official definitions, but let me present them to you in another way:

> **Errors** are mistakes made by the programmer in implementing the software design.

---

[1] https://en.wikipedia.org/wiki/Defect

[2] https://yourdiectionary.com/error

[3] www.computerhope.com/jargon/e/error.htm

**Defects** are mistakes resulting from unanticipated interactions or behaviors when implementing the software design.

**Bugs** are fictitious scapegoats developers create to shift blame and responsibility from themselves to an unseen, unaccountable entity.

Ouch! Those are some challenging definitions to swallow!

I gravitate toward defects when considering which terminology makes the most sense for developers. It's not as harsh as saying that it's an error that gets management or nontechnical folks all worked up, but it does put the responsibility in the right place. A defect in this sense could be an unintended consequence of an expected result in the application. The defect doesn't have a mind of its own and could have resulted from many sources such as

- Improper requirements specification

- Misunderstanding hardware

- Complex system interactions

- Integration issues

- Programmer error (blasphemy!)

When I lecture on defect management (formerly known as debugging), I often ask the audience how much time they spend on average in their development cycle debugging. The most significant responses are usually around 40–50% of the development cycle, but I always have outliers at 80%! Being able to deliver code successfully requires developers to have the correct mindset. The belief that bugs oppose our every line of code leads to loosey-goosey developed software high in defects.

Isn't it time that we squash the bugs once and for all and start using terminology that is not just accurate but shifts developers' perspectives away from being software victims and instead put the responsibility and control back on the developer?

If you are game to ditch the bugs and take responsibility for the defects you and your team create, let's move forward and look at how we can jump-start development to minimize defects. The phase that we are about to cover is the process that I follow when I am starting to implement a new software project. You can still follow the process if you already have existing legacy code.

The end goal is to decrease how much time you spend fixing software defects. If you are like the average developer, spending 40% of your time and 4.5 months per year, set a goal to decrease that time. Could you imagine if you could just reduce that time in half? What would you do with an extra 2.25 months every year? Refactor and improve that code you've wanted to get? Add new features your customers have been asking for? Or maybe, you'll stop working on weekends and evenings and start to live a more balanced life? (LOL.)

# The Defect Minimization Process

The defect minimization process isn't about implementing a rigid, unchangeable process. Defect minimization is about starting a project right, using the right tools, and having the right processes in place to catch defects sooner rather than later. Remember, the later a defect is found in the development cycle, the more costly it is to find and fix. The defect minimization process is more about finding defects as soon as they occur rather than letting them stew and burn up development time and budget. The ultimate goal is to not put the bugs in the software in the first place, but when they do find their way in, catch them as quickly as possible.

The process that I follow typically has seven phases. I roughly order the phases as follows:

- Phase 1 – Project Setup

- Phase 2 – Build System Setup

- Phase 3 – Test Harness Configuration

- Phase 4 – Documentation Facility Setup

- Phase 5 – Code Analysis

- Phase 6 – RTOS-Aware Debugging

- Phase 7 – Debug Messages and Trace

## Phase 1 – Project Setup

The first phase is all about getting your project set up. Setting up a project is about much more than just creating and starting to bang out code. The project setup is about getting organized and understanding how you will write your code before writing a single line of it!

The first part of a project is to set up revision control. Today, that means getting a Git repository set up. Several popular repository vendors are available such as GitHub, GitLab, Bitbucket, and so forth. A repo setup is more than just creating a single repo. Developers need to plan out how multiple repositories will come together. For example, Figure 10-1 shows a project repository with multiple subrepos that hold modular project code so it can be reused in various projects. Git organization can be as simple as using a single repo or as complex as creating a tree of dependencies. I've found it's best to decide how things will be done early to minimize history loss.

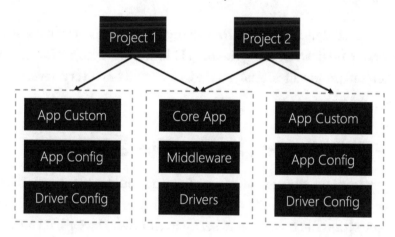

***Figure 10-1.***  *An example "tree" of Git repositories used to maintain multiple products*

---

**Best Practice**   While setting up your Git repos, populate your .gitignore file.

---

In addition to setting up the repo structure, developers in this phase will also organize their projects. Project organization involves identifying the directory structure used for the project and where specific types of files will live. For example, I often have gitignore, dockerfiles, makefiles, and readme files in the root directory. An example organization might look something like Figure 10-2.

*Figure 10-2.*  *A simple project directory structure organization that contains vital elements in a project*

To some degree, it doesn't matter how you organize your project if it's obvious to the developers working on it where things belong. I like having a project folder where I can have IDE-generated project files, analysis files, etc. You'll find every developer has their own opinion.

## Phase 2 – Build System and DevOps Setup

Once the project is set up, the next step is configuring your build system. The build system has two parts: the local build system consisting of the makefiles necessary to build your target(s) and the build system set up to run your DevOps. If you aren't familiar with Embedded DevOps, check out Chapter 7 for details.

Rely on your IDE to build your application. You may not realize that behind the scenes, your IDE is autogenerating a makefile that tells the compiler how to take your source files, generate objects, and then link them together into a final binary image. A lot of modern development is moving away from relying on these autogenerated makefiles or at least removing them from being the center of the build process. With the popularity of Linux and the trending need for DevOps, many developers are moving away from IDEs and returning to working at the command line. There is a significant need to customize the build process with DevOps, and if you just rely on the makefiles generated by your IDE, you will come up short.

A typical project today will require the build system to have several different makefiles. This includes

- An application makefile

- A configuration management makefile

- A makefile for the test harness

- A high-level "command" makefile

The relationship between these makefiles and the functionality they provide can be seen in Figure 10-3.

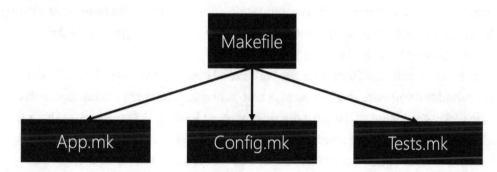

***Figure 10-3.*** *Relationships between various makefiles in a modern embedded application*

Beyond the makefile, developers should also configure their DevOps processes during this phase. Again, there are plenty of tools that developers can leverage today, like GitLab and Jenkins. Besides configuring the CI/CD tool, developers will need to set up a container like Docker to hold their build system and make it easily deployable in their DevOps build pipeline. Appendix C gives a basic example on how to do this. What's excellent about Docker is that we can immediately set up our entire build environment and make it easily deployable to nearly any computer! No more setting up a slew of tools and hoping we don't have a conflict!

The details about how to set up and use DevOps processes will be covered in Chapter 7. At this point, I think it's essential for developers to realize that they don't have to set up their entire DevOps process in one fell swoop. If you are new to DevOps, you can take a phased or agile approach to implement your DevOps. Start by just getting a basic build running in your pipeline. Then build out more sophisticated checks and deployments over time.

Having your DevOps processes set up early can help you catch build issues and bugs much faster than if you don't have one. I can't tell you how many times I've committed code that I thought worked only to have the DevOps system send me a build failed or test failed email. It would be embarrassing to commit code to a master or develop branch to break the branch for all the other developers. So getting DevOps in place early can help save a lot of time and money for defects.

# Phase 3 – Test Harness Configuration

Before ever writing a single line of code, it's highly recommended that developers configure their test harnesses. There are certainly many types of testing that can be done on a system, but unit testing is the first line of defense for developers looking to decrease defects. A unit test harness provides developers with a mechanism for testing the functionality of each function and having those functions' regression tested locally before committing code to the repo and in the DevOps pipeline.

At first, using unit test harnesses for embedded systems can seem like a fool's errand. We're embedded software engineers! Much of our code interacts with hardware that can't be easily simulated in the host environment. At least, that is the excuse that many of us have used for years. If I think back to all the projects I've worked on in my career, probably 70%[4] of the code could have easily been tested in a unit test harness. The other 30% touches the hardware and probably still could have been tested with some extra effort.

Unit testing your software is crucial because it verifies that everything works as expected at the function level. You can even build up unit tests that verify function at the task level if you want to. The ability to perform regression testing is also essential. I can't tell you how many times I've worked on a project where a change in one part of the system broke a piece in a different part. Without the ability to run regression tests, you would never know that new code added broke something already working.

We must discuss testing much, but we will go into deeper detail later. In this phase, remember that you want to get your test harness set up and ready to go. We aren't quite ready yet to start writing tests and developing our application.

# Phase 4 – Documentation Facility Setup

The fourth phase in minimizing defects is configuring how the software will be documented. I have heard many arguments about how code is self-documenting and blah blah. Still, I've rarely encountered a code base where I could pick it up and understand what was going on without documentation or talking to the original developer. Perhaps I'm just dense, but good documentation goes a long way in making sure others understand design and intent. If I know that, I'll reduce the chances of injecting new defects into a code base.

---

[4] I'm making this number up from experience; I have no real evidence of this at this time.

Many tools and techniques are available to developers today to document their code. I may be considered a bit "old school" because I still like to use Doxygen. Today, a lot of open source projects use Sphinx. Honestly, I don't know that it matters which tool you use so long as you are using something that can pull comments from your code and that you can add high-level design details to. Everyone has their own preferences. I tend to still gravitate toward Doxygen because it does a great job of processing inline comments and making great documentation package. Developers can build templates that make it quick and easy to document code too.

During this phase, I will often configure Doxygen using the Doxygen Wizard. I'll configure my build system so that when I commit my code, the documentation is automatically generated, and the HTML version is posted to the project website. Since I reuse as much code as possible, I'll often copy over Doxygen template files for application and driver abstractions that I know I'll need in the project. I'll also set up a main page for the documentation and write any initial documentation that describes the overarching project goals and architectural design elements. (These details will change, but it helps to start with something that can be changed as needed on the fly.)

# Phase 5 – Static Code Analysis

Installing and configuring code analysis tools are key pillars to developing quality software. We want to ensure that with every commit, we are automatically analyzing the code to ensure it meets our coding standard and style guides and following industry best practices for our programming language (see Chapter 15 for practical tools). We also want to ensure that we are collecting our code metrics! The best way to do this is to install code analysis tools.

There are a couple of options for developers in handling their code analysis. First, developers should install all their code analysis tools as part of their DevOps pipelines. Utilizing DevOps in this way ensures that with every commit, the code is being analyzed, and a report is generated on anything that needs to be adjusted. Second, potentially optional, developers can have a duplicate copy of the tools on their development machines to analyze their software before committing it.

To some degree, it's again opinion. I usually like to analyze my code and ensure everything is good to go before I commit it. I don't want to commit code that has defects. However, if I am often committing in a feature branch that I have control over, why not just leverage the DevOps system to analyze my code while I'm working on other activities? For this reason, I will often push as much analysis to the build server as I can. My short cycle commits may not be pristine code, but with feedback from the DevOps tools, I can refactor those changes and have pristine code by the time I'm ready to merge my branch into the develop or master branches.

Again, we must recognize that what gets measured gets managed. This phase is about setting up our code analysis tools, whatever they are, before we start writing our code.

# Phase 6 – Dynamic Code Analysis

In addition to statically analyzing code, developers must also dynamically allocate the code. The same tools developers use for this analysis will vary greatly depending on their overall system architecture and design choices. For example, suppose a team is not using an operating system. In that case, the tools they set up for dynamic code analysis may be limited to stack checkers and custom performance monitoring tools.

Teams using an RTOS may have more tools and capabilities available to them for monitoring the runtime behavior of their software. For example, Figure 10-4 shows a screenshot from Renesas' e$^2$ Studio where a simple, two-task application's behavior is monitored. We can see that the IO_Thread_func has run to completion, while the blinky_thread_func is currently in a sleep state. The run count for these tasks is 79:2. On the flip side, we can see the maximum stack usage for each task.

| Profile | Thread | Stack | MessageQueue | CountingSemaphore | Mutex | EventFlag | MemoryBlockPool | MemoryBytePool | Timer | System | ReadyQueue(No.=Priority) |
|---|---|---|---|---|---|---|---|---|---|---|---|

| No. | Name | Entry | Status | SuspendedFactor(ControlBlock*) | OwnedTX_MUTEX*(top) | Priority | RunCount |
|---|---|---|---|---|---|---|---|
| 1 | Blinky T... | blinky_thread_func | SLEEP | | | 1 | 79 |
| 2 | IO_Thread | IO_Thread_func | COMPLETED | | | 1 | 2 |
| 3 | | | Not created | | | | |
| 4 | | | Not created | | | | |
| 5 | | | Not created | | | | |
| 6 | | | Not created | | | | |
| 7 | | | Not created | | | | |
| 8 | | | Not created | | | | |
| 9 | | | Not created | | | | |
| 10 | | | Not created | | | | |
| 11 | | | Not created | | | | |

| Profile | Thread | Stack | MessageQueue | CountingSemaphore | Mutex | EventFlag | MemoryBlockPool | MemoryBytePool | Timer | Syster |
|---|---|---|---|---|---|---|---|---|---|---|

| No. | Name | Entry | StackPointer | StackStart | StackEnd | StackSize(bytes) | MaxStackUsage(bytes) |
|---|---|---|---|---|---|---|---|
| 1 | Blinky Thread | blinky_thread_func | 1ffe2ba8 | 1ffe2840 | 1ffe2c3f | 1024 | 152 |
| 2 | IO_Thread | IO_Thread_func | 1ffe27e0 | 1ffe2440 | 1ffe283f | 1024 | 128 |
| 3 | | | | | | | |

***Figure 10-4.*** *RTOS-aware debugging using e$^2$ Studio and ThreadX*

Having dynamic analysis tools early can help developers track their system's performance and runtime characteristics early. For example, if, in the ThreadX example, we had been expecting the two tasks to run in lockstep with a ratio of 1:1, the RTOS-aware debugging tab would have shown us immediately that something was wrong. Seeing that the IO_Thread_func had completed might even direct us to the culprit! Stack monitoring can provide the same help. For example, if one of the tasks only had 128 bytes of stack space allocated, but we were using 152, we would see that in the monitor and realize that we had a stack overflow situation.

RTOS-aware debugging is not the only tool developers should use at this stage. Real-time tracing is another excellent capability that can help developers quickly gain insights into their applications. There are several different tools developers might want to set up. For example, developers might want to set up SEGGER Ozone or SystemView to analyze code coverage, memory, and RTOS performance. One of my favorite tools to set up and use early is Percepio's Tracealyzer, shown in Figure 10-5. The tool can be used to record events within an RTOS-based application and then used to visualize the application behavior.

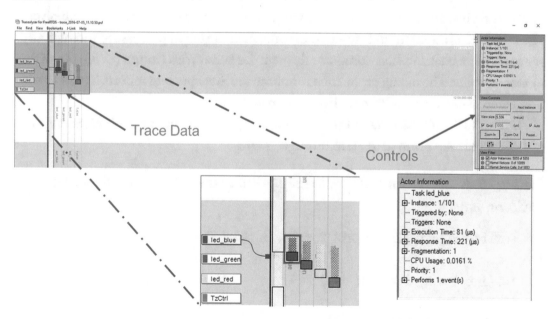

***Figure 10-5.*** *Percepio Tracealyzer allows developers to trace their RTOS application to gain insights into memory usage, performance, and other critical runtime data*

There are quite a few insights developers can gain from these tools. First, early dynamic code analysis can help developers monitor their applications as they build them. Examining the data can then help to establish trends. For example, if I've seen that my CPU usage typically increases by 1–2% per feature but suddenly see a 15% jump, I know that perhaps I have a problem and should investigate the changes I just made.

# Phase 7 – Debug Messages and Trace

The last phase for jump-starting software development to minimize bugs is to set up debug messages and trace capabilities. Finally, phase 7 is about getting data out of the embedded system and onto a host machine where the developer can see and understand how the application behaves. The capabilities used can range from simple printf messages to real-time data graphing.

An excellent first example of a capability set up in this phase is printf through the Instruction Trace Macrocell (ITM), if you are using an Arm part that supports it. The ITM provides 32 hardware-based stimulus channels that can be used to send debug information back to an IDE while minimizing the number of CPU clock cycles involved. The ITM is a hardware module internal to some Arm Cortex-M processors and typically sends the data back to a host computer through the Serial Wire Output (SWO) pin of the microcontroller. The debug probe then monitors the SWO pin and reports it back to the host where it can be decoded and displayed in an IDE.

Listing 10-1 shows an example of how a developer would utilize the ITM_SendChar CMSIS API to map printf to the ITM through _write.

***Listing 10-1.*** An example of how to map printf to the ITM in an Eclipse-based environment

```
int _write(int32_t file, uint8_t *ptr, int32_t len)
{
    for(int i = 0; i < len; i++)
    {
        ITM_SendChar((*ptr++));
    }
}
```

In addition to setting up printf, developers would also ensure that they have their assertions set up and enabled at this stage. As we saw earlier, assertions can catch defects the moment they occur and provide us with useful debug information. Other capabilities can be enabled at this point as well. For example, some microcontrollers will have data watchpoints or can leverage the serial wire viewing to sample the program counter (PC) and perform other readings on the hardware. These can all help a developer understand their application and the state of the hardware, making it more likely that it is found and fixed immediately as a defect occurs.

# When the Jump-Start Process Fails

There are going to be times when you are not able to prevent a defect in your code. A requirement is going to be missing or changed. You'll have a foggy mind day and inject a defect into your code. Embedded software today is just too complex, and the human mind can't keep track of everything. When defects occur, developers must have troubleshooting skills to help them root out the defect as quickly as possible.

When it is time to roll up your sleeves and troubleshoot your system, there are eight categories of troubleshooting techniques that you have at your disposal that you can see in Figure 10-6. Starting at the top of the figure, you have the most straightforward troubleshooting technique, which involves using breakpoints. Then, moving clockwise, techniques become more advanced and provide more insights into the system.

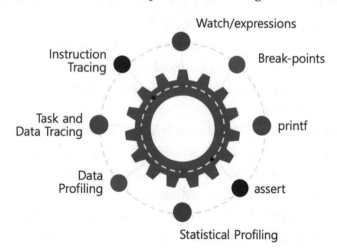

***Figure 10-6.*** *The technique categories used to find and remove defects from an embedded system*

Let's briefly define what each of these troubleshooting techniques is:

**Watch/expressions** – An essential technique used to watch memory and the result of calculations. This technique is often used with breakpoints.

**Breakpoints** – One of the least efficient troubleshooting techniques, but one of the most used. Breakpoints allow the application execution to be halted and the developer to step through their code to evaluate how it is executing. This technique often breaks the real-time performance of an application while in use.

**Printf** – Using the printf facility to print "breadcrumbs" about the application behavior to understand its behavior. Using printf will often break real-time performance, but it depends on the implementation.

**Assertions** – A technique for verifying assumptions about the state of a program at various points throughout. Great for catching defects the moment they occur.

**Statistical profiling** – A method for understanding code coverage and application behavior. Statistical profiling periodically samples the PC to determine what code areas are being executed and how often. It can provide general ideas about performance, code coverage, and so forth without sophisticated instruction tracing tools.

**Data profiling** – A technique that leverages the capabilities of the microcontroller architecture and debugger technology to read memory during runtime simultaneously. The memory reads can be used to sample variables in near real time and then plot them or visualize them in the most effective way to troubleshoot the system.

**Task and data tracing** – A technique that uses an event recorder to record task change and other events in an RTOS. Task and data tracing can visualize CPU utilization, task switching, identify deadlocks, and understand other performance metrics.

**Instruction tracing** – A technique for monitoring every instruction executed by a microprocessor. The method helps debug suspected compiler errors, trace application execution, and monitor test coverage.

Now that we understand a little bit about what is involved in each of the technique categories, it's a good idea to evaluate yourself on where your skills currently are. Look through each category in Figure 10-6. For each category, rank yourself on a scale from zero to ten, with zero being you know very little about the technique and ten being that you are a master of the method. (Note: This is not a time to be humble by giving yourself a nine because you can continually improve! If you've mastered the technique, give yourself a 10!)

Next, add the sum for each category to have a total number. Now let's analyze our results. If your score ranges between 0 and 40, you are currently stumbling around in the dark ages. You've probably not mastered many techniques and are less likely to understand the advanced techniques. You are crawling out of the abyss if your score is between 40 and 60. You've mastered a few techniques but probably are not fully utilizing the advanced techniques. Finally, if your score is between 60 and 80, you are a fast, efficient defect squasher! You've mastered the techniques available to you and can quickly discover defects in your code.

If your score is a bit wanting, you do not need to get worried or upset. You can quickly increase your score with a little bit of work. First, start by reviewing how you ranked yourself in each category. Identify the categories where you ranked yourself five or less. Next, put together a small plan to increase your understanding and use of those techniques over the coming months. For example, if you don't use tracing, spend a few lunch periods reading about it and then experimenting with tracing in a controlled environment. Once you feel comfortable with it, start to employ it in your day-to-day troubleshooting.

# Final Thoughts

As developers, we must move beyond the thinking that bugs are crawling into our code and giving us grief. The reality is that there are defects in our code, and we're most likely the ones putting them there! In this chapter, we've explored a few ideas and a process you can follow to help you prepare for the inevitable defects that will occur in your code but will help you discover them as quickly as possible and remove them.

Changing how you think about bugs will 100% remove all bugs from your code! Instead, there will be only defects. Using the processes we've discussed, though, will allow you to catch defects quickly. As a result, you'll be able to decrease the time you spend troubleshooting your system, which should reduce stress and help developers become more effective. The benefits of having a software development process to minimize defects include

- Working fewer hours to meet deadlines

- More time to focus on product features

- Improved robustness and reliability

- Meeting project deadlines

- Detailed understanding of how the system is behaving and performing

## ACTION ITEMS

To put this chapter's concepts into action, here are a few activities the reader can perform to start decreasing how much time they spend debugging their code:

- Calculate what it costs you and the company you work for each year to debug your code:

    - What percentage of your time do you spend debugging?

    - Convert the percentage to hours spent.

    - Calculate your fully loaded salary and calculate your average hourly rate.

    - What are the total time and the total dollar amount?

- Based on how much time you spend each year debugging, what can you do to decrease the time by 10%? 20%? 50%? Put an action plan in place.

- What do you think about bugs? Do you need to change how you think about bugs and take more responsibility for them?

- Review the seven phases of the Jump-Starting Software Development to Minimize Defects process. Which phases do you need to implement before writing any additional code?

# PART III

# Development and Coding Skills

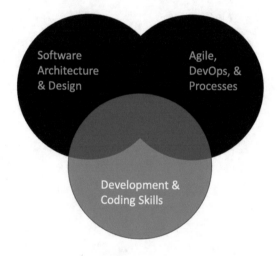

# CHAPTER 11

# Selecting Microcontrollers

Selecting the right microcontroller for the job can be a stressful and disconcerting endeavor. Take a moment to perform a microcontroller search at Digi-Key, Mouser, or your favorite parts distributor. While I write this, Digi-Key has over 90,000 microcontrollers available in their catalog from more than 50 different manufacturers! I suspect that there are still far more than this; it's just that Digi-Key does not have them on their line card.

The sheer number of microcontrollers available is overwhelming. So how does one go about finding a microcontroller that fits their application? That, dear reader, is precisely what we are going to discuss in this chapter.

## The Microcontroller Selection Process

When I first started to develop embedded software professionally in the early 2000s (yes, I was a maker in the mid-1990s before being a maker was a thing), selecting a microcontroller was simple; we let the hardware engineer do it! Microcontroller selection was left to the hardware engineer. At some point, in the design process, there would be an excellent unveiling meeting where the software engineers would learn about the processor they would be writing code for.

Software developers, at least in my experience, had little to no input into the microcontroller selection process. If we were lucky, the hardware designers would ask for feedback on whether the part was okay for software. In many cases, the unveiling of the hardware wasn't just schematics or a block diagram; it was the meeting where prototype hardware was provided to the developers. The modern selection process is quite different and emphasizes the software ecosystem (at least it should!). Failure to consider the software needs often results in a poor solution with a high chance of never making it to market.

© Jacob Beningo 2022
J. Beningo, *Embedded Software Design*, https://doi.org/10.1007/978-1-4842-8279-3_11

The modern microcontroller selection process contains seven significant steps:

1) Create a block diagram of the hardware

2) Identify all the system data assets

3) Perform a threat model and security analysis (TMSA)

4) Review the software model and architecture

5) Research microcontroller and software ecosystems

6) Evaluate development boards

7) Make the final microcontroller selection

These steps may seem obvious or entirely foreign, depending on how experienced you are in selecting processors. Let's examine each of these steps in more detail so that you can choose the right microcontroller on your next project. On the other hand, you may find that some of these steps are pretty familiar and correspond to the design philosophies we discussed in Chapter 1.

# Step #1 – Create a Hardware Block Diagram

The first step toward selecting a microcontroller, and probably for any step in developing an embedded system, is to create a hardware block diagram. The block diagram helps constrain the design and tells the engineer what they are dealing with. A software architecture can't be developed without one, and the microcontroller can't be selected without one either!

A block diagram serves several purposes in microcontroller selection. First, it identifies the external components connected to the microcontroller. These components could be anything from simple GPIO devices like push buttons or relays to more complex analog and digital sensors. Second, it identifies the peripherals needed to communicate with those external components. These could be peripherals like I2C, SPI, USART, USB, etc. Finally, the block diagram provides a minimum pin count that is required. This pin count can be used in online part searches to filter for parts that don't provide enough I/O.

---

**Best Practice**    Don't forget to include a few spare I/O for future hardware expansion and for debugging such as for toggling for timing measurements.

---

An early analysis of the hardware diagram can help to identify potential issues and shortcomings in the system. The analysis can also help to determine if the design is too complex or if there could be other business-related issues that could prevent the product from being completed successfully. An example block diagram for a smart thermostat can be seen in Figure 11-1. Notice that the block diagram segments the different external components based on the peripheral that will be used to interact with it. I will often create diagrams like this and add the pin count required for each to understand the I/O requirements. For example, Figure 11-1 currently has identified at least 23 I/O pins needed to connect all the external devices.

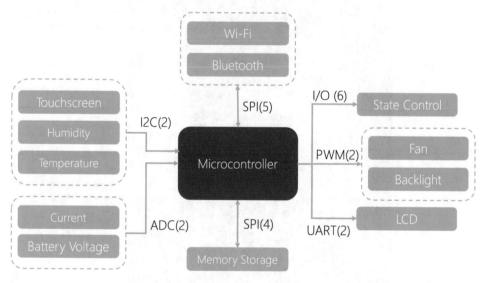

***Figure 11-1.*** *An example block diagram identifies significant system components and bus interfaces used to identify microcontroller features and needs*

Peripherals connected to external components aren't the only peripherals to consider when creating the block diagram. There are internal peripherals that can be just as important. For example, I've often found it helpful to develop a microcontroller-centric block diagram like that shown in Figure 11-2. This diagram identifies how many pins each peripheral uses in addition to internal peripheral needs. Internally, a microcontroller may need features such as

- Direct memory access (DMA)

- Cryptographic accelerators

- Watchdog timers

- Arm TrustZone

- Memory protection unit (MPU)

- Enhanced debug capabilities like the ETM

It's helpful to place these onto a block diagram to ensure that they are not overlooked when it is time to research and select a microcontroller. Otherwise, there may be too much focus on the external peripherals, and essential design features may be overlooked.

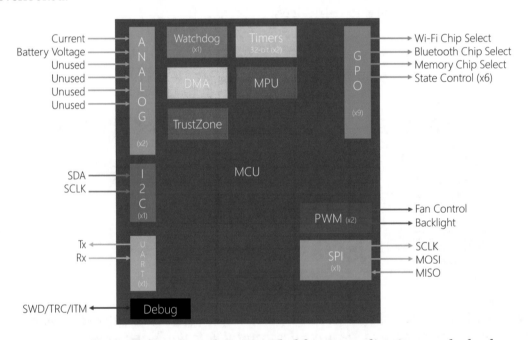

***Figure 11-2.*** *The peripherals and pins needed for an application can be broken down into a simple block diagram that shows the microcontroller requirements in a single, easy-to-digest image*

Once we understand the external and internal peripheral sets, we have identified important constraints on the microcontroller selection process. We know what hardware features our microcontroller needs, but we still don't know how much processing power the microcontroller needs. To understand the processing needs, we need to identify our data assets.

# Step #2 – Identify All the System Data Assets

Well, here we are again! We once again need to identify the data assets in our system! I hope this theme is helping you to understand just how important it is in embedded software and system design to focus on the data. As you should recall from Chapter 1, data dictates design! Selecting a microcontroller is part of the design, even if it is the hardware component of the design.

We want to start to identify the data assets at this point in the microcontroller selection process because the data will help a developer properly scope and size their software. The software required to achieve the desired system behavior will dictate whether a team can use a resource-constrained Arm Cortex-M0+, Cortex-M23, or something with more performance like an Arm Cortex-M7, Cortex-M33, or Cortex-M55 is required. Identifying the data assets used in the system can then be fed into a threat model security analysis (TMSA) which we perform in Step #3.

Identifying data assets can be a straightforward process. I often start by just creating a list of data. Then, I look at the system's inputs and outputs and make a bullet list in a word document. Once I have decomposed the system into the data assets, I create a simple table that lists the data and then try to fill in some operational parameters. Parameters that I often list include

- The type of data asset

- Data size

- Sample rate

- Processing requirements

At this stage, it's important to note that there is more to data assets than simply the operational inputs and outputs. The firmware itself is a data asset! Any certificates, public/private keys, and so forth are all data assets! You want to make sure that you identify each data asset. We often use terminology such as data in-flight (or in-transit) to indicate data that moves into, out of, and within the system. Data at rest is data that doesn't move but can still be more vulnerable to someone interested in attacking the system.

Table 11-1 shows an abbreviated example of what a data asset table might look like. The table lists the data assets that I've identified for the smart thermostat, the type of data it is, and several other parameters that would be pretty useful in scoping the system.

Note that this is not an activity in figuring out software stacks that will be used, digital filtering that would be applied, and so forth. It's just to identify the data assets and get a feel for how much processing power and software may be required.

***Table 11-1.*** *This is a sample data asset table. It has been abbreviated to minimize the space that it takes up. However, it provides a general overview of what a developer would want to list to identify data assets in their system*

| Data Asset | Asset Type | Data Size | Sample Rate | Processing |
|---|---|---|---|---|
| Analog | Sensor | 32 bytes | 1 kHz | Digital notch filter |
| Digital | Sensor | 128 bytes | 1 kHz | Running average – 5 sample |
| Firmware | IP | 256 Kbyte | – | See design documentation |
| Keys | Keys | 128 bytes | – | Secure storage |
| Device ID | Data | 128 bits | – | Secure storage |

# Step #3 – Perform a TMSA

Once the system data assets have been listed, it's time for the team to perform a threat model and security analysis (TMSA). The idea behind a TMSA is that it allows a team to evaluate the security threats that their system may face. A key to performing that analysis is to first list the system data assets, which was exactly what we did in Step #2. Next, we perform a TMSA, and do so early in the design cycle, because it provides us with the security objectives we need to design our software and select a microcontroller.

It's essential to recognize that security cannot be added to a system at the end of the development cycle. Like quality, it must be built into the system from the beginning. The TMSA will provide the security requirements for the system. If security is not needed, then the TMSA will be very fast indeed! Honestly, though, every system has some level of security. At a minimum, most businesses don't want a competitor to be able to pull their firmware from the device so they can reverse-engineer it. They often need secure firmware updates so a system can't have other firmware placed on it. When someone tells me that they don't need security, it just indicates that they don't know what they need and that I'll have to work extra hard to educate them.

The details about designing secure firmware and performing a TMSA can be found in Chapter 3. I won't rehash the details here, but if you are skipping around the book and want to understand the details, it may be worthwhile to pause and go back and read that chapter. Let's discuss the fourth step in the microcontroller selection process for the rest of us.

# Step #4 – Review the Software Model and Architecture

From the very beginning, software designers are hard at work figuring out what the software architecture looks like and how the software will behave. It's critical that the software be designed before the final microcontroller is selected! Without the software architecture, the chosen processor could be far too powerful for the application at hand, or, worse, it's possible that it won't have enough processing power for the application!

Now, you don't necessarily want to perfectly pair your microcontroller with the software you need to run. Instead, you generally want to have a little more processing power, memory, and capability than you need. The reason for this is that products often scale and grow over time. So if you pick the perfectly sized processor at launch and then try to add new features over the next several years, you may end up with a slow system or something that causes customers to complain about.

However, we need to figure out the right processing power, memory, and capabilities based on the software we need to run. We shouldn't just pick a processor because we like it or think it should work. Instead, we need to evaluate our software model and architecture and select the right processor based on the software needs. Yes, software dictates what hardware we need, NOT the other way around. In the "old days," that's exactly how things used to be done.

Typically, at this stage, I identify the software stacks and components that I need to meet the system's objectives. For example, if I know I have an Ethernet stack or USB, I want to list that out. That may require a specialized part, or it may lead me to decide I want to use multiple processors or even a multicore processor. The goal is to get a feel for what is needed to run the various software stacks and the stacks and middleware that will be required. Once we understand that, we can start researching different microcontrollers and the software ecosystems surrounding them!

# Step #5 – Research Microcontroller Ecosystems

I feel that over the past several years, the differentiation between different microcontrollers hasn't come from the physical hardware but the software ecosystems surrounding them. The hardware is standard and agnostic. Most vendors have moved to an Arm Cortex-M core, although some proprietary cores still exist, and RISC-V has been slowly gaining traction in some circles. The differences in hardware between vendors, though, are not terribly dramatic. Even though 32-bit microcontrollers have been coming to dominate the industry, sometimes the best match will be a little 8-bit part like a Microchip PIC processor. In our analysis, we shouldn't immediately rule these out.

Most GPIO peripherals have the same or very similar features. USARTs, SPI, PWM, and many other peripherals also provide similar capabilities. The registers to configure them are often different, but it doesn't matter whether I go with microcontroller Vendor A or Vendor B. In fact, the hardware capabilities are so standard that industry-wide hardware abstraction layers (HALs) have popped up to ease migration between vendors and software portability. (Note that I am exaggerating a bit; there are plenty of parts that are designed for specialized applications, but there are a lot of general-purpose parts out there too.)

The fundamental difference between microcontrollers is in the ecosystem that the vendor provides! In a world where we are constantly working with short delivery times, limited budgets, and so forth, the microcontroller's ecosystem can be the difference between whether we deliver on time with the expected quality or are over budget over time shipping junk. The ecosystem is the differentiator and probably should be one of the most significant factors we use as developers to select a microcontroller.

The ecosystem encompasses everything that supports the microcontroller, including

- Driver and configuration tools

- Real-time operating systems (RTOS)

- Middleware stacks

- Modeling and code generation tools

- Compilers and code optimizers

- Programming, trace, and debug tools

- The developer community and forums that support developers

The ecosystem can even include application examples and anything that makes the developers' job more manageable.

I once had a customer who came to me with his hardware already selected and developed. He just wanted someone to write the software. Upon examination of the hardware, I discovered the customer had chosen a microcontroller that was only available with a six-month lead time, was not supported by significant tool manufacturers, and had no example code or supporting forums because no one was using the part. Furthermore, they had selected their microcontroller poorly! As you can imagine, everything in that project was an uphill battle, and there was no one to turn to when things didn't work as expected.

Don't put yourself in that situation! Instead, select widely adopted, supported, available microcontrollers that have a rich ecosystem.

# Step #6 – Evaluate Development Boards

One of my favorite parts of the microcontroller selection process is purchasing development boards for evaluation. At this point in the project, we've usually gone through the due diligence of the design and are ready to start working on the low-level software. In some cases, the high-level business logic of the application has already been developed, and we are just missing the embedded processor. In other cases, we are evaluating and starting with the low-level firmware to prove any high-risk areas and help the hardware team develop their piece in parallel.

Evaluating development boards can also fill in design gaps. For example, it's good to put blocks in a design for MQTT, Wi-Fi, and cloud connectivity modules, but no team will write these themselves. Instead, the selected ecosystem and vendor will dictate these design blocks. In addition, running example software and software stacks on the development board can help pin down what modules, tasks, and timing are required. These elements, once known, can require adjustments to the design to ensure the system works as expected.

When evaluating development boards, there are generally a few tips I like to follow. First, I recognize that the development board microcontroller is likely far more powerful than the microcontroller I will select. An evaluation board is usually the fastest and best processor in that chip family. The processor will have the most RAM, ROM, and so forth. The quicker I can build a rapid prototype of my application on the evaluation board, the better idea I can get about which part in the microcontroller family will best fit my application.

Chances are that nearly an unlimited number of vendors could supply a microcontroller for the desired application. So the first thing I do is pick two potential vendors I already have experience with. I'll then look for a third that I don't have experience with, but that looks like they have an ecosystem that fits our needs. I'll then look through their offerings to identify several microcontrollers that could work and build up a list like Table 11-2 that lists out the development boards that are available for evaluation.

***Table 11-2.***   *This example table lists development boards that might be evaluated for an IoT application. Again, this is truncated for space but would include additional details like the MCU part #, price, lead times, and checkboxes for whether the board met specific requirements*

| Vendor | Development Board | Clock Speed (MHz) | RAM (KB)/ROM (KB) | Features |
| --- | --- | --- | --- | --- |
| STM | L4S51-IOT01A | 120 | 640/2048 | Wi-Fi, sensors, etc. |
| STM | IDW01M1 | – | 64/512 | Wi-Fi, sensors, etc. |
| Microchip | SAM-IoT | 20 | 6/48 | Wi-Fi, security, etc. |

An important point to note is that I don't look at the price of the evaluation boards. They will range from $20 to $600. I decide which one has everything I need to make my evaluation. Suppose that is the $20 board, awesome. If it's the $600 board, that's perfectly fine. I don't let the price get in the way. Sure, if I can save $500, that's great, but I'd rather spend the $500 if it will shave days off my evaluation or, more importantly, help me to select the correct part. Choosing the wrong part will cost more than $500, so I'd instead do it right. (Too many developers get hung up on the price for boards, tools, training, etc. It may seem like a lot to the developer, but these costs don't even make the decimal place on the balance sheet to the business. So don't cut corners to save a buck in the short term and cost the business tens of thousands later.)

Once the selected development boards arrive, spend a few days with each to get the main components and examples up and running. Evaluate how well the base code and configurations fit your application. Use trace tools to evaluate the architectural components that are already in place. Explore how the existing code could impact your architecture. Take some time to run out to the forums and ask questions and see how quickly you can get a response. Get the contact information for the local Field Application Engineer (FAE) and introduce yourself and ask questions again. Again, check and see how quickly they respond. (Don't base too much of your decision on response time, though. If you are only building 1000 units, the forum will be where you get most of your support. The FAE will do their best, but they often are pulled in many directions and must respond to the customers by building millions or hundreds of thousands of units first.)

## Step #7 – Make the Final MCU Selection

The last step is to make the final selection on the microcontroller, which one seems to fit best for the application. In most cases, this is not a life-or-death decision. Many systems can be built successfully no matter which microcontroller is selected. What is really on the line is whether the microcontroller and ecosystem will match the product and business needs. Unlike at school or college, there isn't necessarily a right or wrong answer. It's finding the best fit and then going for it!

Our discussion on selecting a microcontroller has been high, with some tips and tricks scattered throughout. You might wonder how you can be certain that the selected part will be a good fit for your product and the company you work for. If you are working on a team, you'll discover that every engineer will have their own opinion on which part should be used. In many cases, personal bias can play a massive role in the microcontroller selection process. Let's now examine a process I use myself and my clients to help remove those biases and select the right microcontroller for the application.

## The MCU Selection KT Matrix

Whether or not we want to admit it, our personal bias can play a significant role in our decision-making processes. For example, if I have used STM32 microcontrollers in the past with ThreadX successfully, I'm going to be more biased toward using them. If I am an open source junky, I may push the team to avoid using a commercial RTOS like uC OS II/III or VxWorks, even if they are the right solution for the project. When selecting a microcontroller or making any major team decision, we need to try to remove personal bias from the decision. Now don't get me wrong, some personal bias can be good. If I'm familiar with the STM32 parts, that can be an asset over using a part I am completely unfamiliar with. However, we do want to remove unmanaged bias that doesn't benefit the microcontroller selection process. This is where a KT Matrix can make a big difference.

A KT Matrix, more formally known as the Kepner Tregoe matrix, is a step-by-step approach for systematically solving problems, making decisions, and analyzing potential risks.[1] KT Matrixes limit conscious and unconscious biases that steer a decision away from its primary objectives. The KT Matrix has been used to make organizational decisions for decades, but while it is a standard tool among project managers, embedded engineers rarely have been exposed to it. We're going to see precisely how this tool can be used to help us select the right microcontroller for our application.

## Identifying Decision Categories and Criterions

The idea behind a KT Matrix is that we objectively identify the decision we want to make. Therefore, we start by defining the problem. For example, we want to decide which microcontroller seems to be the best fit for our application. To do this, we need to identify the criteria used to make the decision and determine the weights each criterion has on our decision-making process. For example, I may decide to evaluate several categories in my decision-making, such as

- Hardware

- Features

- Cost

- Ecosystem

- Middleware

- Vendor

- Security

- Experience

These are the high-level categories that I want to evaluate. Each category will have several criteria associated with it that help to narrow down the decision. For example, the middleware category may include

- RTOS

- File system

---

[1] www.valuebasedmanagement.net/methods_kepner-tregoe_matrix.html

- TCP/IP networking stacks
- USB library
- Memory management

When I identify each criterion, I can apply weight to how important it is to my decision. I will often use a scale of one to five, where one is not that important to the decision and five is very important. I'm sure you can see that the weight for each criterion can easily vary based on opinion. We all know that from a business perspective, management will push the cost to be a rating of four or five. To an engineer, though, the price may be a two. The KT Matrix is powerful in that we can get the stakeholders together to decide on which categories and criteria will be evaluated and how much weight each should have!

## Building the KT Matrix

Once the categories and criteria have been decided, a KT Matrix can be developed using a simple spreadsheet tool like Excel or Google Sheets. When building the KT Matrix, there are several columns that we want to make sure the spreadsheet has, such as

- Criterion
- Criterion weight (1–5)
- Microcontroller option (at least 3)

Each microcontroller column is also broken up to contain a column for each person who will help decide whether the microcontroller fits the application or not. Each person rates on a scale from one to five how well the criterion meets the product's needs. These are then all weighted and summed together for each team member. Figure 11-3 shows an abbreviated example of how the microcontroller KT Matrix looks using just a few categories.

| | Criteria | Weight | Microcontroller #1 | | | | | | Microcontroller #2 | | | | | |
|---|---|---|---|---|---|---|---|---|---|---|---|---|---|---|
| | | | Rating 1 | Rating 2 | Rating 3 | Rating 4 | Rating 5 | Weighted Rating Total | Rating 1 | Rating 2 | Rating 3 | Rating 4 | Rating 5 | Weighted Rating Total |
| Hardware | 32-bit Architecture | 4 | | | | | | 0 | | | | | | 0 |
| | Processor speed | 4 | | | | | | 0 | | | | | | 0 |
| | Instruction set | 5 | | | | | | 0 | | | | | | 0 |
| | Minimial interrupt latency | 5 | | | | | | 0 | | | | | | 0 |
| | Loweset energy consumption | 5 | | | | | | 0 | | | | | | 0 |
| | Part Availability | 5 | | | | | | 0 | | | | | | 0 |
| | Memroy footprint / speed | 4 | | | | | | 0 | | | | | | 0 |
| Features | Best Real-time trace capabilities (ITM, ETM) | 3 | | | | | | 0 | | | | | | 0 |
| | Memory protection unit (MPU) | 4 | | | | | | 0 | | | | | | 0 |
| | FPU | 4 | | | | | | 0 | | | | | | 0 |
| | Driver and middleware configuration tools | 5 | | | | | | 0 | | | | | | 0 |
| | Safety certifications | 5 | | | | | | 0 | | | | | | 0 |
| | Hardware accelerated cryptography | 5 | | | | | | 0 | | | | | | 0 |
| | Multicore | 3 | | | | | | 0 | | | | | | 0 |
| | RTOS Support | 3 | | | | | | 0 | | | | | | 0 |
| | Wireless Connectivity | 4 | | | | | | 0 | | | | | | 0 |
| Cost | Lowest upfront licensing costs | 5 | | | | | | 0 | | | | | | 0 |
| | Lowest royalty cost per unit | 3 | | | | | | 0 | | | | | | 0 |
| | Smallest tool investment | 4 | | | | | | 0 | | | | | | 0 |
| | Lowest training investment | 5 | | | | | | 0 | | | | | | 0 |
| | Lowest cost of middleware (price and integration effort vs quality) | 5 | | | | | | 0 | | | | | | 0 |
| | Least open source ( minimize new IP release ) | 3 | | | | | | 0 | | | | | | 0 |

*Figure 11-3.*  *An example microcontroller selection KT Matrix shows how a team can evaluate how well a microcontroller fits their application unbiasedly*

A more detailed example, and one that is more readable, can be downloaded and modified by the reader from beningo.com using the link: `https://bit.ly/3cL1P2X`. Now that the matrix is in place, let's discuss how we perform our analysis.

# Choosing the Microcontroller

What I love most about the KT Matrix is that the decision-making process is reduced to nothing more than a numerical value! At the end of the day, the microcontroller that has the highest numerical value is the microcontroller that best fits the application for the decision makers! (After all, the results would differ for a different team.) Getting the numerical value is easy.

First, each team member reviews the KT Matrix and the microcontrollers. They go through the criterion and rate on a scale from one to five how well that criterion meets the team's needs. Next, once each team member has put in their ranking, each criterion is weighted and summed for each microcontroller. Remember, we applied weight to each criterion based on how important it was to our decision. For example, if cost is weighted as four, we sum each decision maker's cost rating and multiply by four. We do this for each criterion. Once all the criteria have been summed and weighted, we sum up each microcontroller's weight column. Finally, the microcontroller with the largest weighted sum is our choice! An example can be seen in Figure 11-4.

| | Criteria | Weight | Microcontroller #1 | | | | | | Microcontroller #2 | | | | | |
|---|---|---|---|---|---|---|---|---|---|---|---|---|---|---|
| | | | Rating 1 | Rating 2 | Rating 3 | Rating 4 | Rating 5 | Weighted Rating Total | Rating 1 | Rating 2 | Rating 3 | Rating 4 | Rating 5 | Weighted Rating Total |
| Hardware | 32-bit Architecture | 4 | 3 | 3 | 3 | 3 | 3 | 60 | 2 | 2 | 2 | 2 | 2 | 40 |
| | Processor speed | 4 | 2 | 2 | 2 | 2 | 2 | 40 | 1 | 1 | 1 | 1 | 1 | 20 |
| | Instruction set | 5 | 2 | 1 | 1 | 1 | 2 | 35 | 1 | 2 | 2 | 2 | 1 | 40 |
| | Minimial interrupt latency | 5 | 1 | 2 | 2 | 1 | 1 | 35 | 3 | 1 | 1 | 3 | 2 | 50 |
| | Loweset energy consumption | 5 | 1 | 1 | 1 | 1 | 1 | 25 | 2 | 2 | 2 | 2 | 2 | 50 |
| | Part Availability | 5 | 1 | 2 | 1 | 1 | 1 | 30 | 2 | 3 | 3 | 3 | 3 | 70 |
| | Memroy footprint / speed | 4 | 3 | 3 | 3 | 3 | 3 | 60 | 2 | 2 | 2 | 2 | 2 | 40 |
| Middleware | File system best meets system requirements | 4 | 2 | 1 | 2 | 2 | 1 | 32 | 3 | 2 | 3 | 3 | 1 | 48 |
| | TCP/IP stack best meets system requirements | 4 | 2 | 1 | 2 | 2 | 1 | 32 | 3 | 2 | 3 | 3 | 1 | 48 |
| | USB stack best meets system requirements | 4 | 2 | 1 | 2 | 2 | 1 | 32 | 3 | 2 | 3 | 3 | 1 | 48 |
| | Graphics stack best meets system requirements | 4 | 2 | 1 | 2 | 2 | 1 | 32 | 3 | 2 | 3 | 3 | 1 | 48 |
| | Middleware requires minimal integration effort | 4 | 2 | 1 | 2 | 2 | 1 | 32 | 3 | 2 | 3 | 3 | 1 | 48 |
| | Additional 3rd party tools integrated seamlessly | 3 | 1 | 2 | 1 | 2 | 1 | 21 | 2 | 3 | 2 | 3 | 2 | 36 |
| Engineer | Maxmize professional growth potential | 2 | 2 | 2 | 1 | 3 | 1 | 18 | 1 | 1 | 3 | 2 | 3 | 20 |
| | Least amount of stress to implement | 2 | 2 | 3 | 1 | 1 | 3 | 20 | 1 | 2 | 3 | 3 | 2 | 22 |
| | Most fun / interesting | 1 | 2 | 3 | 3 | 1 | 2 | 11 | 3 | 1 | 1 | 2 | 3 | 10 |
| | Minimized labor intensity | 3 | 1 | 2 | 3 | 1 | 3 | 30 | 2 | 3 | 1 | 2 | 1 | 27 |
| | Least deadline constrained to get up to speed | 2 | 2 | 1 | 2 | 1 | 3 | 18 | 3 | 2 | 3 | 2 | 1 | 22 |
| | Most internal resources available | 3 | 1 | 2 | 3 | 3 | 3 | 36 | 2 | 3 | 1 | 1 | 1 | 24 |
| Security | Security Certified RTOS | 5 | 2 | 2 | 1 | 3 | 1 | 45 | 3 | 3 | 2 | 1 | 2 | 55 |
| | Supports Arm TrustZone | 4 | 1 | 1 | 2 | 1 | 1 | 24 | 2 | 2 | 3 | 2 | 2 | 44 |
| | Supports TF-M | 5 | 1 | 1 | 1 | 2 | 2 | 35 | 2 | 2 | 2 | 3 | 3 | 60 |
| | Secure OTA / Bootloader support | 3 | 2 | 2 | 1 | 2 | 2 | 27 | 1 | 1 | 2 | 3 | 3 | 30 |
| | Total | 198 | 98 | 94 | 101 | 101 | 95 | 1852 | 104 | 113 | 109 | 116 | 102 | 2059 |
| | | | Microcontroller #1 | | | | | | Microcontroller #2 | | | | | |

**Figure 11-4.**  *An example microcontroller selection KT Matrix that the team members have evaluated. In this example, we can see that microcontroller #2 best fits the application and team needs*

Looking closely at Figure 11-4, the reader can see that the weighted sum for microcontroller #1 is 1852, while the weighted sum for microcontroller #2 is 2059. In this example, microcontroller #2 has a higher numeric value and is the better fit for our application and team.

The KT Matrix is an excellent tool because it lets us make a decision objectively! The selection process is not determined by a single engineer with the strongest personality or management pushing down on costs. At the end of the day, selecting the right microcontroller for the job can make or break budgets, time to market, and the entire business!

# Overlooked Best Practices

There are several best practices associated with microcontroller selection that teams can very easily overlook. Failure to adhere to these best practices won't end the world, but it can make things more difficult. If the project is more complicated, deadlines will likely be missed, budgets will be consumed, and other nasty things may happen.

The first overlooked best practice is to size the microcontroller memory, RAM, and ROM so that there are more than the current application needs. Sizing the memory early can be really tough. Designers can look for microcontroller families that are pin compatible that allow different size memories to be selected. Final microcontroller selection can then be pushed off toward the end of development when the true memory needs are known.

The reason for oversizing the memory is to leave room for future features. I can't remember a single product I've worked on, consulted on, or advised on that did not add features to the product after the initial launch. Customers always ask for more, and businesses always want to keep adding value to their offerings. During microcontroller selection, developers need to anticipate this scope creep!

The next overlooked best practice is to select a microcontroller that has extra pins that are unused. Products evolve, and having to choose a new microcontroller, rewrite firmware, and update a PCB can be expensive and time-consuming! Instead, select a microcontroller that leaves room to add new hardware features in the future. For example, these pins can be used to add new sensors, or they could also be used during development to debug the application code.

Another overlooked best practice is to select a microcontroller that either has a built-in bootloader or that is sized for you to add a bootloader. Bootloaders are fantastic for providing in-field update capabilities. But, again, so many products evolve after launch that it's essential to provide a mechanism to update the firmware. I can't tell you how many times I have had a client tell me that they don't need a bootloader, only to call me within a month after launch telling me they need help making one (despite my advice to have one from the start!).

Using a bootloader can also add constraints to the design. It's hard to design a robust bootloader, especially a secure one. In many cases, to do it right, you need to trade off risk by having twice the memory so that you can keep a backup copy. The price of the microcontroller can be much larger due to these memory needs. Bootloaders aren't trivial and should be looked at early in the development cycle as well.

The last best practice for selecting a microcontroller isn't about choosing a microcontroller! But it can save some headaches for developers needing to upgrade their microcontroller. When laying out a PCB, leave extra space around the microcontroller. Then, if the microcontroller ever goes end of life, gets a long lead time, or needs to be upgraded, a developer may be able to save time and budget on the upgrade if the entire PCB doesn't need to be redesigned. This trick saved me several times, especially during the COVID era, when microcontrollers became very hard to acquire in the supply chain.

# Final Thoughts

Selecting the right microcontroller can be a tough job! There are so many choices and considerations that the selection process can be dizzying. In this chapter, we've seen that we can follow a simple process to identify and evaluate the microcontrollers that may best fit our applications. In addition, designers can't forget that the main purpose of selecting a microcontroller is to reduce the risk of failure. Businesses at the end of the day are looking to go to market and make a profit! Being too stingy or trying to overoptimize and use a small part can be a disaster.

Choose carefully. Consider the options. Evaluate future needs and availability and then find the microcontroller that is right for your application.

---

## ACTION ITEMS

To put this chapter's concepts into action, here are a few activities the reader can perform to start applying microcontroller selection best practices to their application(s):

- Work through the seven steps to select a microcontroller for your design or a fictitious design.

    - How is this process different from the one you used in the past?

    - Does/did it help you select a microcontroller that was a closer fit for your application?

- List the top three challenges that you encounter when you select a microcontroller. Then, what can you do going forward to alleviate or minimize these challenges?

- Download the RTOS Selection KT Matrix and examine how the KT Matrix works. (Hint: You don't need to use all the criteria I have put there! You can remove them, add more, or do whatever you need to select your microcontroller!)

- In the RTOS Selection KT Matrix, explore the results tab. You'll notice that I break out how much weight each category has on the decision-making process. Why do you think this is important to do?

These are just a few ideas to go a little bit further. Carve out time in your schedule each week to apply these action items. Even small adjustments over a year can result in dramatic changes!

# CHAPTER 12

# Interfaces, Contracts, and Assertions

An essential feature of every object, module, package, framework, and so forth is that it has a well-defined and easy-to-use interface. Perhaps most important is that those interfaces create a binding contract between the interface and the caller. If the caller were to break that contract, a bug would be injected into the code. Without a software mechanism to verify that the contract is being met, the bugs will not just be injected but also run rampant and perhaps become challenging to find and detect.

This chapter will look at developing contracts between interfaces and the calling code. Creating the contract will allow us to introduce and explore the use of assertions available in both C and C++. Assertions in real-time systems can potentially be dangerous, so once we understand what they are and how we can use them to create an interface contract, we will also explore real-time assertions.

---

**Note** I wrote extensively in my book *Reusable Firmware Development* about how to design interfaces and APIs. For that reason, I'll briefly introduce interfaces and dive into contracts and assertions. For additional details on interfaces, see *Reusable Firmware Development*.

---

## Interface Design

An interface is how an object, person, or thing interacts with another item. For example, developers will often work with several different types of interfaces, such as an application programming interface (API) or a graphical user interface (GUI). These interfaces allow someone to request an action to be performed, and the interface provides a mechanism to request, perform, and respond to that action.

© Jacob Beningo 2022
J. Beningo, *Embedded Software Design*, https://doi.org/10.1007/978-1-4842-8279-3_12

The interfaces we are most interested in for this chapter are APIs. APIs are the functions and methods we use as developers to expose features in our application code. For example, Listing 12-1 shows an example interface for a digital input/output (DIO) driver. It includes interactions such as initialization, reading, and writing. The driver also has an interface to extend the API's functionality through low-level interactions like reading and writing registers and defining callback functions for managing interrupt service routine behavior.

***Listing 12-1.*** An example digital input/output interface used to interact with a microcontroller's general-purpose input/output peripheral. (Note: TYPE is replaced with uint8_t, uint16_t, or uint32_t depending on the microcontroller architecture.)

```
void          Dio_Init(DioConfig_t const * const Config);
DioPinState_t Dio_ChannelRead(DioChannel_t const Channel);
void          Dio_ChannelWrite(DioChannel_t const Channel,
                               DioPinState_t const State);
void          Dio_ChannelToggle(DioChannel_t const Channel);
void          Dio_ChannelModeSet(DioChannel_t const Channel,
                                 DioMode_t const Mode);
void          Dio_ChannelDirectionSet(DioChannel_t Channel,
                                      PinModeEnum_t Mode);
void          Dio_RegisterWrite(uint32_t const Address,
                                TYPE const Value);
TYPE          Dio_RegisterRead(uint32_t const Address);
void          Dio_CallbackRegister(DioCallback_t Function,
                      TYPE (*CallbackFunction)(type));
```

Interfaces can come in many shapes and sizes and can be found at all layers of a system, such as in the drivers, middleware, operating system, and even the high-level application code. The interface abstracts low-level details the developer doesn't need to know and packages them behind an easy-to-use interface. The best interfaces are designed and implemented to create a binding contract between the interface and the caller. Personally, I prefer to limit the interface for any module, package, or object to around a dozen or so methods if possible. I find that doing so makes it easier to remember the interface. (Most humans can only retain information in chunks of 7–12 items, as we've discussed before. So if you keep any given interface to around 7–12 methods, it's more likely you can commit it to memory and therefore minimize the time needed to review documentation.)

An interesting technique that developers should be interested in but that I often see a lack of support for in many designs and programming languages is Design-by-Contract (DbC).

# Design-by-Contract

The idea behind Design-by-Contract programming is that every interface should be clearly defined, precise, and verifiable. This means that every interface should have clearly stated

- Preconditions
- Side effects[1]
- Postconditions
- Invariants[2]

By formally defining these four items, the interface creates a fully specified contract that tells the developer

- What needs to happen before calling the interface for it to operate properly

- What the expected system state will be after calling the interface

- The properties that will be maintained upon entry and exit

- Changes that will be made to the system

Design-by-Contract creates the obligations, the contract, that a developer using the interface must follow to benefit from calling that interface.

---

[1] The interface modifies, interacts, or changes the state of the system outside the interface scope.

[2] A set of assertions that must always be true during the lifetime of the interface.

# Utilizing Design-by-Contract in C Applications

At first glance, Design-by-Contract might seem like it is yet another overweight process to add to an already overloaded development cycle. However, it turns out that using Design-by-Contract can be pretty simple and can improve your code base documentation and decrease the defects in the software, which has the added benefit of improving software quality. Design-by-Contract doesn't have to be super formal; instead, it can be handled entirely in your code documentation. For example, look at the Doxygen function header for a digital input/output (DIO) initialization function shown in Listing 12-2.

As you can see from the example, the comment block has a few additional lines associated with it that call out the function's preconditions and postconditions. In this case, for Dio_Init, we specify that

- A configuration table that specifies pin function and states needs to be populated.

- The definition for the number of pins on the MCU must be greater than zero.

- The definition for the number of pin ports must be greater than zero.

- The GPIO clock must be enabled.

If these conditions are all met, when a developer calls the Dio_Init function and provides the specified parameters, we can expect that the output from the function will be configured GPIO pins that match the predefined configuration table.

***Listing 12-2.*** The Doxygen documentation for a function can specify the conditions in the Design-by-Contract for that interface call

```
/*********************************************************
* Function : Dio_Init()
*//**
* \b Description:
*
* This function is used to initialize the Dio based on the
* configuration table defined in dio_cfg module.
*
```

```
* PRE-CONDITION: Configuration table needs to populated
* (sizeof > 0) <br>
* PRE-CONDITION: NUMBER_OF_CHANNELS_PER_PORT > 0 <br>
* PRE-CONDITION: NUMBER_OF_PORTS > 0 <br>
* PRE-CONDITION: The MCU clocks must be configured and
* enabled.
*
* POST-CONDITION: The DIO peripheral is setup with the
* configuration settings.
*
* @param[in]  Config is a pointer to the configuration
* table that contains the initialization for the peripheral.
*
* @return         void
*
* \b Example:
* @code
*    const DioConfig_t *DioConfig = Dio_ConfigGet();
*    Dio_Init(DioConfig);
* @endcode
*
* @see Dio_Init
* @see Dio_ChannelRead
* @see Dio_ChannelWrite
* @see Dio_ChannelToggle
* @see Dio_ChannelModeSet
* @see Dio_ChannelDirectionSet
* @see Dio_RegisterWrite
* @see Dio_RegisterRead
* @see Dio_CallbackRegister
******************************************************************/
```

Documentation can be an excellent way to create a contract between the interface and the developer. However, it suffers from one critical defect; the contract cannot be verified by executable code. As a result, if a developer doesn't bother to read the documentation or pay close attention to it, they can violate the contract, thus injecting a bug into their source code that they may or may not be aware of. At some point, the bug will rear its ugly head, and the developer will likely need to spend countless hours hunting down their mistake.

Various programming languages deal with Design-by-Contract concepts differently, but for embedded developers working in C/C++, we can take advantage of a built-in language feature known as assertions.

# Assertions

The best definition for an assertion that I have come across is

"An *assertion* is a Boolean expression at a specific point in a program that will be true unless there is a bug in the program."[3]

There are three essential points that we need to note about the preceding definition, which include

- An assertion evaluates an expression as either true or false.

- The assertion is an *assumption* of the state of the system at a specific point in the code.

- The assertion is validating a system assumption that, if not true, *reveals a bug* in the code. (It is not an error handler!)

As you can see from the definition, assertions are particularly useful for verifying the contract for an interface, function, or module.

Each programming language that supports assertions does so in a slightly different manner. For example, in C/C++, assertions are implemented as a macro named `assert` that can be found in assert.h. There is quite a bit to know about assertions, which we will cover in this chapter, but before we move on to everything we can do with assertions, let's first look at how they can be used in the context of Design-by-Contract. First, take a moment to review the contract that we specified in the documentation in Listing 12-2.

---

[3] Sadly, I committed this definition to memory and have been unable to find the original source!

Assertions can be used within the Dio_Init function to verify that the contract specified in the documentation is met. For example, if you were to write the function stub for Dio_Init and include the assertions, it would look something like Listing 12-3. As you can see, for each precondition, you have an assertion. We could also use assertions to perform the checks on the postcondition and invariants. This is a bit stickier for the digital I/O example because we may have dozens of I/O states that we want to verify, including pin multiplexing. I will leave this up to you to consider and work through some example code on your own.

***Listing 12-3.*** An example contract definition using assertions in C

```
void Dio_Init(DioConfig_t const * const Config)
{
    assert(sizeof(Config) > 0);
    assert(CHANNELS_PER_PORT > 0);
    assert(NUMBER_OF_PORTS > 0);
    assert(Mcu_ClockState(GPIO) == true);

    /* TODO: Define the implementation */
}
```

# Defining Assertions

The use of assertions goes well beyond creating an interface contract. Assertions are interesting because they are pieces of code developers add to their applications to verify assumptions and detect bugs directly. When an assertion is found to be true, there is no bug, and the code continues to execute normally. However, if the assertion is found to be false, the assertion calls to a function that handles the failed assertion.

Each compiler and toolchain tend to implement the assert macro slightly differently. However, the result is the same. The ANSI C standard dictates how assertions must behave, so the differences are probably not of interest if you use an ANSI C–compliant compiler. I still find it interesting to look and see how each vendor implements it, though. For example, Listing 12-4 shows how the STM32CubeIDE toolchain defines the assert macro. Listing 12-5 demonstrates how Microchip implements it in MPLAB X. Again, the result is the same, but how the developer maps what the assert macro does if the result is false will be different.

**Listing 12-4.** The assert macro is defined in the STM32CubeIDE assert.h header

```
/* required by ANSI standard */
#ifdef NDEBUG
# define assert(__e) ((void)0)
#else
# define assert(__e) ((__e) ? (void)0 : __assert_func \
                 (__FILE__, __LINE__, __ASSERT_FUNC, #__e))
```

Looking at the definitions, you'll notice a few things. First, the definition for macros changes depending on whether NDEBUG is defined. The controversial idea here is that a developer can use assertions during development with NDEBUG not defined. However, for production, they can define NDEBUG, changing the definition of assert to nothing, which removes it from the resultant binary image. This idea is controversial because you should ship what you test! So if you test your code with assertions enabled, you should ship with them enabled! Feel free to ship with them disabled if you test with them disabled.

**Listing 12-5.** The assert macro is defined in the MPLAB X assert.h header

```
#ifdef NDEBUG
#define    assert(x) (void)0
#else
#define assert(x) ((void)((x) || (__assert_fail(#x, __FILE__,\
                            __LINE__, __func__),0)))
#endif
```

Next, when we are using the full definition for assert, you'll notice that there is an assertion failed function that is called. Again, the developer defines the exact function but is most likely different between compilers. For STM32CubeIDE, the function is __assert_func, while for MPLAB X the function is __assert_fail.[4] Several key results occur when an assertion fails, which include

---

[4] Sigh. The devil is always in the details. If you use assert, you literally must check your compiler implementation.

- Collecting the filename and line number where the assertion failed

- Printing out a notification to the developer that the assertion failed and where it occurred

- Stopping program execution so that the developer can take a closer look

The assertion tells the developer where the bug was detected and halts the program so that they can review the call path and memory states and determine what exactly went wrong in the program execution. This is far better than waiting for the bug to rear its head in how the system behaves, which may occur instantly or take considerable time.

---

**Caution**    It is important to note that just stopping program execution could be dangerous in a real-time application driving a motor, solenoid, etc. The developer may need to add extra code to handle these system activities properly, but we will discuss this later in the chapter.

---

# When and Where to Use Assertions

The assert macro is an excellent tool to catch bugs as they occur. That also preserves the call stack and the system in the state it was in when the assertion failed. This helps us to pinpoint the problem much faster, but where does it make sense to use assertions? Let's look at a couple proper and improper uses for assertions.

First, it's important to note that we can use assertions anywhere in our program where we want to test an assumption, meaning assertions could be found just about anywhere within our program. Second, I've found that one of the best uses for assertions is verifying function preconditions and postconditions, as discussed earlier in the chapter. Still, they can also be "sprinkled" throughout the function code.

As a developer writes their drivers and application code, in every function, they analyze what conditions must occur before this function is executed for it to run correctly. They then develop a series of assertions to test those preconditions. The preconditions become part of the documentation and form a contract with the caller on what must be met for everything to go smoothly.

A great example of this is a function that changes the state of a system variable in the application. For example, an embedded system may be broken up into several different system states that are defined using an enum. These states would then be passed into a function like System_StateSet to change the system's operational state, as shown in Listing 12-6.

***Listing 12-6.*** A set function for changing a private variable in a state machine

```
void System_StateSet(SystemState_t const State)
{
    SystemState = State;
}
```

What happens if only five system states exist, but an application developer passes in a state greater than the maximum state? In this case, the system state is set to some nonexistent state which will cause an unknown behavior in the system. Instead, the software should be written such that a contract exists between the application code and the System_StateSet function that the state variable passed into the function will be less than the maximum state. If anything else is passed in, that is a defect, and the developer should be notified. We can do this using assert, and the updated code can be seen in Listing 12-7.

***Listing 12-7.*** A set function for changing a private variable in a state machine with an assertion to check the preconditions

```
void System_StateSet(SystemState_t const State)
{
    assert(State < SYSTEM_STATE_MAX);
    SystemState = State;
}
```

Now you might say that the assertion could be removed and a simple if statement used to check the parameter. This would be acceptable, but that is error handling! In this case, we are constraining the conditions under which the function executes, which means we don't need error handling but a way to detect an improper condition (bug) in the code. This brings us to an important point; **assertions should NOT be used for error handling!** For example, a developer should NOT use an assertion to check that a file exists on their file system, as shown in Listing 12-8. In this case, creating an error handler to create a default version of UserData.cfg makes a lot more sense than signaling the developer that there is a bug in the software. There is not a bug in the software but a file that is missing from the file system; this is a runtime error.

*Listing 12-8.* An INCORRECT use of assertions is to use them for error handling

```
FileReader = fopen("UserData.cfg", 'r');

assert(FileReader != NULL);
```

# Does Assert Make a Difference?

There is a famous paper that was written by Microsoft[5] several years ago that examined how assertion density affected code quality. They found that if developers achieved an assertion density of 2–3%, the quality of the code was much higher, which meant that it had fewer bugs in the code. Now the paper does mention that the assertion density must be a true density which means developers aren't just adding assertions to reach 2–3%. Instead, they are adding assertions where it makes sense and where they are genuinely needed.

# Setting Up and Using Assertions

Developers can follow a basic process to set up assertions in their code base. These steps include

1.  Include <assert.h> in your module.

2.  Define a test assertion using assert(false);.

3.  Compile your code (this helps the IDE bring in function references).

4.  Right-click your assert and select "goto definition."

5.  Examine the assertion implementation.

6.  Implement the assert_failed function.

7.  Compile and verify the test assertion.

---

[5] http://laser.cs.umass.edu/courses/cs521-621.Fall10/documents/ISSRE_000.pdf

I recommend following this process early in the development cycle. Usually, after I create my project, I go through the process before writing even a single line of code. If you set up assertions early, they'll be available to you in the code base, which hopefully means you'll use them and catch your bugs faster.

The preceding process has some pretty obvious steps. For example, steps 1–4 don't require further discussion. We examined how assertions are implemented in two toolchains earlier in the chapter. Let's change things up and look at steps 5–7 using Keil MDK.

## Examining the Assert Macro Definition

After walking through steps 1–4, a developer would need to examine how assertions are implemented. When working with Keil MDK, the definition for assert located in assert.h is slightly different from the definitions we've already explored. The code in Figure 12-1 shows the definitions for assert, starting with the NDEBUG block. There are several important things we should notice.

```
54 ⊟#ifdef NDEBUG
55 │#   define assert(ignore) ((void)0)
56 │#   define __promise(e) ((__ARM_PROMISE)((e)?1:0))
57 │#else
58 ⊟#   if defined __DO_NOT_LINK_PROMISE_WITH_ASSERT
59 ⊟#       if defined __OPT_SMALL_ASSERT && !defined __ASSERT_MSG && !defined __STRICT_ANSI__ && !(__AEABI_PORTABILITY_LEVEL !
60 │#           define assert(e) ((e) ? (void)0 : __CLIBNS abort())
61 │#       elif defined __STDC__
62 │#           define assert(e) ((e) ? (void)0 : __CLIBNS __aeabi_assert(#e, __FILE__, __LINE__))
63 │#       else
64 │#           define assert(e) ((e) ? (void)0 : __CLIBNS __aeabi_assert("e", __FILE__, __LINE__))
65 ┤#       endif
66 │#       define __promise(e) ((__ARM_PROMISE)((e)?1:0))
67 │#   else
68 ⊟#       if defined __OPT_SMALL_ASSERT && !defined __ASSERT_MSG && !defined __STRICT_ANSI__ && !(__AEABI_PORTABILITY_LEVEL !
69 │#undef __promise
70 │#           define assert(e) ((e) ? (void)0 : __CLIBNS abort(), (__ARM_PROMISE)((e)?1:0))
71 │#       else
72 │#           define assert(e) ((e) ? (void)0 : __CLIBNS __aeabi_assert(#e, __FILE__, __LINE__), (__ARM_PROMISE)((e)?1:0))
73 ┤#       endif
74 │#       define __promise(e) assert(e)
75 ┤#   endif
76 │#endif
```

*Figure 12-1.*  *The assert.h header file from Keil MDK defines the implementation for the assert macro*

First, we can control whether the assert macro is replaced with nothing, basically compiled out of the code base, or we can define a version that will call a function if the assertion fails. If we want to disable our assertions, we must create the symbol NDEBUG. This is usually done through the compiler settings.

Next, we also must make sure that we define \_\_DO\_NOT\_LINK\_PROMISE\_WITH\_ ASSERT. Again, this is usually done within the compiler settings symbol table. Finally, at this point, we can then come to the definition that will be used for our assertions:

```
define assert(e) (e ? (void)0 : __CLIBNS __aeabi_assert("e", \
                                    __FILE__, __LINE__))
```

As you can see, there is a bit more to defining assertions in Keil MDK compared to (GNU Compiler Collection) GCC-based tools like STM32CubeIDE and MPLAB X. Notice that the functions that are called if the assert fails are similar but again named differently.[6] Therefore, to define your assertion function, you must start by reviewing what our toolchain expects. This is important because we will have to define our assert_failed function, and we need to know what to call it so that it is properly linked to the project.

## Implementing assert_failed

Once we have found how the assertion is implemented, we need to create the definition for the function. assert.h makes the declaration, but nothing useful will come of it without defining what that function does. There are four things that we need to do, which include

- Copy the declaration and paste the declaration into a source module[7]

- Turn the new declaration into a function definition

- Output something so that the developer knows the assertion failed

- Stop the program from executing

For a developer using Keil MDK, their assertion failed function would look something like the code in Listing 12-9.

---

[6] I wish the function name was included in the standard so that assert failed functions could be more easily ported!

[7] Yes, we must be careful with copy/paste to ensure that we don't inject any bugs! For this operation, the risk is minimal.

***Listing 12-9.*** The implementation for the "assert failed" function in Keil MDK

```
void __aeabi_assert(const char *expr, const char *file,
                            int line)
{
    Uart_printf(UART1, "Assert failed in %s at line %d\n",
                                        file, line);
}
```

In Listing 12-9, we copied and pasted the declaration from assert.h and turned it into a function definition. (Usually, you can right-click the function call in the macro, which will take you to the official declaration. You can just copy and paste this instead of stumbling to define it yourself.) The function, when executed, will print out a message through one of the microcontrollers' UARTs to notify the developer that an assertion failed. A typical printout message is to notify the developer whose file the assertion failed in and the line number. This tells the developer exactly where the problem is.

This brings us to an interesting point. You can create very complex-looking assertions that test for multiple conditions within a single assert, but then you'll have to do a lot more work to determine what went wrong. I prefer to keep my assertions simple, checking a single condition within each assertion. There are quite a few advantages to this, such as

- First, it's easier to figure out which condition failed.

- Second, the assertions become clear, concise documentation for how the function should behave.

- Third, maintaining the code is more manageable.

Finally, once we have notified the developer that something has gone wrong, we want to stop program execution similarly. There are several ways to do this. First, and perhaps the method I see the most, is just to use an empty while (true) or for(;;) statement. At this point, the system "stops" executing any new code and just sits in a loop. This is okay to do, but from an IDE perspective, it doesn't show the developer that something went wrong. If my debugger can handle flash breakpoints, I prefer to place a breakpoint in this function, or I'll use the assembly instruction __BKPT to halt the processor. At that point, the IDE will stop and highlight the line of code. Using __BKPT can be seen in Listing 12-10. (Great, Scott! Yes, there are still places where it makes sense to use assembly language!)

*Listing 12-10.* The complete implementation for the "assert failed" function in Keil MDK includes using an instruction breakpoint to stop code execution and notify the developer

```
void __aeabi_assert(const char *expr, const char *file,
                            int line)
{
    Uart_printf(UART1, "Assert failed in %s at line %d\n",
                                            file, line);

    __asm("BKPT");
}
```

## Verifying Assertions Work

Once the assertion is implemented, I will always go through and test it to ensure it is working the way I expect it to. The best way to do this is to simply create an assertion that will fail somewhere in the application. A great example is to place the following assertion somewhere early in your code after initialization:

```
assert(false);
```

This assertion will never be true, and once the code is compiled and executed, we might see serial output from our application that looks something like this:

"Assertion failed in Task_100ms.c at line 17"

As you can see, when we encounter the assertion, our new assert failed function will tell us that an assertion failed. In this case, it was in the file Task_100ms.c at line number 17. You can't get more specific about where a defect is hiding in your code than that!

## Three Instances Where Assertions Are Dangerous

Assertions, if used properly, have been proven to improve code quality. Still, despite code quality improvements, developers need to recognize that there are instances where using assertions is either ineffective or could cause serious problems. Therefore, before using assertions, we must understand the limits of assertions. There are three specific instances where using an assertion could cause problems and potentially be dangerous.

# Instance #1 – Initialization

The first instance where assertions may not behave or perform as expected is during the system initialization. When the microcontroller powers up, it reads the reset vector and then jumps to the address stored there. That address usually points to vendor-supplied code that brings up the clocks and performs the C copy down to initialize the runtime environment for the application. At the same time, this seems like a perfect place to have assertions; trying to sprinkle assertions throughout this code is asking for trouble for several reasons.

First, the oscillator is starting up, so the peripherals are not properly clocked if an assertion were to fire. Second, at this point in the application, printf will most likely not have been mapped, and whatever resource it would be mapped to would not have been initialized. Stopping the processor or trying to print something could result in an exception that would prevent the processor from starting and result in more issues and debugging than we would want to spend.

For these reasons, it's best to keep the assertion to areas of the code after the low-level initializations and under the complete control of the developer. I've found that if you are using a library or autogenerated code, it's best to leave these as is and not force assertion checks into them. Instead, in your application code, add assertions to ensure that everything is as expected after they have "done their thing."

# Instance #2 – Microcontroller Drivers

Drivers are a handy place to use assertions, but, again, we need to be careful which drivers we use assertions in. Consider the case where we get through the start-up code. One of the first things many developers do is initialize the GPIO pins. I often use a configuration table that is passed into the Gpio_Init function. If I have assertions checking this structure and something is not right, I'm going to fire an assertion, but the result of that assertion will go nowhere! Not only are the GPIO pins not initialized, but printf and the associated output have not yet been mapped! At this point, I'll get an assertion that fails silently or, worse, some other code in the assertion that fails that then has me barking up the wrong tree.

A silent assertion otherwise is not necessarily a bad thing. We still get the line number of the failed assertion and information about what the cause could be stored in memory. The issue is that we don't realize that the assertion has fired and just look confounded at our system for a while, trying to understand why it isn't running. As we discussed earlier, we could set an automatic breakpoint or use assembly instructions to make it obvious to us. One of my favorite tricks is to dedicate an LED as an assert LED that latches when an assertion fires. We will talk more about it shortly.

## Instance #3 – Real-Time Hardware Components

The first two instances we have looked at are truthfully the minor cases where we could get ourselves into trouble. The final instance is where there is a potential for horrible things to happen. For example, consider an embedded system that has a motor that is being driven by the software. That motor might be attached to a series of gears lifting something heavy, driving a propulsion mechanism, or doing other practical work. If an operator was running that system and the motor suddenly stopped being driven, a large payload could suddenly give way! That could potentially damage the gearing, or the payload could come down on the user![8]

Another example could be if an electronic controller were driving a turbine or rocket engine. Suppose the engine was executing successfully, and the system gained altitude or was on its way to orbit, and suddenly an assertion was hit. In that case, we suddenly have an uncontrolled flying brick! Obviously, we can't allow these types of situations to occur.

For systems that could be in the middle of real-time or production operations, our assertions need to have more finesse than simply stopping our code. The assertions need to be able to log whatever data they can and then notify the application code that something has gone wrong. The application can then decide if the system should be shut down safely or terminated or if the application should just proceed (maybe even attempt recovery). Let's now look at tips for how we can implement our assertions to be used in real-time systems.

# Getting Started with Real-Time Assertions

A real-time assertion is

"An assertion in a *real-time* program that, if evaluated as false, will identify a bug without breaking the real-time system performance."

---

[8] This is an example that Jean Labrosse brought to my attention several years ago while we were chatting at Embedded World. I had done some webinars or wrote a blog at the time, and it led to some ideas and topics that we will discuss later in the chapter.

As discussed earlier, you want to be able to detect a bug the moment that it occurs, but you also don't want to break the system or put it into an unsafe state. Real-time assertions are essentially standard assertions, except that they don't use a "BKPT" instruction or infinite loop to stop program execution. Instead, the assertion needs to

1) Notify the developer that an assertion has failed

2) Save information about the system at the failed assertion

3) Put the system in a safe state so the developer can investigate the bug

There are many ways developers can do this, but let's look at four tips that should aid you when getting started with real-time assertions.

# Real-Time Assertion Tip #1 – Use a Visual Aid

The first and most straightforward technique developers can use to be notified of an assertion without just halting the CPU is to signal with a visual aid. In most circumstances, this will be with an LED, but it is possible to use LED bars and other visual indicators. The idea here is that we want to print out the message, but we also want to get the developers' attention. Therefore, we can modify the assert failed function like Listing 12-11, where we remove the BKPT instruction and instead place a call to change the state of an LED.

*Listing 12-11.* When the assertion occurs, a modification to the assert function turns on an LED, LED_Assert

```
void __aeabi_assert(const char expr, const char *file,
                                             int line)
{
   Uart_printf(UART1, "Assertion failed in %s at line %d\n",
                                             file, line);

   // Turn on the LED and signal an assertion.
   LED_StateWrite(LED_Assert, ON);
}
```

The latched LED shows that the assertion has occurred, and the developer can then check the terminal for the details. While this can be useful, it isn't going to store more than the file and line, which makes its use just a notch up from continuously watching the terminal. Unfortunately, as implemented, we also don't get any background information about the system's state. You only know which assertion failed, but if you crafted the assertion expression correctly, this should provide nearly enough information to at least start the bug investigation.

# Real-Time Assertion Tip #2 – Create an Assertion Log

Assertions usually halt the CPU, but if we have a motor spinning or some other reason we don't want to stop executing the code, we can redirect the terminal information to a log. Of course, logs can be used in the development, but they are also one of the best ways to record assertion information from a production system without a terminal. If you look at the DevAlert product from Percepio, they tie assertions into their recording system. They then can push the information to the cloud to notify developers about issues they are having in the field.[9]

There are several different locations to which we could redirect the assertion log. The first location that we should consider, no matter what, is to log the assertion data to RAM. I will often use a circular buffer that can hold a string of a maximum length and then write the information to the buffer. I'll often include the file, line number, and any additional information that I think could be important. I sometimes modify the assertion failed function to take a custom string that can provide more information about the conditions immediately before the assertion was fired. Finally, we might log the data doing something like Listing 12-12.

*Listing 12-12.* An assert function can map where the output from assert will be placed

```
void __aeabi_assert(const char expr, const char *file,
                                            int line)
{
#if ASSERT_UART == TRUE
    Uart_printf(UART1, "Assertion failed in %s at line %d\n",
                                            file, line);
```

---

[9] https://percepio.com/devalert/

```
#elif
    Log_Append(ASSERT, Assert_String, file, line);
#endif

    // Turn on the LED and signal an assertion.
    LED_StateWrite(LED_Assert, ON);
}
```

You'll notice in Listing 12-12 that I've also started adding conditional compilation statements to define different ways the assertion function can behave. For example, if ASSERT_UART is true, we just print the standard assertion text to the UART. Otherwise, we call Log_Append, which will store additional information and log the details in another manner. Once the log information is stored in RAM, there would be some task in the main application that would periodically store the RAM log to nonvolatile memory such as flash, an SD card, or other media.

## Real-Time Assertion Tip #3 – Notify the Application

Assertions are not designed to be fault handlers; a real-time assertion can't just stop the embedded system. Of course, we may still want to stop the system, but to do so, we need to let the main application know that a defect has been detected and that we need to move into a safe mode as soon as possible (at least during development). We can create a signaling mechanism between the assertion library and the main application.

For example, we may decide that there are three different types of assertions in our real-time system:

1)   Critical assertions that require the system to move into a safe state immediately.

2)   Moderate assertions don't require the system to stop, but the developer is immediately notified of an issue so that the developer can decide how to proceed.

3)   Minor assertions don't require the system to be stopped and may not even need to notify the application. However, these assertions would be logged.

We might add these assertion severity levels into an enum that we can use in our code. For example, Listing 12-13 demonstrates the enumeration.

***Listing 12-13.*** An example assertion severity level enumeration

```
typedef enum
{
    ASSERT_SEVERITY_MINOR,
    ASSERT_SEVERITY_MODERATE,
    ASSERT_SEVERITY_CRITICAL,
    ASSERT_SEVERITY_MAX_COUNT
}AssertSeverity_t
```

Our assert failed function would then require further modifications that would allow it to notify the main application. We might even need to move away from the C standard library assert functions. For example, we might decide that the developer should specify if the assertion failed and what severity level the assertion is. We may want to redefine our assertion failed function to look something like Listing 12-14.

***Listing 12-14.*** A custom assertion that can specify the severity level of the assertion failure

```
void assert_failed(const char expr, const char *file, int
                             line, AssertSeverity_t Severity)
{
#if ASSERT_UART == TRUE
    Uart_printf(UART1, "Assertion failed in %s at line %d\n",
                                                file, line);
 #elif
    Log_Append(ASSERT, Assert_String, file, line);
#endif

    // Turn on the LED and signal an assertion.
    LED_StateWrite(LED_Assert, ON);

    App_Notify(Severity)
}
```

In the modified assertion function, we allow an assertion severity to be passed to the function and then have a custom function named App_Notify that passes that severity to a function that will behave how the application needs it to based on the severity level. After all, each application may have its requirements for handling these things. So, for example, App_Notify can decide if the assertion is just logged or if some custom handler is executed to put the system into a safe state.

## Real-Time Assertion Tip #4 – Conditionally Configure Assertions

Assertions are a simple mechanism for detecting defects, but developers can create as sophisticated a mechanism as they decide will fit their application. If you plan to use assertions in both development and production, it can be helpful to create a series of conditions that determine how your assertions will function. For example, you might create conditions that allow the output to be mapped to

- UART
- Debug Console
- Serial Wire Debug
- A log file
- Etc.

There can also be conditions that would disable assertion functionality altogether or the information that would gather and be provided to the log. There may be different capabilities that a developer wants during development vs. what they would like in production. What's important is to think through what you would like to get from your assertion capabilities and design the most straightforward functionality you need. The more complex you make it, the greater the chances that something will go wrong.

## A Few Concluding Thoughts

Interfaces can be designed to create a contract between the interface designer and the developer using that interface. The designer can carefully craft what the developer would expect from their interface by defining preconditions and postconditions. These conditions can then be verified by using executable instructions and assertions.

Assertions provide you a mechanism to verify the interface contract but also verify at any point in an application that developer assumptions are met. Assertions can be a powerful tool to detect bugs the moment they occur. A potential problem with assertions arises if you build real-time embedded systems. You can't have a system just stop! The result could damage the machine, the operator, or the user. To prevent an issue, real-time assertions can keep the system safe while still catching the bug in the act.

Designers and developers do need to be careful when they start to use real-time assertions. The inherent temptation to turn them into fault and error handlers is an entirely different exercise. Faults and errors are used to manage runtime issues, while assertions are meant to catch bugs. Be careful to make sure that you understand and adhere to the differences!

---

**Caution**    Don't allow your assertion functions to become runtime fault and error handlers!

---

## ACTION ITEMS

To put this chapter's concepts into action, here are a few activities the reader can perform to start finding and using the right tools for their job:

- Review the interfaces in your application code. Ask yourself the following questions:

  - Are they consistent?

  - Memorable?

  - Does a contract exist to enforce behavior and use?

- Review Listings 12-2 and 12-3. If you have a microcontroller with two ports with eight I/O channels, what might the postcondition contract look like using assertions? (Hint: You'll need a for loop and the Dio_ChannelRead() function.)

- In your development environment, review how the assert macro is implemented. Identify differences between its implementation and what we've seen in this chapter.

- Enable assertions and write a basic assertion failed function.

- Pick a small interface in your code base and implement a contract using assertions. Then, run your application code and verify that it is not violating the contract!

- Explore the use of real-time assertions. Build out some of the real-time assertion capabilities we explored in this chapter.

- Going forward, make sure that you define enforceable contracts on your interfaces.

- Jack Ganssle has also written a great article on assertions that can be found at

  - `www.ganssle.com/item/automatically-debugging-firmware.htm`

These are just a few ideas to go a little bit further. Carve out time in your schedule each week to apply these action items. Even small adjustments over a year can result in dramatic changes!

# CHAPTER 13

# Configurable Firmware Techniques

By the age of 40, I had the opportunity to review and write code for over 150 firmware projects. One thing that I've noticed about firmware projects is that developers rarely write their code in a configurable manner, even if it is warranted. Don't get me wrong; I've seen a lot of progress toward layering architectures, managing coupling, cohesion, and trying to maximize reuse. Unfortunately, however, I've seen a lot to be desired regarding configuration.

I've seen many projects with similar elements and behaviors yet entirely written from scratch for each project. The use of configurations could simplify so many embedded projects, yet many developers are stuck recreating the wheel repeatedly. The projects I see that use configurable firmware techniques often leverage copy and paste and the core mechanism to prepare configuration files. Obviously, this is not very desirable since copy and paste can inject bugs into a code base due to a developer not updating a line of code. (Interruptions in the office, a phone call, or just the mind drifting can cause a slip up.)

This chapter will explore configurable firmware techniques I've been using in my code for over a decade. I've blogged or spoken about some of these techniques, but this chapter will go further behind the curtain and demonstrate how you can write configurable firmware that is also autogenerated. I've used these techniques to write code that is not just configurable but also created at an accelerated rate. Due to the maturity of my scripts and techniques, they don't easily fit into this chapter, so I've streamlined them to give you the core idea, the technique you can use for your firmware.

You are going to learn in this chapter how to write code that is configurable through configuration tables. These tables allow you to write reusable, scalable, and configurable code for many applications. Those tables can also be generated using configuration files and scripts, resulting in your application's C modules. Let's get started!

© Jacob Beningo 2022
J. Beningo, *Embedded Software Design*, https://doi.org/10.1007/978-1-4842-8279-3_13

# Leveraging Configuration Tables

The one technique that has the potential to revolutionize the way that you write your embedded software is configuration tables. If you were to look at the code that I write, you'd discover that my code is full of patterns that ease system configuration and management that looks like a table. For example, I might have a table like Listing 13-1 that would be passed to a digital input/output initialization function as a pointer. The initialization function would read this table and then initialize the peripheral registers based on the data stored in the table.

*Listing 13-1.* An example digital input/output table

```
static Dio_Config_t const Dio_Config[] =
{
// Name,     Resister, Drive, Filer,    Dir,    State,  Function
   {PORTA_1, DISABLED, HIGH, DISABLED, OUTPUT, HIGH, FCN_GPIO},
   {PORTA_2, DISABLED, HIGH, DISABLED, OUTPUT, HIGH, FCN_GPIO},
   {SWD_DIO, PULLUP,   LOW,  DISABLED, OUTPUT, HIGH, FCN_MUX7},
   {PORTA_4, DISABLED, HIGH, DISABLED, OUTPUT, HIGH, FCN_GPIO},
   {PORTA_5, DISABLED, HIGH, DISABLED, INPUT,  HIGH, FCN_GPIO},
   {PORTA_6, DISABLED, HIGH, DISABLED, OUTPUT, HIGH, FCN_GPIO},
   {PORTA_7, DISABLED, HIGH, DISABLED, OUTPUT, HIGH, FCN_GPIO},
   {PORTA_8, DISABLED, HIGH, DISABLED, OUTPUT, HIGH, FCN_GPIO},
};
```

You'll notice that each row in the table represents a GPIO pin, and each column represents a feature. The table is populated with the initialization data for each feature on each pin. For example, the PORTA_1 pin would have its pull-up/pull-down resistor disabled. The drive strength would be set to high. No filtering would be enabled. The pin would be set as an output with an initial value of high (Vcc) and configured with the GPIO function. Other pins might have functions like MISO, MOSI, CLK, SDA, SCL, etc. The configuration table is nothing more than an array of structures. The structure definition would look something like Listing 13-2.

Configuration tables have a lot of advantages associated with them. First, they are human-readable. I can quickly look at the table and see what is going on. Next, they are easily modified. For example, if you wanted to go in and change PORTA_7 to initialize in the LOW state, you could just change the value in the table, recompile, and make the change! If you had a new microcontroller with more pins, you could just add new rows to the table! If the new microcontroller has fewer pins, then you remove rows. The idea is the same with the features that are represented as columns.

The last advantage I want to mention that is important and that I rarely see exploited is that configuration tables can also be generated automatically using configuration scripts. However, we'll talk a bit more about that later in the chapter. Nevertheless, it's a technique that I think you'll find helpful and may forever change how you write your embedded software.

***Listing 13-2.*** The structure definition for a GPIO peripheral that is populated in an array to initialize a microcontroller's pins

```
/**
 * Defines the digital input/output configuration table
 * elements that Dio_Init uses to configure the Dio
 * peripheral.
 */
typedef struct
{
    DioChannel_t Channel;   /**< The I/O channel                */
    uint8_t Resistor;       /**< DISABLED,PULLUP,PULLDOWN       */
    uint8_t DriveStrength;      /**< HIGH or LOW                */
    uint8_t PassiveFilter;  /**< ENABLED or DISABLED            */
    uint8_t Direction;      /**< OUTPUT or INPUT                */
    uint8_t State;          /**< HIGH or LOW                    */
    uint8_t Function;       /**< Mux Function                   */
}DioConfig_t;
```

Despite the numerous advantages of using configuration tables, there are some disadvantages that you should also be aware of. First, configuration tables will take up additional flash space. This is because we're creating an array of structures, and every element of our structure and array will be stored in flash. Extra flash space usage isn't a big deal if you are working on a microcontroller with 128 kilobytes or more flash. However, every byte may count if you work on a resource-constrained device with only 32 kilobytes.

Another disadvantage is that the initialization functions are written to loop through the configuration table and then set the specific bits in the peripheral register. Looping through the table can burn unwanted CPU cycles compared to just setting the register to the desired value. Obviously, these extra CPU cycles are only used during start-up, and even on a slower processor running at 48 MHz, it has a minimal impact.

I've found that the benefits of using configuration tables outweigh the disadvantages. Configuration tables can be used for much more than just initializing a peripheral. Let's look at a few familiar places where they can make a big difference.

# An Extensible Task Initialization Pattern

Real-time operating systems (RTOS) have steadily found their way into more and more embedded applications. I've noticed that many 32-bit applications start immediately with an RTOS rather than going the bare-metal route. Using an RTOS makes excellent sense in many systems today because it provides a real-time scheduler, synchronization tools, memory management, and much more. However, one exciting thing that I have noticed when I review designs and work with teams in the industry is that they often design their task initialization code in a way that is not very reusable.

When they want to create an RTOS task, many developers will go to their main function and follow the process necessary to complete a task for their specific RTOS. For example, if you are using FreeRTOS, you'll find a handful to several dozen code segments that look like Listing 13-3. There is nothing wrong per se with initializing a task this way, but it's not very scalable. Sometimes, finding what tasks are being created in an application can result in a witch hunt. (I've seen applications where tasks are scattered throughout, which is excellent for reuse since they are all modularized but difficult if you want to analyze the code and see how many tasks are in the system.)

*Listing 13-3.* Applications commonly have code blocks scattered throughout main and the code base to initialize all the application tasks

```
xTaskCreate(Task_LedBlink,              // Task Pointer
            "TaskLedBlink",             // Task Name
            configMINIMAL_STACK_SIZE,   // Stack Depth
            (void * const) 0,           // Data to Pass
            2,                          // Task Priority
            (TaskHandle_t * const)0 );  // Task Handle
```

The problem I have with creating a bunch of these statements to create all the application tasks is multifold. First, if I want to add a new task, I must duplicate the code that is using xTaskCreate. Chances are, I'm going to copy and paste, which means I'll likely inject a bug into the code base because I'll get distracted (squirrel!) and forget to update some parameters like the task priority. Second, if I want to remove a task from the system, I must hunt down the task I'm looking for and remove the xTaskCreate code block by either deleting it or conditionally compiling it out. The final problem that I'll mention is that, from a glance, I have no idea how many tasks might be in an application. Sure, I can use grep or do a search for xTaskCreate and see how many results come up, but that feels clunky to me.

# Defining a Task Configuration Structure

A more exciting solution to creating tasks is to create a configuration table that initializes tasks during start-up. The configuration table would contain all the parameters necessary to initialize a task as a single row in the table. If a developer wants to add a new task, they go to the table and add the task. If they want to delete a task, they go to the table and delete the task. If they want to know how many tasks are in the application, they go to the table and count how many rows are in it! Finally, if they want to port the code, they don't have to rewrite a bunch of task creation code. Instead, they update the table with the tasks in the system!

The implementation of the task configuration table is relatively simple. First, a developer creates a structure that contains all the parameters necessary to initialize a task. For FreeRTOS, the structure would look something like Listing 13-4. Looking at each structure member, you'll notice that they directly correspond to the parameters passed into the xTaskCreate function. You'll also notice that we use const quite liberally to ensure that the pointers can't accidentally be incremented or overwritten once they are assigned.

---

**Caution**   Listing 13-4 shows how the structure would look for a dynamically allocated task in FreeRTOS. Dynamic memory allocation uses malloc, which is not recommended for real-time embedded systems. An alternative approach would be to use the statically allocated API, but I will leave that as an exercise for the reader.

---

**Listing 13-4.** A typedef structure whose members represent the parameters necessary to initialize a task using FreeRTOS's xTaskCreate API

```
/**
 * Task configuration structure used to create a task configuration
 * table. Note: this is for dynamic memory allocation. We should
 * create all the tasks up front dynamically and then never allocate
 * memory again after initialization or switch to static allocation.
 */
typedef struct
{
    TaskFunction_t const TaskCodePtr;       // Pointer to task function
    const char * const TaskName;            // String task name
    const configSTACK_DEPTH_TYPE StackDepth;    // Stack depth
    void * const ParametersPtr;             // Parameter Pointer
    UBaseType_t TaskPriority;               // Task Priority
    TaskHandle_t * const TaskHandle;        // Pointer to task handle
}TaskInitParams_t;
```

Different RTOSes will require different TaskInitParams_t structure definitions. For example, the structure in Listing 13-4 will only work for FreeRTOS. My structure will differ if I use Azure RTOS, formerly known as ThreadX. For example, Listing 13-5 demonstrates what the structure would look like for Azure RTOS. FreeRTOS requires six parameters to initialize a task, whereas Azure RTOS requires ten. Note, this doesn't make FreeRTOS superior! Azure RTOS provides additional features and control to the developer to fine-tune how the created task behaves. For example, you can create a task but not automatically start it based on how the TaskAutoStart parameter is configured.

---

**Tips and Tricks**   When possible, it's a good idea for Arm processors to limit the number of parameters passed to a function to six. Passing more than six parameters can result in a small performance hit due to the passed parameters not all fitting in the CPU registers. If you have six or more parameters, pass a pointer instead.

---

***Listing 13-5.*** A typedef structure whose members represent the parameters necessary to initialize a task using Azure RTOS's tx_thread_create API

```
/**
 * Task configuration structure used to create a task configuration
 * table.
 */
typedef struct
{
    TX_THREAD * const TCB;          // Pointer to task control block
    char * TaskName;                // String task name
    void * const TaskEntryPtr;      // Pointer to the task function
    void * const ParametersPtr;     // Parameter Pointer
    uint8_t ** const StackStart;    // Pointer to task stack location
    ULONG const StackSize;          // Task stack size in bytes
    UINT TaskPriority;              // Task Priority
    UINT PremptThreshold;           // The preemption threadshold
    ULONG TimeSlice;                // Enable or Disable Time Slicing
    UINT  TaskAutoStart;            // Enable task automatically?
}TaskInitParams_t;
```

Suppose you are working in an environment where you want to have just a single configuration structure for any RTOS you might be interested in using. In that case, you could design your configuration structure around CMSIS-RTOSv2 (or whatever the current API version is). CMSIS-RTOSv2 is designed to be an industry standard for interacting with an RTOS. Clearly, it acts like an operating system abstraction layer (OSAL) or a wrapper. The most common features of an RTOS are exposed, so there may be cool features that developers can't access, but I digress a bit. Designing a configuration structure for CMSIS-RTOSv2 using their osThreadNew API would look like Listing 13-6.

The TaskInitParams_t configuration structure looks much more straightforward than the other two we have seen. It is simpler because osThreadNew takes a pointer to an attribute structure that contains most of the task's configuration parameters. The structure osThreadAttr_t includes members like

- Name of the thread

- Attribute bits (detachable or joined threads)

- Memory control block

- Size of the memory control block

- Memory for stack

- Size of the stack

- Thread priority

- TrustZone module identifier[1]

Personally, I prefer the method of just passing a pointer to the osThreadAttr_t variable. It's quick, efficient, readable, and configurable. It's also entirely scalable because the API doesn't need to change if the data does (again, I digress …).

***Listing 13-6.*** A typedef structure whose members represent the parameters necessary to initialize a task using CMSIS-RTOSv2's osThreadNew API

```
/**
 * Task configuration structure used to create a task configuration
 * table.
 */
typedef struct
{
    osThreadFunc_t TaskEntryPtr;    // Pointer to the task function
    void * const ParametersPtr;     // Parameter Pointer
    osThreadAttr_t const * const Attributes;   // Pointer to attributes
}TaskInitParams_t;
```

Once you've decided how you want to define your configuration structure, we are ready to define our task configuration table.

## Defining a Task Configuration Table

Now that we have defined TaskInitParams_t, we can create an array of TaskInitParams_t where each element in the array contains the parameters necessary to initialize our task. For example, let's say that I am using FreeRTOS and that my system has two tasks:

---

[1] Yes! The interface accounts for Arm's TrustZone thread context management, but we shouldn't be surprised because CMSIS was designed by Arm. Still a useful feature though.

1.  A blinky LED task that executes every 500 milliseconds

2.  A telemetry task that runs every 100 milliseconds

I can define an array named TaskInitParameters as shown in Listing 13-7. You can easily see that each element of the array, represented by a row, contains the parameters necessary to initialize a task in FreeRTOS. Each parameter needed is aligned as a column. Quite literally, our TaskInitParameters array is a configuration table!

Looking at Listing 13-7, you might wonder where I got some of these values. For example, where is Task_Led defined? In my mind, Task_Led is a public function that manages the LED task and is defined in a LED module like task_led.h and task_led.c. The values TASK_LED_STACK_DEPTH, TASK_TELEMETRY_STACK_DEPTH, TASK_LED_PRIORITY, and TASK_TELEMETRY_PRIORITY are a different story.

***Listing 13-7.*** A typedef structure whose members represent the parameters necessary to initialize a task using CMSIS-RTOSv2's osThreadNew API

```
/**
 * Task configuration table that contains all the parameters necessary to
   initialize
 * the system tasks.
 */
TaskInitParams_t const TaskInitParameters[] =
{
//  Task Pointer,       Task String Name,            Stack Size,
    Parameter,  Task Priority,      Task Handle
//                          Pointer
    {&TaskLed,          "Task_LedBlinky",   TASK_LED_STACK_DEPTH,
    NULL,   TASK_LED_PRIORITY       , NULL},
    {&TaskTelemetry,    "Task_Telemetry",   TASK_TELEMETRY_STACK_DEPTH,
    NULL,   TASK_TELEMETRY_PRIORITY , NULL}
};
```

I mentioned earlier that I hate having to hunt through a project to find tasks; this includes trying to find where the task priority and memory definitions are. Typically, I will define a module named task_config.h and task_config.c. The task_config.h header file would contain the definitions for all the task configurations in the application! That's right; everything is in one place! Easy to find, easy to change.

There are several different ways that we can define those four configuration values. One way is to use macros, as shown in Listing 13-8. Using macros might appeal to you, or this may be completely horrid depending on your experience and programming philosophy. I look at it as a quick and dirty way to do it that is easy to read, although not really in line with OOP beliefs. However, when using C, it fits into the design paradigm well enough.

***Listing 13-8.*** Task configuration parameters that may need adjustment can be defined in task_config.h as macros

```
/*
 * Defines the stack depth for the tasks in the application
 */
#define TASK_LED_STACK_DEPTH            (512U)
#define TASK_TELEMETRY_STACK_DEPTH      (512U)

/*
 * Defines the stack depth for the tasks in the application
 */
#define TASK_LED_PRIORITY               (15U)
#define TASK_TELEMETRY_PRIORITY         (30U)
```

---

**Caution**    FreeRTOS defines task priority based on the most significant numeric number. A task with priority 30 has a higher priority than 15. Priority setting is backward compared to most RTOSes, so be careful when setting priorities! (Or use an OSAL that manages the priority behind the scenes.)

---

Another way to define these configuration parameters, and one you may like a bit better, is to define an enum. First, examine Listing 13-9. You can see that we have created two different enumerations: one for the stack depth and a second for the task priorities. Notice that we must assign the value for each item in the enum, which may overlap! Same values in an enum may or may not bother your software OCD (if you have it) or your desire for perfection.

*Listing 13-9.* Defining task priorities and stack depth as enums

```
/*
 * Defines the stack depth for the tasks in the application
 */
enum
{
    TASK_LED_STACK_DEPTH = 512U,
    TASK_TELEMETRY_STACK_DEPTH = 512U,
};

/*
 * Defines the stack depth for the tasks in the application
 */
enum
{
    TASK_LED_PRIORITY         = 15U,
    TASK_TELEMETRY_PRIORITY   = 30U,
};
```

These are just a few C-compatible options for defining the priority and stack depth. The goal here is to keep these items grouped. For example, if I were to perform a rate monotonic analysis (RMA), I may need to adjust the priorities of my tasks. Having all the tasks' priorities, stacks, and even periodicity values in one place makes it easy to change and less error-prone.

# Initializing Tasks in a Task_CreateAll Function

With the configuration table for the tasks created in the application completed, the only piece of code currently missing is a task initialization function. You can name this function however you like; however, my preference is to name it Task_CreateAll. Task_CreateAll, obviously from its name, creates all the tasks in the application. So let's look at how we can do this.

First, I usually put the Task_CreateAll function into the task_config module. I do this so that the task configuration table can be locally scoped using static and not exposed to the entire application. You could put Task_CreateAll in a tasks.c module and create an interface in task_config that returns a pointer to the task configuration table. There's a part of me that likes it, but for now, we will keep it simple for brevity's sake.

The interface and prototype for the function are then super simple:

```
void Task_CreateAll(void);
```

You could make the interface more complex using something like

```
TaskErr_t Task_CreateAll(TaskInitParams_t const * const
                                             TaskParams);
```

I like the second declaration myself, especially the idea of returning a TaskErr_t so that the application can know if all the tasks were created successfully. There's little danger in passing a pointer to the configuration structure, which is stored in flash early in the boot cycle. The chances of memory corruption or bit flips several microseconds after start-up are highly improbable in most environments. However, to simplify the interface, you may want to just use

```
TaskErr_t Task_CreateAll(void);
```

You can decide which version works best for you. For now, we will explore the interface that I listed first, which returns void and has no parameters. (It's the simplest case, so let's start there.)

Once the interface has been defined, we can write the function, shown in detail in Listing 13-10. The function starts by determining how many tasks are required by the application. The number of tasks in the application can be calculated by taking the size of the TaskInit Parameters array and dividing it by the size of the TaskInitParams_t structure. I've often seen developers create a macro that must be updated with the number of tasks in the system. Manual calculations are error-prone and almost an ensured way to create a bug. I prefer to let the compiler or runtime calculate this for me.

After calculating the number of tasks in the table, Task_CreateAll just needs to loop through each task and call xTaskCreate using the parameters for each task. A for loop can be used to do this. Afterward, the function is complete! You can see the results in Listing 13-10.

A quick review of Listing 13-10 will reveal a simple bit of code written once to manage the creation of all tasks in our embedded application. In this simple function, we've refactored potentially hundreds of lines of code that would usually be written to initialize a bunch of tasks throughout our application. We've also moved code that was likely scattered throughout the application into one central location. The Task_CreateAll function that we have written could use a few improvements, though.

First, the void return means that the application code cannot know whether we successfully created all the tasks. It is assumed that everything will go perfectly well. Second, our example uses the dynamic task creation method xTaskCreate. If we did not size our heap correctly, it's possible that the task creation function could fail! If xTaskCreate fails, we will never know because we are not checking the return value to see if it is pdPASS (which says the function ran successfully). When I write production code, I can generally get away with a task creation function that doesn't check the return value if we statically allocate tasks. You'll know immediately if you cannot create the task at compile time.

***Listing 13-10.*** The Task_CreateAll function loops through the task configuration table and initializes each task in the table, creating a reusable, scalable, and configurable solution

```
void Task_CreateAll(void)
{
    // Calculate how many rows there are in the task table. This is
    // done by taking the size of the array and dividing it by the
    // size of the type.
    const uint8_t TasksToCreate = sizeof(TaskInitParameters) /
                                  sizeof(TaskInitParams_t);

    // Loop through the task table and create each task.
    uint8_t TaskCount = 0;
    for(TaskCount = 0; TaskCount < TasksToCreate; TaskCount++)
    {
        xTaskCreate(TaskInitParameters[TaskCount].TaskCodePtr,
                    TaskInitParameters[TaskCount].TaskName,
                    TaskInitParameters[TaskCount].StackDepth,
                    TaskInitParameters[TaskCount].ParametersPtr,
                    TaskInitParameters[TaskCount].TaskPriority,
                    TaskInitParameters[TaskCount].TaskHandle);
    }
}
```

# Autogenerating Task Configuration

The coding technique we've been exploring can easily be copied and pasted into any number of projects. The problem with copy and paste is that there seems to be a sizeable statistical probability that the developer will overlook something and inject a bug into their code. Even if a bug is not injected into the code, I've found that if I copy the task_config module into new projects, I tend to comment out large blocks of code which may live that way for months which is not ideal. The optimal way to manage the task configuration table and the function to initialize all the tasks is to autogenerate the code.

Code generation has been around in embedded systems for decades, but I've generally only seen it exercised by microcontroller vendors as part of their toolchains. For example, if you look at STM32CubeIDE, it integrates with STM32CubeMx, which allows developers to select their driver and middleware configurations. The tool then generates the source code based on those settings. Of course, developing such tools is time-consuming and expensive. However, you can still take advantage of these techniques on a much more limited basis to improve the configurability of your projects. At a minimum, a code generator will contain four main components: a configuration, templates, the generator, and the output code module. These four components and their interactions can be seen in Figure 13-1.

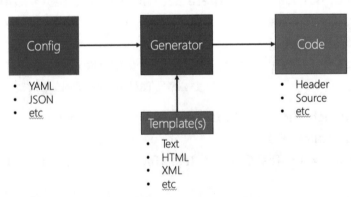

***Figure 13-1.*** *The high-level process for creating autogenerated configuration modules*

The configuration file stores all the configuration information to generate the code. For example, our task code configuration file would contain information such as the task and all the parameters necessary to configure that task. (We will look more closely at how these files look in the next section.) Several different formats are available to use, such as YAML or JSON. (These formats are even interchangeable using JSON to YAML and YAML to JSON converters.)

Templates are used to create easily updatable files with replaceable text to create the output module. For example, if you have a template to organize how a C header and source file are organized, these could be used as templates. Templates could be text files, an HTML file, XML, etc. Each file type has its advantages and disadvantages, depending on the end goal. We're looking to generate C modules, so plain text files are sufficient.

The generator takes in the configuration file(s) and template(s) and then generates the final code module(s). The generator can be written in nearly any language, such as C++, Java, or Python. However, my personal preference is to write it in Python because almost everyone these days knows Python. If you don't, there are so many online resources that nearly any coding problem can be solved with a quick web search.

At the end of the day, we're looking to generate a C header and source code in the most scalable way possible. In addition, we want to minimize the number of errors generated and generate clean and human-readable code. To understand the entire process, let's build a simple code generator in Python for our task configuration module.

## YAML Configuration Files

YAML, Yet Another Mark-up Language, is a file format that lends itself well to configuration files. YAML is minimalistic and follows the Python style of indentation. YAML only allows spaces to be used for formatting and can be used to create lists and maps of data in nearly any data type that a software engineer may want to use. Each YAML file starts with a "---" to denote the start of a YAML document. Each file may have more than a single document included.

For our task configuration module example, we need the YAML file to store the parameter information required to initialize a task. We can do this by creating a dictionary of X tasks where we include a key and a value for the configuration parameter we want to configure. For example, suppose I had a task named "Task Telemetry" with a periodicity of 100 milliseconds, a stack size of 4096, and a priority of 22. In that case, I might save the configuration in a YAML file that looks something like Listing 13-11.

***Listing 13-11.*** The YAML configuration for a task named Task_Telemetry. All the details necessary to initialize the task are included

```
---
 Task1:
   TaskName: "Task_Telemetry"
   TaskEntryPtr: "Task_Telemetry"
   PeriodicityInMS: 100
   ParametersPtr: 0
   StackSize: 4096
   TaskPriority: 22
   TaskHandle: "NULL"
```

The great thing about a YAML configuration file is that if I need to add or remove tasks, I don't have to modify my code! All I need to do is add the new task to the configuration and run my Python generator! For example, if I wanted to add two more tasks to my configuration, my YAML file might become something like Listing 13-12.

***Listing 13-12.*** The YAML configuration file now includes three tasks: Task_ Telemetry, Task_RxMessaging, and Task_LEDBlinky

```
---
 Task1:
   TaskName: "Task_Telemetry"
   TaskEntryPtr: "Task_Telemetry"
   PeriodicityInMS: 100
   ParametersPtr: 0
   StackSize: 4096
   TaskPriority: 22
   TaskHandle: "NULL"
 Task2:
   TaskName: "Task_RxMessaging"
   TaskEntryPtr: "Task_RxMessaging"
   PeriodicityInMS: 0
   ParametersPtr: 0
   StackSize: 4096
```

```
      TaskPriority: 1
      TaskHandle: "NULL"
  Task3:
      TaskName: "Task_LedBlinky"
      TaskEntryPtr: "Task_LedBlinky"
      PeriodicityInMS: 500
      ParametersPtr: 0
      StackSize: 256
      TaskPriority: 5
      TaskHandle: "NULL"
```

YAML files are a fantastic option for storing configuration data. Configuration files also provide you with a flexible way of configuring source code for different projects and product SKUs without having to maintain "extra" code in your source repository. Instead, it's all one common code base, and the build system chooses which configuration to generate, build, and deploy.

## Creating Source File Templates

Templates provide a mechanism from which you can outline how a header and source module will be organized once the final code modules are generated. Templates can be generated in several different ways. For example, a developer could build a template in XML, text, or even an HTML file. The format that we choose depends on the result. An HTML template might make the most sense if you were to autogenerate a report. For C/C++ modules, it's far easier to just use a simple text file.

To generate a module, there are two file templates that we would want to use: a header template file, t_header.h, and a source header file, t_source.c. These files would contain all the typical comments that one would expect in a header and outline the structure of the files. However, the main difference between the template and the final file is that instead of containing all the code, the template files have tags that are replaced by the desired code blocks.

Tags are special text that the Python template library can detect and use for string replacements. For example, you might in your template files have a tag like ${author} which is where the name of the author is supposed to go. Our Python code will specify that the author is Jacob Beningo and then search out all the ${author} tags and replace

it with Jacob Beningo. In the header file, we would include tags such as ${author},
${date}, ${task_periodicity}, ${task_priorities}, and ${task_stacks}.
Listing 13-13 shows how an example header file template looks.

The source module template is very similar to the header source template, except for
having different tags. The source module has tags for ${author}, ${date}, ${includes},
and ${elements}. The header files point to the individual task modules that a developer
would write, and the elements are our configuration table. The source module template
also includes the Task_CreateAll function that allows us to initialize all the tasks in the
configuration table. Listing 13-14 shows how an example source file template looks.
Note: I would normally also include sections for Configuration Constants, Macros,
Typedefs, and Variables for each template, but I've omitted them to try and get the code
to fit on as few pages as possible.

***Listing 13-13.*** An example template header file for task_config.h with string
replacement tags

```
/***********************************************************
* Title              :    Task Configuration
* Filename           :    task_config.h
* Author             :    ${author}
* Origin Date        :    ${date}
***********************************************************/
/** @file task_config.h
 *   @brief This module contains task configuration parameters.
 *
 *   This is the header file for adjusting task configuration
 *   such as task priorities, delay times, etc.
 */
#ifndef TASK_CONFIG_H_
#define TASK_CONFIG_H_
/***********************************************************
* Includes
***********************************************************/

/***********************************************************
* Preprocessor Constants
***********************************************************/
```

```
${task_periodicity}
${task_priorities}
${task_stacks}
/***************************************************************
* Function Prototypes
***************************************************************/
#ifdef __cplusplus
extern "C"{
#endif

void Task_CreateAll(void);

#ifdef __cplusplus
} // extern "C"
#endif

#endif /*TASK_CONFIG_H_*/

/*** End of File ******************************************/
```

**Listing 13-14.** An example template source file for task_config.c with string replacement tags

```
/***************************************************************
* Title                 :    Task Configuration
* Filename              :    task_config.c
* Author                :    ${author}
* Origin Date           :    ${date}
* See disclaimer in source code
***************************************************************/
/***************************************************************
* Doxygen C Template
* Copyright (c) 2013 - Jacob Beningo - All Rights Reserved
***************************************************************/
/** @file task_config.c
 *  @brief This module contains the task configuration and
 *  initialization code.
 */
```

```c
/***************************************************************
* Includes
***************************************************************/
#include "task_config.h"        // For this modules definitions
#include "FreeRTOS.h"           // For xTaskCreate
${includes}
/***************************************************************
* Module Preprocessor Macros, Typedefs, etc
***************************************************************/
/**
 * Task configuration structure used to create a task
 * configuration table. Note: this is for dynamic memory
 * allocation. We create all the tasks up front dynamically
 * and then never allocate memory again after initialization.
 */
typedef struct
{
    TaskFunction_t const TaskCodePtr;
    const char * const TaskName;
    const configSTACK_DEPTH_TYPE StackDepth;
    void * const ParametersPtr;
    UBaseType_t TaskPriority;
    TaskHandle_t * const TaskHandle;
}TaskInitParams_t;

/***************************************************************
* Module Variable Definitions
***************************************************************/
/**
 * Task configuration table that contains all the parameters
 * necessary to initialize the system tasks.
 */
TaskInitParams_t const TaskInitParameters[] =
{
//      Pointer to Task,    Task String Name,  Stack Size,  Param Ptr,
        Priority, Handle
```

```
${elements}
};

/*****************************************************************
 * Function Definitions
 *****************************************************************/
/*****************************************************************
 * Function : Task_CreateAll()
 *//**
 * @section Description Description:
 *
 *   This function initializes the telemetry mutex and then
 *   creates all the application tasks.
 *
 * @param               None.
 * @return              None.
 *
 * @section Example Example:
 * @code
 *     Task_CreateAll();
 * @endcode
 *****************************************************************/
void Task_CreateAll(void)
{
    // Calculate how many rows there are in the task table.
    // This is done by taking the size of the array and
    // dividing it by the size of the type.
    const uint8_t TasksToCreate = sizeof(TaskInitParameters) /
                                  sizeof(TaskInitParams_t);

    // Loop through the task table and create each task.
    for(uint8_t TaskCount = 0; TaskCount < TasksToCreate; TaskCount++)
    {
        xTaskCreate(TaskInitParameters[TaskCount].TaskCodePtr,
                    TaskInitParameters[TaskCount].TaskName,
                    TaskInitParameters[TaskCount].StackDepth,
                    TaskInitParameters[TaskCount].ParametersPtr,
```

```
                  TaskInitParameters[TaskCount].TaskPriority,
                  TaskInitParameters[TaskCount].TaskHandle);
    }
}

/*************** END OF FUNCTIONS ****************************/
```

## Generating Code Using Python

The Python code we are about to read in our template and configuration files is written in Python 3.9. Hopefully, it will continue to be compatible with future versions, but we all know that Python changes quite rapidly, and nothing is sacred during those changes. I would also like to point out that the code is literally written in script form and could be refactored and improved dramatically. The tools I've developed to manage a lot of my project configuration would be difficult to include and take up far more pages than any reader probably wants to work through. So, we will look at a simplified version that gets the trick done and demonstrates how this technique can be used.

The Python code aims to read the YAML configuration file and generate strings that will replace the tags in our template files. Those generated strings will complete our C module and should compile and drop into any project using FreeRTOS. A developer should only need to include "task_config.h" and then call Task_CreateAll. (Of course, you must also write the actual task functions we are trying to create, too!)

Figure 13-2 shows the high-level flowchart for the Python script.

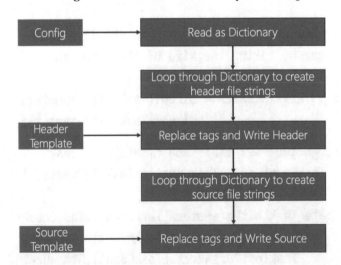

***Figure 13-2.*** *The high-level flowchart for the Python code generator*

The YAML files are pretty easy to read and manipulate in Python. There is a library called PyYAML[2] that has all the methods necessary to use them. The script reads in the entire YAML file and stores it as a dictionary. The dictionary can then be looped through for each task, and the configuration data read and converted into a string that will replace a tag in the template files.

The script's setup and header portion can be seen in Listing 13-15. The source portion of the script can be seen in Listing 13-16. I'm assuming you understand the basics of Python, but just in case, I'll describe a few high-level techniques that are being used here.

*Listing 13-15.* Python code to generate the header file

```python
import yaml
import string
from datetime import date

today = date.today()
Author = "Jacob Beningo"

# mm/dd/YY
Date = today.strftime("%m/%d/%Y")

if __name__ == '__main__':

    stream = open("tasksFRTOS.yaml", 'r')
    dictionary = yaml.safe_load(stream)

    #
    # Prepare the header file first
    #

    # Create a string with the task periodicity macro
    # definitions
    periodicityString = ""
    prioritiesString = ""
    stacksString = ""
    LastTaskName = ""
```

---

[2] https://pyyaml.org/

```
for task in dictionary:
    # get the data to build the period macros
    taskData = dictionary.get(task)
    periodicity = taskData.get('PeriodicityInMS')
    name = taskData.get('TaskName')

    if periodicity != 0:
        tempString = f"#define {name.upper()}_PERIOD_MS"
        temp1String = f"({str(periodicity)}U)\n"
        periodicityString +=
                            f'{tempString:<50}{temp1String:>10}'

    # get the data to build the priorities string
    priority = taskData.get('TaskPriority')
    tempString = f"#define {name.upper()}_PRIORITY"
    temp1String = f"({str(priority)}U)\n"
    prioritiesString +=
                        f'{tempString:<50}{temp1String:>10}'

    # get the data to build the stack string
    stackSize = taskData.get('StackSize')
    tempString = f"#define {name.upper()}_STACK_SIZE"
    temp1String = f"({str(stackSize)}U)\n"
    stacksString += f'{tempString:<50}{temp1String:>10}'

    LastTaskName = str(name)

# Read in the task.h template file
with open("t_task.h") as t_h:
    template = string.Template(t_h.read())
t_h.close()

# Substitute the new strings with the template tags
header_output = template.substitute(author=Author,
                                    date=Date,
                                    task_periodicity=periodicityString,
                                    task_priorities=prioritiesString,
```

```
                                    task_stacks=stacksString
                                )
with open("task_config.h", "w") as output:
    output.write(header_output)
output.close()
```

***Listing 13-16.*** Python code to generate the source file

```
#
# Prepare the C file
#
includeString = ""
content = ""

for task in dictionary:
    taskData = dictionary.get(task)
    TaskName = taskData.get('TaskName')

    includeString += '#include "' + TaskName + '.h"\n'

    EntryFunction = '    { &' + \
                                taskData.get('TaskEntryPtr') +','
    TaskNameStr = '"'+str(taskData.get('TaskName')) + '",'
    StackSize = str(taskData.get('StackSize')) + ","
    Parameters = str(taskData.get('ParametersPtr')) + ","
    Priority = str(taskData.get('TaskPriority')) + ","
    Handle = str(taskData.get('TaskHandle')) + "}"

    content += f'{EntryFunction:<25}{TaskNameStr:<21}{StackSize:^8}
                {Parameters:^14}{Priority:^8}{Handle:>8}'

    if TaskName != LastTaskName:
        content += '\n'

# Read in the task.c template file
with open("t_task.c") as t_c:
    template = string.Template(t_c.read())
t_c.close()
```

```
# Substitute the new strings with the template tags
 source_output = template.substitute(author=Author,
                                     date=Date,
                                         includes=includeString,
                                     elements=content,
                                     )

 with open("task_config.c", "w") as output:
     output.write(source_output)
 output.close()
```

First, the script uses Python f strings to format the strings. You'll notice statements like:

```
f'{tempString:<50}{temp1String:>10}'
```

This is an f string that contains two variables tempString and temp1String. tempString is left justified with 50 spaces, and temp1String is right justified with ten spaces. (Yes, some assumptions are built into those numbers as to what will allow the configuration values to appear properly in columns.)

Next, you'll notice that several statements use the Python template library, such as

```
template = string.Template(t_c.read())
```

and

```
source_output = template.substitute(…)
```

The substitute method is the one doing all the heavy lifting! Substitute is replacing the tags in our template files with the string we generate from looping through the configuration dictionary.

Finally, after looking through all this code, you might wonder how to use it. I've written this example to not rely on having to pass parameters into the command line or anything fun like that. To run the script, all you would need to do is use the following command from your terminal:

```
python3 generate.py
```

I use a Mac with Python 2.7 and Python 3.9 installed, so I must specify which version I want to use, which is why I use python3 instead of just python. If you run the script, you'll discover that the header file output will look like Listing 13-17, and the source file output will look like Listing 13-18. Notice that the tags are no longer in the files and have now been replaced with the generated strings. You'll also notice in the header file that macros for the PERIOD_MS are only generated for two tasks! This is because Task_RxMessaging is event driven and has a period of zero milliseconds.

***Listing 13-17.*** The header file output from the generate.py script

```
/****************************************************************
* Title               :    Task Configuration
* Filename            :    task_config.h
* Author              :    Jacob Beningo
* Origin Date          :    04/16/2022
****************************************************************/
/** @file task_config.h
 *   @brief This module contains task configuration parameters.
 *
 *   This is the header file for adjusting task configuration
 *   such as task priorities, delay times, etc.
 */
#ifndef TASK_CONFIG_H_
#define TASK_CONFIG_H_
/****************************************************************
* Includes
****************************************************************/
/****************************************************************
* Preprocessor Constants
****************************************************************/
#define TASK_TELEMETRY_PERIOD_MS              (100U)
#define TASK_LEDBLINKY_PERIOD_MS              (500U)

#define TASK_TELEMETRY_PRIORITY               (22U)
#define TASK_RXMESSAGING_PRIORITY             (1U)
#define TASK_LEDBLINKY_PRIORITY               (5U)

#define TASK_TELEMETRY_STACK_SIZE             (4096U)
#define TASK_RXMESSAGING_STACK_SIZE           (4096U)
#define TASK_LEDBLINKY_STACK_SIZE             (256U)
```

```
/*************************************************************
* Configuration Constants, Macros, Typdefs, and Variables
*************************************************************/
/*************************************************************
* Function Prototypes
*************************************************************/
#ifdef __cplusplus
extern "C"{
#endif

void Task_CreateAll(void);

#ifdef __cplusplus
} // extern "C"
#endif

#endif /*TASK_CONFIG_H_*/

/*** End of File ***************************************/
```

**Listing 13-18.**  The source file output from the generate.py script

```
/*************************************************************
* Title              :   Task Configuration
* Filename           :   task_config.c
* Author             :   Jacob Beningo
* Origin Date        :   04/16/2022
* See disclaimer in source code
*************************************************************/
/** @file task_config.c
 *  @brief This module contains the task configuration and
initialization code.
 */
/*************************************************************
* Includes
*************************************************************/
#include "task_config.h"
#include "FreeRTOS.h"
```

```
#include "Task_Telemetry.h"
#include "Task_RxMessaging.h"
#include "Task_LedBlinky.h"
/***************************************************************
* Module Preprocessor Constants, Macros, etc
***************************************************************/

/***************************************************************
* Module Typedefs
***************************************************************/
/**
 * Task configuration structure used to create a task
 * configuration table. Note: this is for dynamic memory
 * allocation. We create all the tasks up front dynamically
 * and then never allocate memory again after initialization.
 */
typedef struct
{
    TaskFunction_t const TaskCodePtr;
    const char * const TaskName;
    const configSTACK_DEPTH_TYPE StackDepth;
    void * const ParametersPtr;
    UBaseType_t TaskPriority;
    TaskHandle_t * const TaskHandle;
}TaskInitParams_t;

/***************************************************************
* Module Variable Definitions
***************************************************************/
/**
 * Task configuration table that contains all the parameters
 * necessary to initialize the system tasks.
 */
```

```
TaskInitParams_t const TaskInitParameters[] =
{
 {&Task_Telemetry, "Task_Telemetry",    4096, 0,22,      NULL}
 {&Task_RxMessaging,"Task_RxMessaging",4096, 0, 1,      NULL}
 {&Task_LedBlinky, "Task_LedBlinky",    256, 0, 5,      NULL}
};

/*************************************************************
* Function Prototypes
*************************************************************/
/*************************************************************
* Function Definitions
*************************************************************/
/*************************************************************
* Function : Task_CreateAll()
*//**
* @section Description Description:
*
*   This function initializes the telemetry mutex and then
* creates all the application tasks.
*
* @param              None.
*
* @return             None.
*
* @section Example Example:
* @code
*     Task_CreateAll();
* @endcode
*
* @see
*************************************************************/
void Task_CreateAll(void)
{
    // Calculate how many rows there are in the task table.
    // This is done by taking the size of the array and
```

```
// dividing it by the size of the type.
const uint8_t TasksToCreate = sizeof(TaskInitParameters) /
                              sizeof(TaskInitParams_t);

// Loop through the task table and create each task.
for(uint8_t TaskCount = 0; TaskCount < TasksToCreate; TaskCount++)
{
    xTaskCreate(TaskInitParameters[TaskCount].TaskCodePtr,
                TaskInitParameters[TaskCount].TaskName,
                TaskInitParameters[TaskCount].StackDepth,
                TaskInitParameters[TaskCount].ParametersPtr,
                TaskInitParameters[TaskCount].TaskPriority,
                TaskInitParameters[TaskCount].TaskHandle);
}
}

/************** END OF FUNCTIONS **************************/
```

# Final Thoughts

The ability to create configuration tables is a powerful tool for embedded developers. The ability to automatically generate those files from configuration files is even more powerful. The scalability, configurability, and reusability of embedded software can grow exponentially using the techniques we've looked at in this chapter. The question becomes whether you'll continue to hand-code configurable elements of your software or join the modern era and start writing generated, configurable code.

---

## ACTION ITEMS

To put this chapter's concepts into action, here are a few activities the reader can perform to start using configurable firmware techniques:

- Review your software and identify areas where configuration tables could be used to improve your code's reuse, configurability, and scalability.

- For your RTOS or CMSIS-RTOSv2, create a task_config module. You should include

- - A Task_CreateAll function

  - A TaskInitParams_t structure

  - A TaskInitParamt_t configuration table

- Download the example Python script files from Jacob's GitHub account at `https://github.com/JacobBeningo/Beningo`. Then, run the example to see the default configuration and output.

  - Add a new custom task of your choosing and rerun the script. How did the code change?

  - Change the priorities on several tasks. Rerun the script. How did the code change?

- Modify the Python script to autogenerate the code you created for your RTOS.

- Carefully consider where else you can leverage configurable firmware techniques in your code base.

- If you are serious about building a configurable code based on the techniques in this chapter, consider reaching out to Jacob at jacob@beningo.com for additional assistance.

# CHAPTER 14

# Comms, Command Processing, and Telemetry Techniques

While there is no such thing as an absolute in embedded systems or life, most embedded systems require the ability

- To communicate with the external world

- To receive and process commands

- To report the system's status through telemetry

To some degree, these are necessities of every system. However, the complexity of each can vary dramatically. For example, some systems only have a simple USART communication interface that is not encrypted and is wrapped in a packet structure. Other systems may communicate over Bluetooth Mesh, require complex encryption schemes, or be encapsulated in several security layers requiring key exchanges and authentication.

This chapter will examine how to build reusable firmware for simple communications and command processing. We will start with developing a packetized communication structure that can be used to send commands to a system. We will then explore how to decode those commands and act on them. Finally, we'll explore how to build reusable telemetry schemes that can be used in real-time applications. The techniques discussed in this chapter will use more straightforward examples that you can then extrapolate into more complex real-world scenarios.

© Jacob Beningo 2022
J. Beningo, *Embedded Software Design*, https://doi.org/10.1007/978-1-4842-8279-3_14

# Designing a Lightweight Communication Protocol

There are several problems that developers often face when it comes to sending and receiving data from their embedded systems. First, they need some mechanism for retrieving the data from their system, whether it's health and wellness data or sensor data. Second, they need some mechanism for commanding their system and any data that tells the system how to carry out those commands. Finally, data transfer to and from an embedded system can take many forms and potential physical interfaces.

A common theme between many physical layer interfaces is that it is up to the developer to design and build their own higher-level application protocol that sits on top of the physical layer. For example, in many instances, a system might have a USART, USB, or Wi-Fi interface that can send/receive data. Still, the developer must decide on the protocol to encode and decode the payload sent over the interface. Even if a standard protocol is selected like CAN or MQTT, these protocols include a payload area where it is up to the developer to define their application data transfer protocol. That protocol defines what all the bits and bytes mean to the application. The need for a higher-level application-specific protocol makes sense because there are a wide variety of systems and requirements, and there is no one solution to fit them all.

The use of communication packets can be an extremely effective method for developers to define their application-specific transfer mechanism because it provides a set of predefined fields that the sender or receiver can decode, validate, and process. In addition, the packet does not have to be a fixed length and can adapt to the system's needs on the fly. A packet can also be assembled and decoded using a state machine that can easily handle real-time streaming and be reset if something does not go as expected. Figure 14-1 shows the general process of packet data entering a state machine and then being converted into a format usable by the application code.

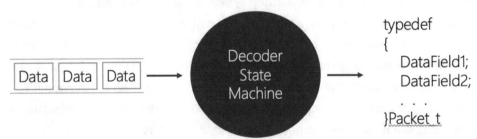

***Figure 14-1.*** *A protocol decoder is often implemented as a state machine that consumes input data, transitions to various decoding states, and then, if successful, results in a populated data structure that the application can use*

The exciting thing about packet protocols is that if you design and use enough of them, you'll discover that there are several requirements that nearly every packet protocol needs despite the industry of the system that is being built. For example, you'll find that

- A packet's integrity needs verification to ensure the packet was received and has not become corrupted.

- The packet must support the system's different operations through commands and requests.

- The packets need to support the ability to transfer data of various sizes to support the system's commands.

There is also the need to define requirements around how the protocol should operate. For example, if the protocol will be used on a slow serial bus such as a UART, it may be necessary to specify that the overhead should be low. Low overhead means the protocol needs to support sending and receiving information using the fewest number of bytes possible to ensure we don't max out the communication bus's bandwidth.

## Defining a Packet Protocol's Fields

When defining a packet protocol to use in your systems, the exact fields that you include will vary dramatically based on your goals and what you want to achieve; however, what you'll find is that you can create a lightweight protocol whose structure remains relatively consistent across a wide range of applications. For example, as a graduate student working on small satellite flight software, we used a simple packet architecture internally to the satellite for point-to-point communication between subsystems. As a result, the communication protocol had the structure seen in Figure 14-2.

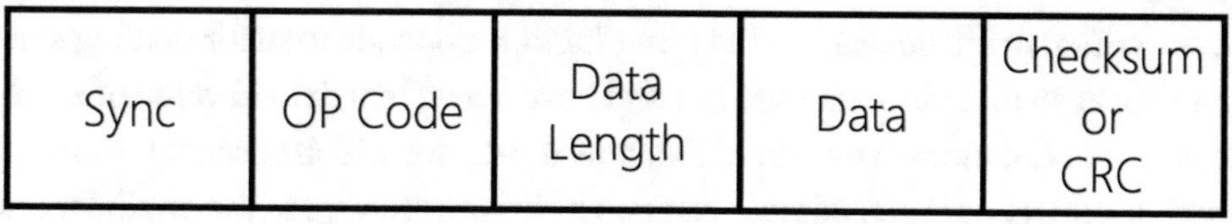

| Sync | OP Code | Data Length | Data | Checksum or CRC |
|------|---------|-------------|------|-----------------|

**Figure 14-2.**  *An example packet protocol definition that supports point-to-point communication[1]*

---

[1] I first published this image in *CubeSat Handbook: From Mission Design to Operations,* Chapter 10, Figure 5.

As you can see, the packet protocol only contains five fields, which could be applied to nearly any application. Before we discuss that in more detail, let's discuss what each field is doing. First, the sync field is used to synchronize the receive state machine. This field is basically a defense against spurious noise from accidentally triggering the processing of a message. You may get a noise spike in noisy environments that tricks the hardware layer into clocking in a byte. Of course, that byte could be anything, so you add a synchronization byte (or two) that must first be received before triggering the receive state machine.

The operational code (opcode) field is used to specify what operation will be performed or to tell the receiver what information can be found in the packet. For example, one might command their system to issue a soft reset by sending an opcode 0x00. Opcode 0x10 might be used to write configuration information for the device and opcode 0x20 to read sensor data. I think you get the idea. Whatever it is that your system does, you create an opcode that is associated with that operation. In smaller systems, I would define the opcode as a single byte, allowing 256 potential commands. However, this field could be expanded to two or more if needed.

Next, the data length field is used to specify how many bytes will be included in the data portion of the packet. In most applications that I have worked on, this is represented by a single byte which would allow up to 255 bytes in the packet. Multiple bytes could be used for larger payloads, though. The data size field generally does not include the sync byte or the checksum. This makes the data length a payload size field that tells how much data to include from the OP code and the data fields. There are many designs for packet protocols, and some include a packet size that covers the entire packet. I've generally designed my packets for data length only because they are more efficient from a performance standpoint.

The data field is then used to transmit data. For example, the opcode 0x10 that writes configuration information would have a data section with the desired configuration data to be written to the device. On the other hand, an opcode 0x20 for reading sensor data may not even include a data section and have a data length field value of zero! Then again, it may contain a data field that tells the device which sensors it wants to get readings back on. How much data for each opcode is entirely up to the engineer and application.

Finally, we have the checksum field. This field could be a checksum, or it could be a Cyclical Redundancy Check (CRC) that is placed on the data packet. I've used both depending on the application and the hardware I was using. For example, I've used

Fletcher16 checksums when working with resource-constrained devices that did not include a CRC hardware accelerator. The Fletcher16 algorithm is nearly as good as a CRC but can be calculated quickly compared to the CRCs calculated in hardware. Optimally, we would use an internal hardware accelerator if it is available.

# A Plethora of Applications

The general format of the packet protocol we've been exploring can be extrapolated to work in various environments. Figure 14-2 shows that it is designed to work on a point-to-point serial interface. The protocol could easily be modified to work in a multipoint serial interface like RS-422 or RS-485 with just a few minor modifications. For example, in multipoint communication, there is usually a need to specify the address of the source node and the destination node. Developers could add an additional section to the packet protocol that defines their applications' address scheme, as shown in Figure 14-3.

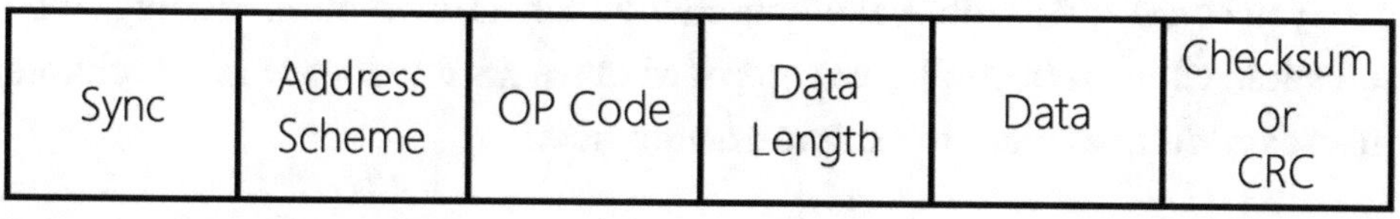

| Sync | Address Scheme | OP Code | Data Length | Data | Checksum or CRC |
|------|----------------|---------|-------------|------|-----------------|

*Figure 14-3.* *An example packet protocol definition that supports multipoint communication by including an addressing scheme*

Developers may even want to modify the packet protocol to be a framed packet. A framed packet will often start and end with the same sync character. If the sync character is included in the packet data, it is replaced with some type of escape character to not interfere with the encoding or decoding algorithm. I've often seen these protocols eliminate the data or packet length fields, but I personally prefer to include them so that I can use them as a sanity check. An example of a framed packet protocol can be seen in Figure 14-4.

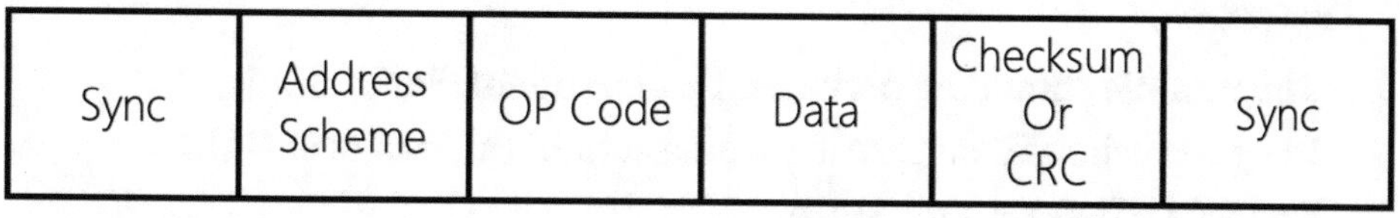

| Sync | Address Scheme | OP Code | Data | Checksum Or CRC | Sync |
|------|----------------|---------|------|-----------------|------|

*Figure 14-4.* *An example packet protocol definition that supports framing for multipoint communication*

Packet protocols don't necessarily have to only be applied to interfaces like USART and USB. Developers can just as quickly adapt them to run on other interfaces like I2C.

In such a system, a developer wouldn't need the sync field because the I2C bus already includes a slave address which is 7 bits, along with a single bit for reading and writing. The general packet protocol could be modified to work on I2C with the modifications shown in Figure 14-5. It's essentially the same protocol again, just with some interface-specific improvements.

| Slave Address (7-bits) | R/W (1-bit) | OP Code (8-bits) | Data Length (8-bits) | Data (x bytes) | Checksum / CRC (16-bits) |
|---|---|---|---|---|---|

***Figure 14-5.*** *An example packet protocol adapted to work on the I2C bus*

As you can imagine, this general format can be used on nearly any physical interface. If needed, you could add additional functionality such as encryption, signing, and authentication, which is beyond the scope of what we want to look at in this chapter. Just note those as additional ideas to explore on your own.

# Implementing an Efficient Packet Parser

We've looked at several different adaptations for lightweight packet protocols. But, first, let's look at what it can take to architect a protocol that is reusable using the protocol that we defined in Figure 14-2.

## The Packet Parsing Architecture

Before we start banging out code, it's helpful to stop and think through how a parsing module would work. There are several characteristics that we would want the module to have:

1. The module should be peripheral independent. For example, the parser should not care if the data stream came in over UART, USART, USB, CAN, I2C, SPI, etc.

2. The module should not depend on any specific circular buffer, message queue, or operating system.

3. The implementation should be scalable and configurable.

Considering these three requirements, we might arrive at a simple data flow diagram for parsing a packet that looks like Figure 14-6.

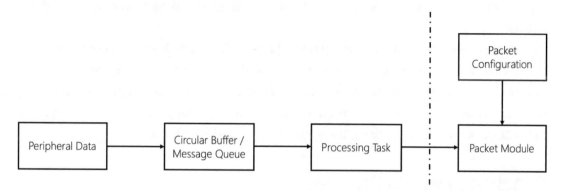

***Figure 14-6.*** *The data flow diagram for packet data starts at the peripheral and makes its way to the packet parser. The dotted line represents the interface to the packet parsing module*

The preceding diagram conveys a lot of information to us. First, the data is received from a peripheral and could be retrieved through either hardware buffers, interrupts, DMA, or any other mechanism in the system. To the architect, we don't care at this stage. Second, once the data is received, it is passed into a buffer where it waits for a task or some other mechanism to consume it. Finally, the task takes the data and uses the packet parser interface to pass the data into the parsing state machine.

In this design, the packet module is abstracted from anything happening on the left side of the dashed line. It doesn't care about how it gets its data, what interface is used, or anything like that. Its only concern is that it is provided with the correct configuration to know how to parse the data and that it is provided a byte stream. We aren't going to go into details on the generic configuration during this post.

The packet parser must have two public functions to parse the packet. These public functions include

```
bool Packet_DecodeSm(uint8_t const Data);
PacketMsg_t const * const Packet_Get(void);
```

As you can see, the primary function for our packet parser is the Packet_DecodeSm function which just takes a single byte of data and returns true if that byte completes a packet whose checksum is verified. (This example will not return checksum errors or other status information, but it can be easily added by extending the packet interface.)

339

I abbreviate StateMachine as Sm in my code. The second function, Packet_Get, can be called to get a pointer to the packet so it can be read and processed by the application. As you can see, the packet is kept as constant data along with the pointer for where the data is.

Notice that the user of the packet module has no clue how the packet parser looks. The state machine that parses the data is a black box. This is an example of encapsulation, an excellent object-oriented design technique that can be applied even in C. In fact, with this interface, we could use any number of parsing algorithms, but as long as the user can call this interface, the application code doesn't care.

## Receiving Data to Process

Typically, we might start to now code up our packet module, but I want to instead write the code that will use the packet module to demonstrate how the module will be used first. For example, I will use an STM32 development board and leverage their STM32Cube HALs. I will also assume we will use a UART to receive the packet data.

The first step is to set up the callback function for a UART receive interrupt. An example of how the interrupt callback function can be defined can be seen in Listing 14-1.

***Listing 14-1.*** Defining a UART callback function to receive packet data

```
void HAL_UART_RxCpltCallback(UART_HandleTypeDef *huart2)
{
    HAL_UART_Transmit(&huart2, (uint8_t *)aRxBuffer, ONE_BYTE, DELAY_MAX);

    CBUF_Push(RxDataBuffer, aRxBuffer[0]);

    HAL_UART_Receive_IT(&huart2, (uint8_t *)aRxBuffer, 1);
}
```

You'll notice that this callback uses HAL_UART_Transmit to repeat the sender's character. It also uses cbuf.h, an open source circular buffer implementation, to push the character into a circular buffer. The peripheral is then placed into a receive state to wait for the next character.

Setting up the circular buffer occurs during the system start-up in the initialization routine. How cbuf works is beyond the scope of the book, but in case you want to try this yourself, the code to initialize the circular buffer can be found in Listing 14-2.

***Listing 14-2.*** Example code for initializing cbuf

```
// initialize the circular buffer
CBUF_Init(RxDataBuffer);

// Initialize the buffer to all 0xFFFF
for(i = 0; i < myQ_SIZE; i++)
{
    CBUF_Push(RxDataBuffer, 0xFF);
}

// Clear the buffer
for(i = 0; i < myQ_SIZE; i++)
{
    CBUF_Pop(RxDataBuffer);
}
```

With these two code snippets in place, Listings 14-1 and 14-2, we can fill out some application code to receive the data from the circular buffer and then feed it to the packet parsing module. The rate at which a system decodes data packets can be designed in many ways. First, it could be completely event driven. As part of our ISR, we could use a semaphore to tell a task that there is data to process. Second, we could define a periodic task that processes the circular buffer regularly. Which one is correct? It depends on the application's latency needs. For this example, the code could be placed in a 100-millisecond periodic task, as shown in Listing 14-3.

***Listing 14-3.*** Example task code for reading the circular buffer data, feeding the data into the packet parser, and detecting if a valid packet is ready

```
char ch;
bool PacketReady = false;
PacketMsg_t const * const Packet = Packet_Get();

// While the buffer is not empty
while(!CBUF_IsEmpty(RxDataBuffer) && PacketReady == false)
```

```
{
    ch = CBUF_Pop(RxDataBuffer);

    PacketReady = Packet_DecodeSm(ch);
}

if(PacketReady == true)
{
    // Process Packet
    Command_Process(Packet);

    PacketReady = false;
}
```

Now that we have the application pieces in place, let's look at the packet parsing module itself.

## Packet Decoding As a State Machine

So far, we have designed the interface and application code that feeds the packet parsing module. The application code that feeds the module is entirely independent, and the parser only cares about receiving a data byte and nothing about the physical interface. We are now at the point where we are feeding data into an empty stub. Then, we can define a simple state machine to implement the packet parser where each state represents an element from our packet protocol. For example, compare the packet design in Figure 14-2 with the state machine design defined in Figure 14-7. You'll notice that the only difference is that we now have defined transitions between the various states.

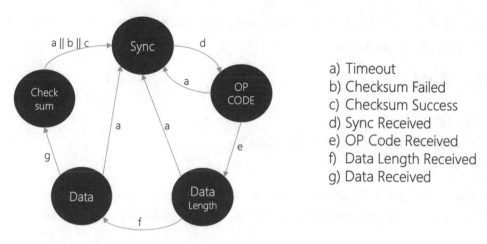

a) Timeout
b) Checksum Failed
c) Checksum Success
d) Sync Received
e) OP Code Received
f) Data Length Received
g) Data Received

***Figure 14-7.*** *An example packet decoding state machine based on the packet protocol fields. Not receiving a byte in a given period resets the state machine as a timeout error*

Each of the states defined in the state machine can be represented by an enumeration member, as shown in Listing 14-4.

***Listing 14-4.*** The state machine from Figure 14-7 converted into an enum

```
typedef enum
{
    SYNC,               /**< Looks for a sync character  */
    OP_CODE,            /**< Receives the op code        */
    DATA_LENGTH,        /**< Receives the data size       */
    DATA,               /**< Receives the packet data     */
    CHECKSUM,           /**< Receives the checksum        */
    PACKET_STATE_END    /**< Defines max state allowed    */
} PacketRxState_t;
```

There are many different ways we could write our state machine in C. However, a method that I am particularly fond of is using tables. Tables allow us to easily specify the function that should be executed for each state and provide a place to easily add, remove, or maintain the states. To start, we define a structure that contains the information required for each state in the state machine, as shown in Listing 14-5.

**Listing 14-5.** A typedef defines the information needed to execute a state in the packet parsing state machine

```
typedef struct
{
    PacketRxState_t State;
    (bool)(*State)(uint8_t const Data);
}PacketSm_t;
```

We can now define the state machine in table form using the enumeration members and the functions that should be called to execute each state, as shown in Listing 14-6.

**Listing 14-6.** Defining the table that maps states to their respective function

```
static PacketSm_t const PacketSm[]=
{
    {SYNC,          Packet_Sync        },
    {OP_CODE,       Packet_OpCode      },
    {DATA_LENGTH,   Packet_DataLength },
    {DATA,          Packet_Data        },
    {CHECKSUM,       Packet_Checksum    }
};
```

Notice in Listing 14-6 that we define our array of PacketSm_t to be constant and static. We define the table as const for several reasons. First, we don't want the state machine to be modified during runtime. Second, declaring the table const should force the compiler to store the table in flash rather than RAM. This is good because our table contains function pointers, and we don't want to open an attack surface by allowing a buffer overrun, etc., to modify what the state machine is pointing to. (Most embedded compilers, by default, put const into flash, not RAM. Usually, you must force it into RAM using a compiler switch, but this is compiler dependent.) Finally, we also define the table as static to limit the scope of the table to the packet module so that no external module can use it or link to it.

Now, with the state machine table set up, you are ready to implement the interface that will use the state machine. Listing 14-7 shows how simple the state machine code can be. First, Packet_DecodeSm performs a simple runtime sanity check that the PacketSmResult variable that tracks the state machine state is within the correct boundaries. We perform this runtime check rather than an assertion because if memory corruption ever occurs on PacketSmResult, we don't want to dereference an unknown memory location and treat it like a function pointer. To prevent this, we wrap a simple boundary condition check, and if that fails, we reset the state machine and set an error.

***Listing 14-7.*** An example Packet_DecodeSm implementation that uses the PacketSm table and some basic runtime error handling

```
bool Packet_DecodeSm(uint8_t const Data)
{
    bool PacketSmResult = false;

    if(PacketState < PACKET_STATE_END)
    {
        PacketSmResult = (*PacketSm[PacketState].State)(Data);
    }
    else
    {
        Packet_SmReset();
        PacketError = BOUNDARY_EROR;
    }

    return PacketSmResult;
}
```

The key to the Packet_DecodeSm function and running the state machine is dereferencing the function pointer. For example, Listing 14-7 contains a line of code that looks similar to the following:

```
(*PacketSm[PacketState].State)(Data);
```

This line is going to the PacketState element in the PacketSm array and dereferencing the State function pointer. So, if PacketState is SYNC, then the Packet_Sync is executed from the PacketSm table. If PacketState is DATA, then the Packet_Data function is executed, and so on and so forth. There are several advantages to this state machine setup, such as

- A new state can be added by adding a new row to the StateSm table.

- Likewise, a state can be deleted by adding a new row to the StateSm table.

- All states are executed through a single point in the packet module rather than through complex and convoluted switch statements.

- Each state function can easily be remapped based on the application needs.

Looking at the code that we have defined so far, it's apparent that we are missing the definitions for each function in our state machine. Defining each function goes above and beyond what we want to explore in this chapter. At this point, you have a framework for decoding packets. Implementing the remaining details in the state machine functions depends on your packet protocol definition. Before we move on to command decoding, we should examine the last state in the state machine and explore how we validate the packet.

## Validating the Packet

PacketDecodeSm accepts a character and proceeds through the various decoding states based on the character data it receives. The state machine progresses until it has received the checksum. Once the checksum has been received, the checksum state will run a checksum validation function to ensure that the received packet matches the received checksum. Checksum validation can be performed through a Packet_Validate function. Packet_Validate could use a software-based checksum like a Fletcher16 algorithm or a microcontroller hardware-based CRC calculator. It's completely application dependent. An example of using Fletcher16 can be seen in Listing 14-8.

***Listing 14-8.*** An example Packet_Validate function that uses a Fletcher16 checksum

```
bool Packet_Validate(void)
{
    uint16_t ChecksumRx = 0;
    bool ChecksumValid = false;

    // Verify the checksum before executing the command
    ChecksumRx = (Packet.Checksum[0] << 8);
    ChecksumRx |= Packet.Checksum[1];
```

```
// Calculate checksum of packet
// The magic number 3 is to add in the sync, opcode and
// data length to the calculation
ChecksumValid = Fletcher16_CalcAndCompare((char*)&Packet,
                              Packet.Length + 3, ChecksumRx);

    return ChecksumValid;
}
```

The reader can take the example function in Listing 14-6 and modify it so that the Fletcher16_CalcAndCompare function is replaced with the correct function at compile time. Optimally, this function would be assigned at compile time based on how we want to validate the packet. We can do this through a function pointer, which gives us a degree of polymorphism to the implementation, which is an object-oriented design technique.

You'll recall that the Packet_DecodeSm returns a bool. Once the Packet_Validate function validates that a complete packet exists, it returns true to the higher-level application. The application will see that Packet_DecodeSm is valid and can retrieve the completed packet from memory for processing. Let's now look at how we take the validated packet and process the associated command and payload.

# Command Processing and Execution

Command processing is a core feature in many embedded systems. The ability to command the system and change its state is fundamental. However, parsing and executing commands can turn application code into a giant ball of bud! Let's explore several methods that can be used to parse a packet to execute a command that is contained within it.

## Traditional Command Parsers

When it comes to parsing command packets or any type of messaging protocol in C, I often see developers use two parsing methods:

- If/else if/else statements
- Switch statements

These C constructs can be effective for protocols with only a few commands, but as the number of commands in the system approaches a dozen or more, they become inefficient. For example, consider a bootloader where we have four commands that can be executed:

- Bootloader exit

- Erase device

- Program device

- Query device

We could quickly write an if/else if/else statement to determine which OP Code was sent in the packet using code similar to that shown in Listing 14-9. In the example, five commands are supported, creating five potential branches in our packet decoder. Notice that the variable Packet is a pointer to the validated packet memory location. Therefore, we need to use -> rather than dot notation to access the OP Code. Overall, it is not too bad for just a few commands using if/else if/else statements. Our command parser, in this case, has five independent branches, therefore a McCabe Cyclomatic Complexity of only 5; therefore, the complexity of the command parser is also relatively low, testable, and manageable. (Recall, we discuss McCabe's Cyclomatic Complexity in Chapter 6.)

***Listing 14-9.*** An example packet decoding method that uses if/else/else if statements to identify the command and execute the command function

```
if(Packet->OP_CODE == BOOT_EXIT)
{
  Bootloader_Exit(Packet);
}
else if(Packet->OP_CODE == ERASE_DEVICE)
{
  Bootloader_Erase(Packet);
}
else if(Packet->OP_CODE == PROGRAM_DEVICE)
{
  Bootloader_Program(Packet);
}
else if(Packet->OP_CODE == QUERY_DEVICE)
```

```
{
  Bootloader_DeviceQuery(Packet);
}
else
{
  Bootloader_UnknownCommand(Packet);
}
```

One potential problem with the technique used in Listing 14-9 is that the code looks through every possible command that could be received until it finds one that matches or decides that the command is invalid. This is a bit inefficient because if we receive the last message in our statement, we're wasting clock cycles working through a bunch of if/else if cases. A more efficient way to implement the parser would be to use a switch statement like the one shown in Listing 14-10.

***Listing 14-10.*** An example packet decoding method that uses switch statements to identify the command and execute the command function

```
switch(Packet->OP_CODE):

  case BOOT_EXIT:

        Bootloader_Exit(Packet);

        break;

  case ERASE_DEVICE:

        Bootloader_Erase(Packet);

        break;

  case PROGRAM_DEVICE:

        Bootloader_Program(Packet);

        break;

  case QUERY_DEVICE:

        Bootloader_DeviceQuery(Packet);

        break;
```

```
default:

    Bootloader_UnknownCommand(Packet);

    break;
```

We could argue that a good compiler will reduce both code sets to identical machine instructions. However, we might only improve human readability and maintainability at this point. The great thing about the switch statement is that the compiler will generate the most efficient code to execute the statement when it is compiled, which usually results in a jump table. Instead of checking every case, the compiler creates a table that allows the value to be indexed to determine the correct case in just a few clock cycles.

The problem with switch statements is that they can become complex and challenging to maintain as they grow to over a dozen or so commands. For example, I've worked on flight software for several different small satellite missions, and these systems often have several hundred possible commands. Can you imagine the pain of managing a switch statement with several hundred entries? Spoiler alert, it's not fun! In addition, the complexity of our function will shoot to the moon because of all the possible independent branches that the code could go through.

When it comes down to it, we need a more efficient technique and one that is less complex and easier to maintain than just using switch statements. This is where a command table can come in handy.

## An Introduction to Command Tables

A command table is an array of a structure that contains all information necessary to find a command that needs to be executed and the function that should be executed for that command. Creating one is quite simple. First, define a command structure that contains a human-readable command name and a function pointer to the function that should be executed. This can be done using code like that shown in Listing 14-11.

***Listing 14-11.*** Defining a command structure that contains all the information necessary to execute a system command

```
typedef struct
{
    Command_t Command;
    void (*function)( CommandPacket_t * Data);
}CommandRxList_t;
```

You may notice in Listing 14-11 that we are once again following a similar pattern. We once again have a human-readable value as one structure member and then a function pointer as the other member. In addition, there is an undefined Command_t in the declaration. Command_t is an enumeration that contains all the commands we would expect to receive in our system. For example, for the bootloader, I may have an enumeration that looks like Listing 14-12.

***Listing 14-12.*** The bootloader command list as a typedef enum

```
/**
* Defines the commands being received by the bootloader.
*/
typedef enum Command_t
{
    BOOT_ENABLE,        /**< Enter bootloader */
    BOOT_EXIT,          /**< Exit bootloader */
    ERASE_DEVICE,       /**< Erase application area of memory */
    PROGRAM_DEVICE,     /**< Program device with an s-record */
    QUERY_DEVICE,       /**< Query the Device */
    END_OF_COMMANDS     /**< End of command list */
};
```

With the enumeration and the structure defined, we can create a command table containing all the commands supported by the system and map those commands to the function that should be executed when the command is received. The bootloader command table then becomes something like Listing 14-13.

***Listing 14-13.*** An example command table that lists the system command and maps it to the command to execute

```
/**
 * Maps supported commands to a command function.
 */
CommandRxList_t const CommandList[] =
{
    {BOOT_EXIT,           Command_Exit          },
    {ERASE_DEVICE,        Command_Erase         },
    {PROGRAM_DEVICE,      Command_Program       },
    {QUERY_DEVICE,        Command_Query         },
    {END_OF_COMMANDS,     NULL                  },
};
```

There are several advantages to using a table like Listing 14-12 to manage commands which benefit the developer, including

- Humans can easily read through the table quickly, which gives an "at a glance" look at the commands in the system.

- If a command needs to be added, a developer just needs to insert a new row into the table.

- If a command needs to be removed, a developer can just remove that row from the table.

- If a command operational code needs to change, it can be updated in the enumeration, and the command table does not need to change.

- The command table can be generated by a script using a system configuration file similar to what we saw in Chapter 13 with the task table generation.

There are additional advantages one could also think up, such as minimizing complexity, but at this point, our time is best spent looking at how we can execute a command from the command list.

# Executing a Command from a Command Table

Once a command table has been implemented, executing a command from the table can be in several different ways. First, and most preferably, the Command_t enum should be sequential without any gaps and start at 0x00. If this is done, executing the command can be done using the code in Listing 14-14. You'll notice that this code looks very similar to how we executed the state machine!

**Listing 14-14.** A function for executing the OP Code, a command, from a received data packet

```
void Command_Process(CommandPacket_t const * const Packet)
{
    if(Packet->OP_CODE < END_OF_COMMANDS)
    {
        (*CommandList[Packet->OP_CODE]->function)(Packet);
    }
}
```

All our Command_Process function needs to do is index into the array and dereference the function pointer stored there! That's it! Of course, we would probably want to add some additional error checking and maybe add some error codes if we go outside the defined bounds, but I think the reader at this point understands it, and I've removed the code for brevity.

If the enum is not sequential or has gaps in the ordering, the code in Listing 14-14 will not work. Instead, the table needs to be searched for a matching operational code, and only when a match is found is the pointer dereferenced. This makes the command parsing much less efficient. However, several methods could be used. While developers may be tempted to jump into algorithms for binary searches and so forth, using a simple loop such as that shown in Listing 14-15 can often be the simplest solution.

**Listing 14-15.** An example command processing function is when the command OP Codes are not sequential and contain "holes" in the command map

```
void Command_Process(CommandPacket_t const * const Packet)
{
    const CommandRxList_t * CmdListPtr;
    uint8_t CmdIndex = 0;
```

```
    // Loop through the command list for a match.
    CmdListPtr = CommandList;

    while((CmdListPtr->function != NULL) ||
          (CmdListPtr->Command != Packet->OpCode))
    {
        CmdListPtr++;
        CmdIndex++;
    }

    // Verify that we found a match and that the function is
    // not NULL
    if(CmdListPtr->function != NULL)
    {
       // Execute the command and parse out the command byte
       (*CmdListPtr->function)(Packet->Data);
    }
}
```

As you can see, we use a while loop to search through the table for a match. This is not as efficient as being able to directly index into the array, but it can still work quite well and is easier to maintain than switch statements. There are also several other ways to write the search code, but it can come down to your personal preference, so use the code provided as an example, not gospel.

# Managing System Telemetry

Let's shift focus from parsing packets and executing commands to discussing telemetry. Once again, nearly every embedded system has some type of telemetry data that it sends back to the user or even the manufacturer. The telemetry could consist of sensor values, health, and wellness data, among other things. However, there is a significant problem that is often injected into systems due to telemetry:

1.    Telemetry has the potential to break dependency models and turn architecture into a giant ball of mud.

Telemetry code has a bad habit of reaching out into the system and touching every module. As a result, the smallest modules in the system suddenly start to depend on the telemetry module so that they can update it with their data. This is something that we don't want to happen. So in this section, let's explore some ideas on how you can successfully manage your telemetry code without breaking the elegance of your design.

# Telemetry As a "Global" Variable

Telemetry data is stored in a structure variable in many systems and declared at global scope. Several problems present themselves when a developer takes this approach with their telemetry. First, every task and module in the system can directly access the telemetry structure. The opportunity for race conditions, data corruption, and so forth dramatically rises. Now you might think in an RTOS-based application that you can just throw a mutex at the structure, but every module also needs to have an RTOS dependency! Before you know it, the dependency map is out of control!

The trick to minimizing the telemetry dependencies and keeping things in control is to limit how the telemetry structure is updated. In an RTOS-based application, the only application code that should be able to update the telemetry is a task. Therefore, supporting code modules like drivers, board support packages, and middleware should be queried and controlled by their respective tasks, and then that task updates the telemetry structure. This limits how data flows into and out of the structure by default.

Limiting telemetry access to tasks can still present developers with a problem. The temptation to create a global variable still exists. If the global variable exists, the chances are higher that someone who isn't supposed to access it will. However, developers can leverage an exciting technique to create the telemetry variable at a high scope with the application and then pass pointers to the telemetry structure into the task creation function! Listing 14-16 shows a task created with the telemetry task receiving a pointer to the telemetry structure through the parameter's pointer.

***Listing 14-16.*** Passing the telemetry structure into a task at creation can limit the scope and access to the telemetry data

```
xTaskCreate(TaskCodePtr,
            TaskName,
            StackDepth,
            &Telemetry,
```

```
    TaskPriority,
    TaskHandle);
```

There are a couple of caveats to passing the telemetry pointer into the task this way. First, the task code will depend on the telemetry module to decode what members are in the structure. That should be okay since there is a limit on how many modules will access it. Next, the task is just receiving a pointer to something. It doesn't know that it is a pointer to telemetry data. In the task function itself, we will need to cast the received pointer into a pointer to the telemetry data. For example, Listing 14-17 shows how the casting would be performed. Finally, there is still a potential for race conditions and the telemetry data to become corrupted due to no mutual exclusion.

***Listing 14-17.*** Casting the pointer to telemetry data in a task to a telemetry pointer

```
void Task_Module(void * pvParameters)
{
    ...

    Telemetry_t * const Telemetry = (Telemetry_t * const) pvParameters;

    ...
}
```

The shared variable problem with multiple tasks being able to access the telemetry structure is not something we can ignore. The chances are that different tasks will access different members, but we can't guarantee that. On top of that, it's possible a task could be updating a telemetry member and be interrupted to transmit the current telemetry, sending out a partially updated and incorrect value. However, there is a simple solution. We can build a mutex into the telemetry structure.

Adding a mutex into the telemetry structure bundles the protection mechanism with the data structure. I like to make the structure's first member a pointer to the telemetry data. In most modern IDEs with autocomplete, when you try to access the telemetry structure, the first autocomplete option that will show up is the mutex! The developer should immediately remember that this is a protected data structure, and the mutex should first be checked. (Note: You could also build a telemetry interface that abstracts this behavior.)

***Listing 14-18.*** Example mutex pointer being added to the telemetry data structure

```
typedef struct __attribute__((packed))
{
    SemaphoreHandle_t * MutexLock;
    ...
}Telemetry_t;
```

# Telemetry As a Service

Another technique developers can use to limit telemetry from turning an elegant architecture into a giant ball of mud is to treat telemetry as a service. Developers can create a telemetry task that treats the telemetry data structure as a private data member. The only way to update the telemetry data is to receive telemetry data updates from other tasks in the system through a telemetry message queue. For example, examine the architectural pattern shown in Figure 14-8. The figure demonstrates how the message queue is available to all the tasks, but the telemetry data and access to Task Telemetry are restricted.

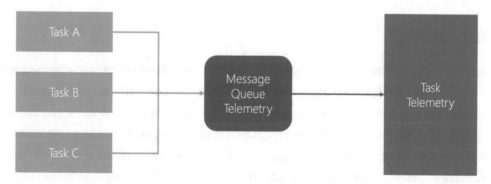

***Figure 14-8.*** *A message queue can be used as the "entry point" to perform updates on a private telemetry data member residing in Task Telemetry*

Using an architectural pattern like Figure 14-8 has some trade-offs and key points to keep in mind, just like anything else in software and engineering. First, there does need to be some type of coordination or identifier between the tasks (Tasks A–C) and Task Telemetry so that Task Telemetry can identify whose telemetry data it is in the message.

The coordination will require each message placed into the queue to have some type of telemetry identifier and fit a specific format. In fact, a simplified version of the packet protocol that we looked at earlier in this chapter could even work.

Next, we need to consider how to properly size the message queue. Each task will send its latest telemetry to the queue many times per cycle. The designer must decide if the message queue is read out once or twice per cycle. If it's just read once, the message queue may need to be slightly larger to maintain all the telemetry message updates. This is a bit inefficient from a memory perspective because there may be multiple duplicate messages that, when processed, will overwrite each other. However, if we process the message queue more often, we're potentially wasting clock cycles that another task could use. It's a trade-off and up to the developer to decide what is most appealing to them.

Finally, if the software changes, the maintenance for this pattern should be minimal. For example, if a new task with telemetry points is added, a new message specifier can be added. Task Telemetry can remain as is with just the added case for the new telemetry message. The same goes for if a task is removed. Generally, the pattern is relatively scalable and should prevent the software architecture from decaying into a giant ball of mud.

# Final Thoughts

Communication, commands, and telemetry are standard features in many embedded systems. Unfortunately, many implementations consider these features as one-off design decisions that result in unnecessary development costs and time expenditures. However, communication interfaces, command processing, and telemetry are common problems that most systems need to solve. In this chapter, we've seen several techniques that developers can leverage to write reusable and easily configurable code for various applications. These techniques should help you to decrease cost and time to market in future projects.

## ACTION ITEMS

To put this chapter's concepts into action, here are a few activities the reader can perform:

- Identify the characteristics that you need in a packet protocol. Does your current protocol meet those needs? Then, create an updated packet protocol that meets your needs.

- What is the difference between a checksum and a CRC? What can be done to improve the efficiency with which a CRC is calculated on a microcontroller?

- Implement an example software checksum and CRC. Compare the execution and memory differences. Implement a hardware-based CRC. How did the memory and execution times change?

- Design and implement a packet parsing state machine for your packet protocol. What techniques can you use to make it more reusable and scalable?

- Experiment with using state tables to transition through the various states of a state machine. How does this compare to other techniques you've used in the past?

- Implement a command parser. What differences do you notice about using a command table vs. using if/else if/else and switch statements?

- Review how you currently implement your telemetry transmission functions. What changes can you make to minimize module dependencies?

# CHAPTER 15

# The Right Tools for the Job

When I teach a class, speak at a conference, or work with a client, one of the most asked questions I get is related to the tools I use to get the job done. Of course, engineers love their tools, and they should! Tools help them get their job done; if you aren't using the right tools, it will most likely take longer and cost more to get the job done. However, I often find that engineers and companies overlook the benefits and value of tools. This chapter will explore the benefits of using the right tools, how to justify getting those tools, and what tools you need to succeed. I'll also share a little about the tools I use personally, but keep in mind it changes and evolves yearly. (You might consider subscribing to my newsletter to stay up to date on new and exciting tools![1])

## The Types of Value Tools Provide

Imagine for a moment that you are building a house or a shed or finishing a basement. If you wanted to do so as fast as possible, with minimal labor, would you choose to frame the structure using a hammer and nails or nail gun and nails? Would you use a screwdriver to install the drywall or a cordless drill? The tools we choose to use can have a dramatic effect on a project's success. When it comes to embedded software engineering worldwide, it would surprise you how many teams use a "hammer and nails."

The tool's value is the first and most important consideration when selecting and purchasing tools. Value is the importance, worth, or usefulness of something.[2] Value comes in four different types:

---

[1] www.beningo.com/insights/newsletter/
[2] www.merriam-webster.com/dictionary/value

© Jacob Beningo 2022
J. Beningo, *Embedded Software Design*, https://doi.org/10.1007/978-1-4842-8279-3_15

- Monetary

- Functional

- Emotional

- Social

Monetary value can be equated to the direct revenue generated or saved from having and using the tool. Economic value can also be calculated from the time saved on a project. Development requires time which requires developers who need monetary payment for their services. If a tool can allow a developer to get more done, that tool provides economic value to the company.

Functional value is something that solves a problem. For example, a compiler tool will solve the problem of cross-compiling the application into instructions that the target processor can execute. A logic analyzer will allow the engineer to see the bus signals. Monetary and functional values are the core values that businesses look at. If a developer wants a tool to make their job easier, they must convince their company using monetary and functional values.

---

**Tips and Tricks**    When requesting a new tool, build up a concise but powerful return on investment statement based on monetary and functional values. If done correctly, you'll rarely be denied the tools you need.

---

Emotional and social values are generally less important to a business but can be important to the engineer. Emotional value provides psychological value, such as a sense of well-being or a tool that decreases stress. For example, I leverage my calendar and project management software to schedule my time and understand my workload. These tools are emotionally valuable because they offload activities I would otherwise have to track and ensure I don't overwork myself mentally.

Social value connects someone to others or gives them social currency among their peers. For example, as a consultant, I often leverage techniques or tools that I use or know about to get companies or engineers interested in my work and services. I have tools and knowledge they don't know about, and I can use that to create a social connection. That interest for me will often lead to engineers taking my courses, signing up for mentoring or advisory services, or companies asking me to help them architect and review their software systems. Social value is less valuable, though, to an engineer in the trenches who are working for their boss. However, it could be essential to a company's customers.

When selecting and requesting the purchase of a new tool, developers need to think about and clearly communicate the value proposition for the tool. To do so, they must learn how to calculate the tool's value.

# Calculating the Value of a Tool

A big problem I often encounter is that companies are more than willing to buy oscilloscopes, spectrum analyzers, and other expensive hardware tools, but the moment someone wants to buy a software tool like a compiler, static analyzer, or linter, they are instantly denied the purchase. There seems to be a belief in business that software tools are supposed to be open source and free, which, to be honest, is a great way to set up for being late on a project, going over budget, and complete failure.

Part of the problem with software developers getting the tools they need to be successful is learning how to talk business to their managers and bosses. A company is only interested in investing in the business and producing a profit. Purchasing a new tool (spending money) must have a clear return on investment (profit). The hardware engineers have something physical that management can understand and wrap their heads around. For software developers, they have nebulous bits and bytes that remind managers of the streaming code of the Matrix. They don't understand software. Software to them is expensive, holds up projects, and is something they wish they didn't need (even though the software is a great product differentiator and a huge business asset).

---

**Caution**    Don't get caught in the trap of believing you can build your own tool faster and cheaper! Developers are notorious for underestimating the value of a tool. I've seen too many teams try to save money by building it themselves, only to spend 10x or more and still have to spend the money to bail themselves out.

---

Developers need to learn how to communicate the value that software tools bring to the company and how the business will profit from it. Developers need to convert the four values discussed in the last section into business needs. For example, a developer might demonstrate that a tool purchase will

- Improve developer efficiency

- Increase system robustness

- Result in shorter development times

- Decrease costs

- Improve scalability and reuse

- Minimize rework

The list can go on and on.

The trick to ensuring you get the tools you need and want is to request the tool in a way that the business understands: return on investment (ROI). Return on investment can be calculated using the following formula:

$$\text{Return on Investment (ROI)} = \text{Value Provided} - \text{Value Invested}$$

The ROI should be positive. The larger the positive value, the greater the investment in the tool is, and the harder it will be for the business to say no! The company may also want to look at the ROI from a percentage standpoint. For example, instead of saying that a purchase will annually save the business $2000, the investment will return 20%. The ROI as a percentage can be calculated using the following formula:

$$\text{ROI \%} = (\text{Value Provided} - \text{Value Invested}) / \text{Value Invested} * 100\%$$

Let's look at an example of how to justify purchasing "expensive" tools costs. One tool that I often get pushed back on being expensive is debugging tools. So let's calculate the ROI of a professional debugger such as a SEGGER J-Link.

---

**Caution**    It's not uncommon for developers to look at a tool's price and decide that it is expensive based solely on a comparison with their salary or standard of living expenses. Don't get caught in this trap! The price is inconsequential for a business if it has a good return on investment! (Look at your salary and ask yourself if you would pay someone that much for what you do!)

---

# The ROI of a Professional Debugging Tool

Compilers and professional debuggers are software tools I often see a lot of resistance to purchasing. I say professional here because nearly every development board today comes with some sort of onboard debugger, and there are many "one-chip"[3] debuggers out there that can be purchased for less than $100. Since these low-cost debuggers are everywhere, getting permission to buy an $800 or $1500 debugging tool becomes an uphill battle.

The reader must recognize that the low-cost tools provide the minimum debugging capability possible to get the job done. The debugger typically can program and erase flash, perform basic debugging, and provide maybe two or three breakpoints. Breakpoint debugging is one of the most inefficient ways to debug an application. Developers need the ability to trace their application in real time, record the trace, and set up data watchpoints to trigger events of interest. These capabilities take a lot of time to develop and are usually absent from onboard debuggers and low-cost programmers. Low-cost programmers are useful for quick evaluation purposes, though.

---

**Lesson Learned**   Early in my career, I accumulated a box of low-cost programmers that probably cost me upward of $3000 over the course of just a few years. I had dozens of them for all sorts of microcontrollers. Each debugger provided the minimal capability possible, making each debug cycle exceptionally painful. One day, out of annoyance of having to buy yet another tool, I decided to invest in a SEGGER J-Link Ultra+. Upon first use, I realized how much my attempt to be frugal had cost money, time, and performance. I was able to discard my box of programmers, and nearly eight years later, I still have not had to purchase a single programming tool. (Well, I did buy a SEGGER J-Trace so I could do instruction tracing, but I don't count that.)

---

Consider the value a professional debugger might present to a developer and the company they work for. On average, a developer spends ~20% of their time on a project debugging software.[4] (When I speak at conferences and survey the audience, the result is

---

[3] A "one-chip" debugger is a debugger that can only program a single family of microcontrollers or just literally one from the family.

[4] https://bit.ly/3AlhA8u, page 38.

closer to 50% or more.)[5] Assuming that the average work year is 2080 hours, the average developer spends 416 hours per year debugging. (Yes, that's 2.6 months yearly when we are incredibly conservative!)

According to glassdoor.com, the average US salary in 2022 for an embedded software engineer is ~$94,000.[6] However, that is not the cost for an employee to the business. Businesses match the social security tax of an employee, ~$6000, and usually cover health care, which can range from $8000 to $24,000 per year (I currently pay ~$15,000/year). Therefore, we will assume health-care costs are only $5000. In fact, we will ignore all the other employee costs and expenditures. In general, it is not uncommon for a business to assume the total cost for an employee is two times their salary!

Adding all this up brings the total to around $105,000 for the average developer or ~$50 per hour. (The exact value is $50.48, but we'll perform our math in integer mathematics as our microcontrollers do.) Of course, we ignore adjustments for holidays, vacations, and the like to keep things simple.

Calculating the tool's value requires us to figure out how much time a developer can save using the new tool. For example, if the tool costs $1000, we need to save 20 hours or decrease our debugging time by ~5% to break even. Personally, I like to break out the value of the improvements the new tool will bring and assign a time value that will be saved per year. For example, the value proposition for the debugger might look something like the following:

- Adds real-time trace capability (saves ~40 hours/year)

- Unlimited breakpoints (saves ~8 hours/year)

- Increases target programming speed (saves ~8 hours/year)

---

[5] This observation was corroborated with Jack Ganssle in addition to (thanks to Jack for the references!)

- *Software Testing in the Real World: Improving the Process*, Addison Wesley, Harlow, England, 1995.
- *Facts and Fallacies of Software Engineering*, Glass, page 90.
- www.cebase.org:444/www/defectreduction/top10/top10defects-computer-2001.pdf
- *Winning with Software*, Watts Humphrey, p. 200, and *The Art of Software Testing* (Myers, Sandler, and Badgett).
- Capers Jones (www.ifpug.org/Documents/Jones-SoftwareDefectOriginsAndRemovalMethods Draft5.pdf).
- www.sei.cmu.edu/library/abstracts/news-at-sei/wattsnew20042.cfm

[6] https://bit.ly/3qMzwpb

We could certainly argue on the justification for the times, but from experience, they are on the conservative side. The ability to trace an application is a considerable saving. While one might argue that programming the device faster or pulling data off more quickly is not a big deal, you're wrong. When I switch from my SEGGER J-Link Ultra+ or my J-Trace, I notice a difference in the time to program and start a debug session. Given how often a developer programs their board daily, these little time frames add up and accrue to measurable time frames. (The caveat here is that if you are using disciplines like TDD, simulation, DevOps, and so forth, then the savings may not be as high, but you'd likely need a more advanced tool anyways for network connections and the build server.)

So far, our conservative estimate would show that we can save 56 hours a year by purchasing the professional debugger. The savings for buying the tool is expected to be

ROI = $2800 (Engineering Labor) – $1000 (Tool Cost) = $1800

ROI % = $1800 / $1000 * 100% = 180%

It's not a considerable amount of money, but it's also freeing up the engineer by 36 hours to focus on something else during the year. For example, it might be enough time to implement a new feature or improve the CI/CD process.

An important point to realize here is that this is just the first year, and it's only for one developer! The first J-Link Ultra+ that I purchased lasted six years before it met an untimely death while connected to a piece of client's hardware. (SEGGER has a fantastic trade-in policy where even though the device was damaged, I could trade it in and purchase a new one at half price!) If we assume that the developer will use the debugger for five years, suddenly, we see a time savings of 250 hours total! Our ROI becomes

ROI = ($50 x 250 hours (Engineering Labor)) – $1000 (Tool Cost)
= $11,500

ROI % = $11,500 / $1000 * 100% = 1150%

I don't think anyone can argue with that math. Who doesn't want a return on investment of 1150% in five years?

What's interesting is that this can still get better. For example, suppose there are five developers in the company, and each gets their own debugger, for over five years. In that case, we're suddenly talking about injecting nearly 6250 more development hours into the schedule! What could your company do with three extra person-years worth of development?

**Tips and Tricks**   Small investments and nearly imperceptible improvements can have a dramatic long-term effect on a development team and business. What minor enhancements do you need to make today to change your long-term results?

# Embedded Software Tools

In the Introduction of the book, we discussed that the most successful teams strike a balance between their software architecture, processes, and implementation. This book is organized around the same concept, and the best approach to discuss the tools we need to develop software is to follow that approach. There are more tools and types than we can explore in this chapter; however, we can look at some of the primary tools developers should have in their toolbox. For those interested, I often post blogs about tools on my website at www.beningo.com. If you go to the blog, there is a category called tools that you can use to access the tool blogs. In addition, the tool reviews and tutorials I have done can also be found here.[7]

**Caution**   Tools must provide an ROI. Some tools and processes related to project management can become tangled cobwebs that prevent rapid progress. Make sure that you dictate the tool's behavior, not the other way around!

# Architectural Tools

Architectural tools help a developer and team design the architecture for their product. We've discussed in Part 1 the importance of software architecture. The architecture is the road map, the blueprints developers use to implement their software system. While there are many architectural tools, three primary categories in this area come to mind: requirements elicitation and management, storyboarding, and UML diagramming and modeling.

---

[7] www.beningo.com/category/tools/

## Requirements Solicitation and Management

I often joke that requirements are the bane of a software developer's existence. Requirements are boring, tedious, and perhaps one of the most important aspects of developing a system. Requirements are a critical key to managing the customers' expectations, and if those expectations aren't managed correctly, even success could be viewed as a failure. (This is why scope creep is so dangerous and needs to be carefully managed.)

Several different types of tools can be used to meet management requirements. The tool selected will be based on the organization's size and traceability requirements. The first tool, and the one that I still see small organizations or teams working with legacy requirements use, is just a simple Excel spreadsheet. Within Excel, teams can create columns for requirement numbers, requirements descriptions, and other status and traceability requirements needs. Excel is low cost and straightforward but does not allow other tools' linking and full design cycle management capabilities.

Commercial tools often used within the embedded systems industry include Jama Software[8] and DOORS.[9] These tools provide customizability and end-to-end traceability and integration. Thankfully, several great web pages describe many of these tools in detail. You can check out this page for more options.[10]

## Storyboarding

One of my favorite ways to manage requirements is to create a series of stories describing the product's use cases. I personally believe a picture is worth 1000 words and several hundred thousand lines of code. Breaking the product requirements into a series of pictures describing the use cases and user needs can dramatically help clarify and simplify requirements. A storyboard allows the entire life cycle of an idea to be expressed in one image by showing the problem, a high-level solution, and the benefit to the user.[11]

There are a lot of great tools available to help developers create storyboards. These tools can be either stand-alone or part of a larger software development life cycle process software. From a pretty picture standpoint, developers can use tools like

---

[8] www.jamasoftware.com/
[9] www.ibm.com/docs/en/ermd/9.7.0?topic=overview-doors
[10] www.guru99.com/requirement-management-tools.html
[11] https://devpost.com/software/storyboard-that-for-jira

- Adobe Photoshop

- Boords Storyboard Creator

- Canva

- Jira

- PowerPoint

- StoryboardThat

My go-to tool for my storyboarding is to use Jira from Atlassian (and it's not the last time I will mention it). Jira allows me to create a series of user stories I can track as part of my software development life cycle processes. You might be saying yes, but that doesn't give you a pretty picture. So instead, for my storyboarding, I create UML use case diagrams.

## UML Diagramming and Modeling

A critical tool when developing a software architecture is to utilize diagramming tools. Developers can use tools like PowerPoint, but it is far more powerful to use UML tools. UML tools provide a common language for developing the software architecture and guiding the software implementation. A really good tool will generate C/C++ code, so the developers don't need to hand-code. Of course, you must be careful with autogenerate tools. In general, the code they generate is not very human-readable, and the code could be tough to debug if there's a bug in the generator.

One popular tool that I often see used is Lucidchart.[12] Lucidchart is an online UML diagramming tool. Just like with many good tools, they provide professional templates that can be used to jump-start a design. The significant advantage of this tool is that it is great for distributed teams. Everything is hosted online, meaning developers worldwide can collaborate and access the same documentation. While this is great, I often consider this a disadvantage. Given the many corporate hacking cases, anything stored online must be considered vulnerable.

My personal go-to tool currently is Visual Paradigm.[13] Many of the UML diagrams in this book were created using it. The tool is potent, and one thing I like about it is that it links diagrams together, making navigating an architecture very easy. There are free

---

[12] www.lucidchart.com/

[13] www.visual-paradigm.com/solution/freeumltool/

editions of the tool along with commercial versions. The commercial versions allow the use of storyboards and can generate C++ code. There's nothing like designing a state machine or class diagram and clicking generate code. Just remember, the generated code most likely won't meet your coding standard and could use features you would like to avoid.

There are many other tools, and I can't do them all justice in print. Guru99 has a list of the best UML tools, which you can find here.[14] There are lots of great tools on there. Some other favorites of mine are Dia, Draw.io, Edraw Max, and Visio. Again, try out several tools. When evaluating tools, I create a little test project to see which tool best fits my workflows and needs.

# Process Tools

I love process tools. Professionals act professionally and use the best tools to complete the job. Tools help me to plan things out and ensure that I have a consistent development process. Honestly, I think with the right tools, teams can consistently do more, deliver more value, and help more people. There are a lot of great process tools out in the world, and there are also many bad ones. In general, at a minimum, teams should have tools for managing revision control, the software development life cycle, quality control, testing, and DevOps.

## Revision Control

When it comes to revision control, obviously, the industry standard today is Git. Mercurial attempted to become the industry standard, but many services have discontinued it. Git has won (for now) and looks like it will be the revision control tool to use for the foreseeable future.

Git repository tools are everywhere. I have repositories on Bitbucket, GitHub, GitLab, and Azure DevOps. Sigh. It's quite a bit, but it also helps me understand the significant services and what they offer. The key is finding a service that provides what you need with the desired security level. For example, many of these services will allow you to set up a personal access token. Unfortunately, I set mine to 30 days, which means they expire, and I must go through the pain of creating new ones monthly. The security is well worth it, though.

---

[14]www.guru99.com/best-uml-tools.html

I think the best tools for revision control aren't necessarily the online repositories but the client-side tools to manage the repositories. Yes, all we need is a terminal, and we can command line bang away Git commands all day long. I do this regularly. However, I find that user interface tools can be beneficial. For example, if I want to view the history and visualize the active branches and merges, I will not get that from a terminal.

---

**Caution**   When using a hosted repository service, I recommend enabling SSH and a personal access token. The personal access token should expire at most 90 days from creation.

---

One tool that I like is Sourcetree.[15] Sourcetree is available for Windows and Mac OS X and provides an excellent visual interface for working with repos. If you are like me and need to manage multiple accounts with several dozen active repos at any given time, a GUI-based tool is handy. I can organize my repos by customer, manage multiple repos, and track commits between employees, contractors, and clients while managing most aspects of the repos. I find that everything I need is unavailable through a button click, but that's why I keep my terminal handy. (Plus, working through the CLI[16] keeps my skills sharp.)

For Windows users, I have used TortoiseGit[17] for years. It provides all the tools one might need to manage a repo, all through a simple right-click of the mouse. Honestly, when I used it, I nearly forgot how to use the Git terminal commands because everything I wanted to do had a button click available. When I switched to exclusively using Mac OS and focused on terminal use only, I had to relearn many commands. I highly recommend no matter what tool you use, you occasionally go a week or two just using the terminal to keep your skills sharp if you decide to use a GUI-based tool.

## Software Development Life Cycle

Software development life cycle tools are the tools that we use to track and monitor progress on a project. They often will provide us with the concept of delivery tracking and visibility. As you might imagine, there are hundreds of tools out there that can be used to manage projects. However, there are several that have gained the most traction.

---

[15] www.sourcetreeapp.com/

[16] CLI = Command-Line Interface.

[17] https://tortoisegit.org/

My favorite tool for managing software projects is to use Atlassian's Jira.[18] A significant benefit is that it is free to use for teams up to around ten (assuming it has not changed since I wrote this). Jira can pretty much do anything that a development team needs. For example, developers can create a development road map, define their backlog, plan sprints, track issues, integrate it into their Git repo, and even manage software deployments. As I write this, Jira is the model tool for managing software development in an agile team.

Another interesting tool that is more on the commercial side is Wrike.[19] I've used it on several projects over the years and found that it provides most tools developers would want for an Agile-based software process.

Commercial tools aren't the only options for SDLC management. In the past, I've also used tools like Trac[20] and Redmine.[21] These tools allow a team to set up their project management server. I used to enjoy deploying and managing my own hosted servers, but over time, with all the cloud-based services that can do this for you, I've moved away from them. However, they do provide similar tools and workflows to other tools. Which one you choose comes down to your own choice and budget.

Every team needs to have an SDLC tool. Even when I operate as a one-person team, I found that trying to skip the use of a tool like these just made it harder for me to plan and get visibility into what it was I was working on. I don't care what tool you use, find the one that works for you and use it. You'll discover that over time you'll become far more efficient; just don't get stuck in the trap of spending too much time managing your tools.

---

**Tip**    Don't let sprint planning consume too much time. Limit your planning to 1.5 hours or less and get back to work!

---

## Testing

The ability to test our code and verify that it meets our customers' needs is perhaps one of the most critical aspects of developing software. Unfortunately, too many development teams and engineers I encounter don't have good testing tools (or

---

[18] www.atlassian.com/software/jira

[19] www.wrike.com/vm/

[20] https://trac.edgewall.org/

[21] www.redmine.org/

processes). The ability to test the software is critical, and the ability to do so in an automated way is perhaps even more important. Testing allows us to verify our assumptions that the code we have written does what it is supposed to. Testing helps us to verify and trace the desired result back to our requirements.

Test harnesses provide developers with a mechanism to perform unit tests, integration tests, and even system-level testing. Testing tools can range from free, open source tools like CppUTest[22] to commercial tools like Parasoft.[23] Personally, I use CppUTest in most of the projects that I work on. CppUTest provides a test harness that can be used with C/C++ applications. It's easy to get up and run, and I often run it within a Docker container integrated as part of DevOps processes.

Following a process like Test-Driven Development, you should develop your tests before writing your production code. (We discussed this in Chapter 8.) Defining test cases doesn't necessarily mean you will create enough test cases to get full coverage. One technique that I often use to verify my test count is McCabe's Cyclomatic Complexity.[24] Cyclomatic Complexity tells a developer how many linearly independent paths there are through a function. Stated another way, Cyclomatic Complexity means the developer needs the minimum number of test cases to cover all their branches in a function. Additional test cases may be required to test boundary conditions on variable values. To measure Cyclomatic Complexity, I often use tools like pmccabe[25] and Metrix++.[26]

One additional tool that can be pretty useful when testing is gcov. gcov is a source code coverage analysis and statement-by-statement profiling tool.[27] Usually, a developer would enable gcov to be run as part of their test harness. The tool will then provide a file that gives the branch coverage as a percentage and a marked-up version of the source file that shows a developer which branches are covered and which ones are not. Branch coverage allows developers to go back and improve their test cases to get full coverage.

---

[22] https://cpputest.github.io/

[23] www.parasoft.com/

[24] https://en.wikipedia.org/wiki/Cyclomatic_complexity

[25] https://people.debian.org/~bame/pmccabe/pmccabe.1

[26] https://metrixplusplus.github.io/metrixplusplus/

[27] https://bit.ly/3wcUKjl

**Caution**    100% code coverage does not mean there are no bugs in the code! It simply means that the test cases result in every branch being executed. It's still possible to be missing test cases that would reveal bugs, so don't let 100% code coverage lure you into a false sense of well-being.

## Continuous Integration/Continuous Deployment

I believe the most important tool for embedded development teams to embrace and deploy effectively are CI/CD tools. The ability to automate software's build, analysis, testing, and deployment is a game changer for nearly all teams. The potential improvement in quality and decreases in cost can be pretty profound. There are also quite a few different tools that can fit almost any team's needs.

The first tool that I would like to highlight is Jenkins.[28] Jenkins is an open source automation server that provides plug-ins to scale building, deploying, and automating projects. The software is a Java-based program that can be installed on any major OS like Windows, Mac OS, and Linux. It's easy to set up and use their plug-in systems; it can be integrated into many other SDLC programs. Jenkins has been quite popular due to its open source nature.

Another CI/CD tool that I've used and like quite a bit is GitLab.[29] GitLab integrates a project's Git repo and CI/CD pipelines that can then be used to build, test, and deploy the software. The tool also comes with some basic SDLC tools, but they are nowhere near as sophisticated as the other SDLC tools discussed in previous sections. I think the most significant advantage of the tool is how easily CI/CD is integrated into it. Developers can easily create a CI/CD pipeline like in Figure 15-1.

---

[28] www.jenkins.io/
[29] https://gitlab.com/

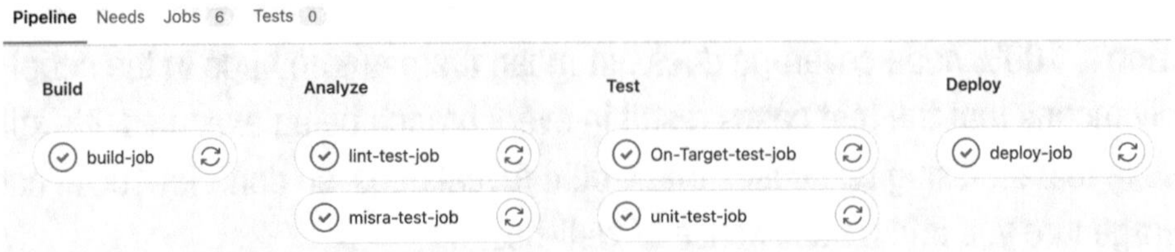

*Figure 15-1.* *An example CI/CD pipeline for an embedded target that leverages the GitLab CI/CD system*

# Implementation Tools

The last primary tool category we will discuss in this book, but probably not the last, is implementation tools. Developers use implementation tools to write and build their software. In addition, various tools are used to develop embedded software, such as compilers, programmers, and code analyzers. Each tool's specific purpose is instrumental in successfully developing a code base.

## IDEs and Compilers

Integrated Development Environments (IDEs) and compilers often go hand in hand. Most microcontroller vendors provide their customers with an IDE that can seamlessly interface to a compiler and do much more. For example, STMicroelectronics provides STM32CubeIDE,[30] which integrates the ability to configure an embedded project, set up peripherals and pin configurations, write code, debug, and deploy it to the target processor. The ecosystem that STMicroelectronics has set up also includes plug-ins with libraries for extended features such as

- Secure Boot and Secure Firmware Update (SBSFU)

- Machine Learning (STM32CubeAI and X-CUBE-AI)

- Motor Control (X-CUBE-MCSDK)

- Sensor and motion algorithms (X-CUBE-MEMS1)

- Display and graphics (X-CUBE-DISPLAY)

- IoT connectivity (X-CUBE-IOTA1)

---

[30] www.st.com/en/development-tools/stm32cubeide.html

The value that these tools can provide is immense! As a junior engineer, I remember the pain I had to go through to set up the clock trees on a microcontroller. It seems that it would take 40 hours or more to review the clock tree properly, identify the registers that had to be modified, and then configure them for the correct frequency. The first attempt often failed, which led to debugging and frustration from just wanting to get the CAN baud rate to 500 kbps, not 423.33 kbps. In addition, each peripheral had to be set up with the correct dividers, so it was not a process that was just done once.

Modern IDEs had taken what took me a week or so when I started my career to about 15 minutes today. Most IDEs now contain a clock tree configurator that allows us to go in, select our clock source, set the frequency, and then magic happens, and our clock tree and peripherals are all configured! An example of this fantastic tool can be seen in Figure 15-2.

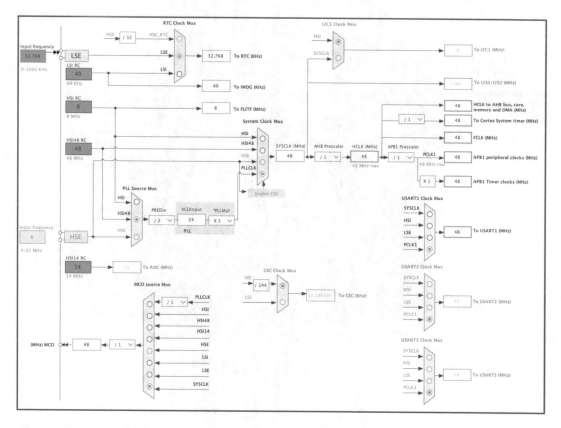

**Figure 15-2.**  *A clock tree configurator example from STM32CubeIDE for an STM32F0XX. Developers can quickly set their clock source, dividers, and peripheral dividers and get to writing code*

The clock tree is not the only tool in these IDEs that makes developers' lives easier; there is also the pin and peripheral configurator! The pin and peripheral configurator again allows developers to avoid digging deep into the datasheet and register map by using a high-level tool instead. In many environments, developers can click individual pins to set the pin multiplexer and even assign project-related names. They can then click a peripheral, use dropdowns to assign peripheral functions and baud rates, and configure many other features. When I started, this entire process could easily take a month, but today, if you spend more than a day, you're probably doing something wrong.

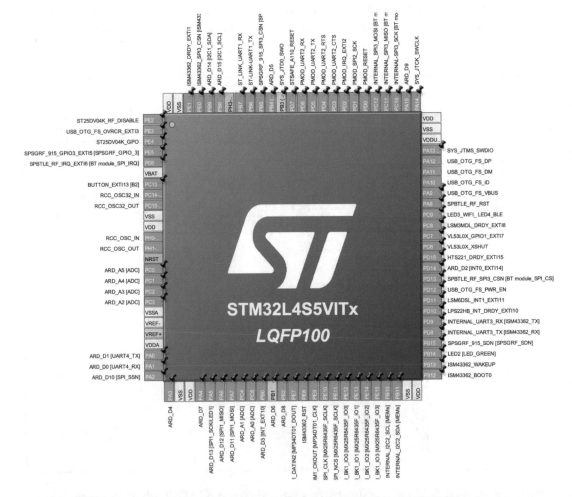

***Figure 15-3.*** *A pin configuration example from STM32CubeIDE for an STM32L4SX. Developers can quickly configure their pins and peripheral functions. In addition, if a development board is used, the tools can automatically configure them for you!*

When considering compiler options, many teams use free compilers like GCC for Arm.[31] However, there are a lot of good commercial compilers available as well. In addition, many IDEs will allow teams to easily integrate compilers such as Keil MDK[32] and IAR Embedded Workbench.[33] In fact, these compiler vendors usually also provide their IDEs. I've worked with all these compilers, and which one I choose depends on the customer I'm working with, the end goals of the project, optimizations needed, and so forth. However, if you carefully examine the data from the Embedded Microprocessor Benchmark Consortium (EEMBC),[34] you'll discover that the compiler chosen can impact code execution.

Before you think I'm pushing STMicroelectronics, I'll just mention that I'm using them as an example. NXP, Microchip, and many other vendors have similar ecosystems and capabilities. It is interesting to note, though, that each has its own spin, libraries, and ways of doing things that often differentiate them from each other. So I would just encourage you to try several out and figure out which fits best for your needs.

Using a compiler-provided IDE or a microcontroller vendor IDE is not the only option for the modern embedded software developer. Microsoft's Visual Studio Code[35] (VSC) has also become popular over the last several years. VSC provides developers with an agnostic development environment that can be used for embedded programming or general application development. The number of plug-ins it has for Arm debugging, code reviews, pair programming, source control, code analysis, and so forth is quite dizzying! The capabilities only seem to be growing too.

There are many benefits to selecting a tool like VSC. For example, VSC provides an integrated terminal, making it very easy in a single environment to issue commands, manage a build, run Docker, and so forth. I've also found that VSC often has far better editor capabilities and runs better than most vendor-supplied IDEs. VSC is also lightweight and cross-platform. I can't tell you how many times I've found vendor IDEs to be wanting in a Mac OS or Linux environment.

---

[31] https://bit.ly/3CRlxmL

[32] www2.keil.com/mdk5

[33] www.iar.com/products/architectures/arm/iar-embedded-workbench-for-arm/

[34] www.eembc.org/products/

[35] https://code.visualstudio.com/

One last point to leave you with is that, on occasion, I will use more than one IDE to develop embedded software. For example, I've occasionally worked with teams where I would work in VSC but maintain an STM32Cube IDE project file or a Microchip MPLab X project file. This allowed developers who are more comfortable with more traditional and noncommand-line tools to make fast progress without having to slow down too much to learn new skills. (Obviously, it's good to know and push yourself to learn new things, but sometimes the resistance is just not worth it!)

## Programmers

Programmers play an essential role in embedded systems; they allow us to push our application images onto our target hardware for production and debugging! Successfully programming a target requires a physical device that can communicate over JTAG/SWD and software that can take the image and manage the physical device. Let's start our discussion with the physical device.

Nearly every development board in existence today comes with some sort of onboard programmer. These programmers are great for evaluating functionality but hardly serve as excellent professional debuggers. The onboard debuggers are often limited in features such as only offering two breakpoints, low clock speeds, and nearly no chance of performing real-time tracing. Advanced features like energy monitoring are also missing. To get professional-level features, developers need a professional programming tool.

Several companies provide fantastic third-party programmers. I mention third party because if you buy a tool like an STLink, you'll only be able to program STMicroelectronics parts. Perhaps only programming one vendor's microcontrollers is acceptable for you, but I've always found it's worth the extra money to be flexible enough to program nearly any microcontroller. The price for programmers can, of course, vary widely depending on the space that you work in. For example, you might pay upward of five figures for Green Hills Softwares' TimeMachine Debugging Suite[36] while paying $700–$3000 for an Arm Cortex-M programmer.

There are several programmers that I've found to be good tools. First, I really like SEGGER's J-Link and J-Trace products. They have become my default programmers for any development that I do. I use a J-Link Ultra+ for most applications unless the ETM port is provided, in which case I then use the J-Trace. Beyond the quality aspect,

---

[36] www.ghs.com/products/timemachine.html

I also like that SEGGER has a no-questions-asked defect policy. For example, I once had customer hardware damage my programmer but could replace it at 50% of the cost through a trade-in program. You can find an overview of the different models here.[37]

The next programmer that I use is Keil's ULINK.[38] I've used several models, but my favorite is the ULINK Plus. The ULINK Plus comes in a small package that travels well and provides a 10 MHz clock. The feature I like the most is that it gives I/O lines for test automation and continuous integration. The ULINK Plus can also be used for power measurements. All around, a pretty cool tool.

There are undoubtedly other programmers out there as well. Other notable programmers I like include IAR's I-Jet[39] and I-Scope,[40] which provide energy monitoring and debugging capabilities. I also like PEmicro's programmers, which can be used for debugging or production.

Beyond the hardware devices, the software is required to program a microcontroller. In many instances, developers will just use the default software built into their IDEs. However, other compelling tools can also be used. For example, if you plan to set up a CI/CD server, you'll want to examine the OpenOCD[41] framework. There is also a scripting framework associated with SEGGER's programming tools.

I've also found that each vendor usually provides some tools and utilities to get a bit more out of their programmers. For example, SEGGER provides a tool called J-Flash and J-Flash Lite,[42] which provide developers with a GUI interface to perform functions such as

- Reading chip memory

- Erasing chip memory

- Programming an application

- Securing the chip

- Modifying option/configuration bytes

---

[37] www.segger.com/products/debug-probes/j-link/models/model-overview/

[38] www2.keil.com/mdk5/ulink

[39] www.iar.com/ijet

[40] www.iar.com/products/architectures/arm/i-scope/

[41] https://openocd.org/

[42] www.segger.com/products/debug-probes/j-link/tools/j-flash/about-j-flash/

These additional software tools can provide developers with a quick and easy bench or production setup that does not require the source code to be "in the loop."

# Code Generators

Code generation can be a sticky subject for developers. In theory, generating a model of your software and then using the model to generate the functional code sounds fantastic! However, generated code doesn't always deliver on that promise. I've seen code generators that create code as expected, but the code is unreadable, slow, and can't be modified by hand. For example, when I started my consulting business in 2009, I used Freescale's CodeWarrior product to generate code that helped me understand how to work with various peripherals and then hand-code the drivers from scratch. Thankfully, these tools have become much better over time.

Before deciding on code generation tools that might fit your development cycle, it's helpful to recognize the pros and cons of their use. First, code generators can bring development teams out of the low-level hardware and help them to focus on high-level application code. Next, developers focus on the application, which can run code in emulators, improve testing and test cases, decrease debug cycles, and much more. Of course, the biggest complaint I often hear is that generated code is larger and runs slower than hand-coded code.

| Pros | Cons |
| --- | --- |
| • Minimal low-level development<br>• High-level models<br>• Can run in emulator<br>• Simplified testing<br>• Shorter debug cycles<br>• Improved scalability<br>• Ease of reuse | • Poorer performance than hand coded software<br>• Larger software image<br>• Tools are considered expensive<br>• Additional training required |

*Figure 15-4.   Leveraging code generation has its pros and cons, which may impact whether a team is willing to adopt it or not. This figure shows a summary comparison of the pros and cons of using code generation*

There are several code generators that I believe embedded developers can readily leverage today to help improve their embedded software and deliver in a reasonable time frame. First, there are code generation tools that nearly every microcontroller vendor provides. For example, STMicroelectronics provides STM32CubeMx,[43] which also integrates into STM32CubeIDE, which can be used to configure and autogenerate drivers and some application code. NXP has MCUExpresso SDK,[44] and Microchip has Harmony.[45] These tools help configure and generate low-level driver code and some middleware. Unfortunately, they don't help with application modeling.

Earlier in the chapter, we discussed several tools that can be used to model an application in UML. These tools are oriented toward developing the software architecture. However, many UML tools will provide mechanisms for generating code in Python, Java, and C/C++. State machines, sequences, and other software constructs can be generated using these tools. Using a modeling tool can provide a huge advantage to developers in that they only need to maintain the model and can generally not care about the generated code as much. This doesn't sit well with me unless the tool generates nice human-readable, efficient code.

Probably the most well-known code generation tool is Matlab.[46] Matlab is mighty and can be used to create a high-level software model, run it in simulation, and then generate the target-specific code. So, for example, if a developer wants to make a digital finite impulse response (FIR) filter, they can model such a filter in Matlab with something like Figure 15-5.

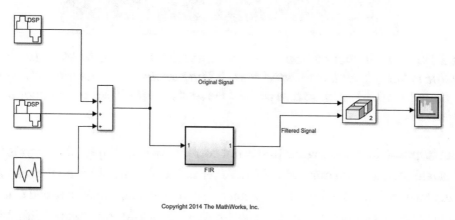

Copyright 2014 The MathWorks, Inc.

***Figure 15-5.*** *Modeling an FIR filter in Matlab*

---

[43] www.st.com/en/development-tools/stm32cubemx.html

[44] https://bit.ly/3qptpa2

[45] www.microchip.com/en-us/tools-resources/configure/mplab-harmony

[46] www.mathworks.com/products/matlab.html

In the example, two frequency signals are fed into the filter and a noise signal. Next, the FIR block performs the filtering based on the FIR filter parameters in that block. Finally, the original and filtered signals are fed into a plotter. When the model is run, the simulated output looks like Figure 15-6.

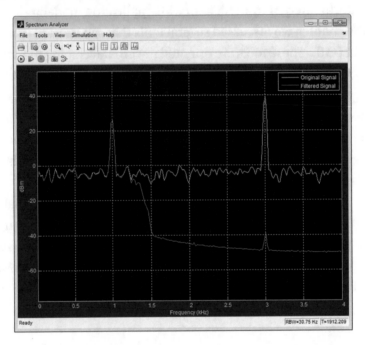

***Figure 15-6.*** *The frequency response of the modeled FIR filter in Matlab. The yellow line is the original, unfiltered signal. The purple line is the filtered response. Notice we get ~-40 dB attenuation at 1.5 kHz and about -80 dB on the signal at 3 kHz*

If the response found in Figure 15-6 is not correct, we can simply adjust our FIR filter block instead of modifying our code and then rerun the filter until we get the response we want. Once we have the desired response, we can use tools like Embedded Coder and the Embedded Coder Support Package for ARM Cortex-M Processors. With the generated code, we can import it into whatever build system we have and integrate on target.

My opinion is that embedded software development's future is in tools that configure, simulate, and generate our software for us. I've believed this, though, since ~2008. There have been huge strides and improvements in the tools since then, but they still are not perfect and can sometimes be cost prohibitive for start-ups and small businesses that are not working in safety-critical environments or have large budgets.

# Analyzers

Analysis tools should be at the heart of every development cycle. Analysis tools provide developers with the metrics they need to ensure that the software is on track. For example, an analysis tool might tell developers how many new lines of code have been added. How many lines of code have been removed? Whether there are any violations in the static analysis.

Analysis tools can help developers remove bugs in their software before they find their way into the field. In addition, they can help developers improve the robustness of their code and software architecture. Finally, an analysis can help developers understand what is in the code base, how it is structured, and the potential problem module developers will need to tiptoe around.

How can developers analyze their software with so many different tools worldwide? There are probably nearly two dozen that I frequently use. Unfortunately, I'm not going to describe them all here, but I will tell my favorites and then try to list out a few for you to further investigate on your own.

The first analysis tool I love and am unsure how I would ever develop without is Percepio's Tracealyzer.[47] Tracealyzer is a runtime analysis tool often used with real-time operating systems like FreeRTOS, ThreadX, Azure RTOS, Zephyr, and others. The tool allows developers to record events in their system like giving or taking a semaphore, context switching to a different task, an interrupt firing, and much more. These events can be captured in either a snapshot mode or a real-time trace for debugging and code performance monitoring.

I was teaching a class one day on real-time operating systems and introducing Tracealyzer to the class. One of my colleagues, Jim Lee, was in the audience. Jim primarily focuses on hardware design but has also been known to write low-level code. Upon explaining Tracealyzer, he best summarized the tool by saying, "So it's an oscilloscope for software developers." It was a simple observation, but it hits the nail right on the head. Hardware developers can't debug and verify their hardware without seeing the signals and understanding what is happening. For software developers, that is what Tracealyzer does for us developers. It allows us to capture and visualize what our software is doing.

---

[47] https://percepio.com/tracealyzer/

Developers can capture information ranging from task context switching, CPU utilization, heap usage, stack usage, data flow, state machine states, performance data, and more. I generally use the tool while developing software to ensure that the software I'm working on is behaving the way I expect it to. I'll also use it to gather runtime performance metrics, which I track with each commit. That way, if I add something that changes the runtime performance, I know immediately.

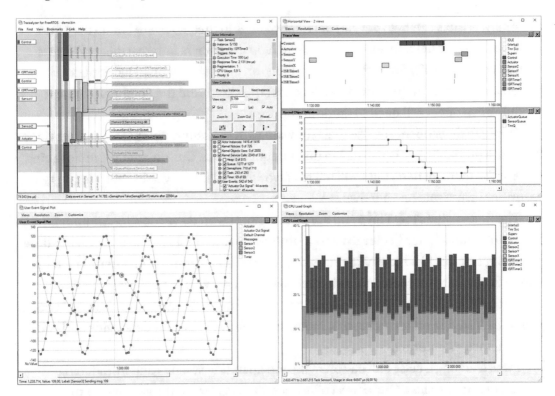

***Figure 15-7.*** *Tracealyzer provides developers with an event recorder and visualization tool to analyze and understand what an embedded system software is doing. This screenshot shows task execution, user event frequency, and CPU utilization. (It can do so much more, though!)*

Other tools out on the market will do similar things to Tracealyzer, but I think it has the largest feature set and is the most user-friendly. There are tools out there like

- SEGGER SystemView[48]

- IAR Embedded Workbench (has a built-in trace capability)

- Green Hills SuperTrace[49]

Another helpful analysis tool category that I have found useful is tools that can monitor test coverage. I will often use gcov[50] to monitor my test case coverage with CppUTest.[51] It's important to note that just because test coverage says 100% doesn't mean there are no bugs. I had a customer I was mentoring whom I was telling this to, and they said, "yea, yea," dismissively. The next week during our call, the first thing out of his mouth was, "So I have 100% test coverage and was still finding bugs, so I see what you mean now." He had the coverage, but not the right tests!

Speaking of testing, there are also a bunch of tools from SEGGER that can be useful for developers looking to hunt bugs. One of my favorites is Ozone.[52] Ozone is a debugger and performance analyzer that can analyze your binary file and monitor code coverage. I've also used the tool in conjunction with my J-Trace to perform instruction tracing. However, I have found that getting instruction tracing set up can be a bit of a nightmare. Once set up, though, it's fantastic.

There are also many metric analysis tools that developers can find to analyze their software. There are all kinds of metrics that developers may want to monitor. For example, one might want to monitor lines of code (LOC), comment density (although there are arguments around this), and assertion density. The tool that I've adopted the most for metrics analysis is Metrix++.[53] Metrix++ can

- Monitor code trends over periods such as daily, weekly, and monthly

- Enforce trends at every code commit

---

[48] https://bit.ly/3JCc5pJ

[49] www.ghs.com/products/probe.html

[50] https://gcc.gnu.org/onlinedocs/gcc/Gcov.html

[51] https://cpputest.github.io/

[52] www.segger.com/products/development-tools/ozone-j-link-debugger/

[53] https://metrixplusplus.github.io/metrixplusplus/

- Automatically review standards in use

- Be configured for various metric classes

Figure 15-8 shows an example of the output from Metrix++ that is monitoring Cyclomatic Complexity. Notice that there is a limit of 15.0 set, and the two functions that are coming up have complexity measurements of 37 and 25. Reports like this can be built into the DevOps process, allowing developers to get results on the quality of their code base continuously. It's pretty cool and, I think, an often overlooked, low-hanging tool that can provide a lot of value.

```
./interprocess/detail/managed_open_or_create_impl.hpp:302: warning: Metric
    Metric name    : std.code.complexity:cyclomatic
    Region name    : priv_open_or_create
    Metric value   : 37
    Modified       : None
    Change trend   : None
    Limit          : 15.0
    Suppressed     : False

./interprocess/streams/vectorstream.hpp:284: warning: Metric 'std.code.com
    Metric name    : std.code.complexity:cyclomatic
    Region name    : seekoff
    Metric value   : 25
    Modified       : None
    Change trend   : None
    Limit          : 15.0
    Suppressed     : False
```

***Figure 15-8.*** *The Metrix++ output for Cyclomatic Complexity on a test project*

Speaking of Cyclomatic Complexity, one of my favorite tools for measuring complexity is pmccabe.[54] The tool only measures complexity and works on the command line in Linux-based systems. It's pretty easy to install. All one needs to do is type

```
sudo apt-get update -y
sudo apt-get install -y pmccabe
```

---

[54] https://people.debian.org/~bame/pmccabe/pmccabe.1

Once installed, it can then be run on a file like packet.c by using

`pmccabe -v packet.c`

The tool provides several different outputs, such as traditional and modified McCabe Cyclomatic Complexity. I'll often use this tool before committing to ensure that the module(s) I'm working on is within my desired parameters (usually <=10, but if it makes sense, I'll occasionally allow a slightly higher value). Figure 15-9 shows an example output.

```
beningo@Beningos-MacBook-Pro Application % pmccabe -v packet.c
Modified McCabe Cyclomatic Complexity
|    Traditional McCabe Cyclomatic Complexity
|        |    # Statements in function
|        |        |    First line of function
|        |        |        |    # lines in function
|        |        |        |        |    filename(definition line number):function
|        |        |        |        |        |
7        7        18       233      67       packet.c(233): Packet_DecodeSm
2        2        4        335      10       packet.c(335): Packet_Sync
1        1        2        381      5        packet.c(381): Packet_Version
1        1        1        422      4        packet.c(422): Packet_SourceAddress
1        1        1        462      4        packet.c(462): Packet_DestinationAddress
1        1        1        502      4        packet.c(502): Packet_MessageID
3        3        7        542      24       packet.c(542): Packet_ProcessTwoByteField
5        5        13       602      59       packet.c(602): Packet_Data
1        1        2        697      5        packet.c(697): Packet_Checksum1
1        1        2        738      6        packet.c(738): Packet_Checksum2
2        2        6        780      20       packet.c(780): Packet_EndSync
1        1        1        834      4        packet.c(834): Packet_Get
3        3        36       878      57       packet.c(878): Packet_Validate
1        1        1        962      4        packet.c(962): Packet_ErrorGet
1        1        1        992      4        packet.c(992): Packet_ErrorClear
1        1        13       998      22       packet.c(998): Packet_Encode
1        1        3        1021     6        packet.c(1021): Packet_ResetSm
```

***Figure 15-9.***  *The results of running pmccabe on a test module named packet.c. Notice that the complexity measurements are all below 10*

There are a lot of additional tools that you might consider helping you analyze your embedded software. I always recommend that developers use a static code analyzer. Wikipedia has a pretty good list of various static analysis tools that you can find here.[55]

---

[55] https://en.wikipedia.org/wiki/List_of_tools_for_static_code_analysis

I've discovered that IAR's C-Stat is also a good one. Years ago, I used to use PC-Lint quite a bit, but their business model changed considerably, and it is not friendly toward start-ups and small teams anymore. Still a good product, though.

The tool landscape is constantly evolving, and it is a good idea to always keep your eye out for new tools that can be used to analyze software and improve it.

# Open Source vs. Commercial Tools

Before we close this chapter, I want to take the opportunity to discuss the differences between open source and commercial tools. I often come across teams who have a mentality that everything should be open source. While this is possible, I find that teams often don't look at the full picture when going this route. So let's quickly discuss the differences.

Open source tools are fantastic. They provide developers with a valuable tool at no cost! The source code is right there, so we can review it, modify it, and understand what the tool is doing. These immediately seem like wins; however, this may be shortsighted in some cases. For example, a developer often writes open source tools with a sole purpose in mind. That developer may not be considering your use cases. They also might not be interested in maintaining the tool in the long term if it's a tool that will evolve and change over time.

I've also found that open source tools tend to lack good documentation. Yes, I'm generalizing because there are some great projects out there; however, I've seen teams adopt an open source tool to regret it later. Sometimes, the tool won't entirely do what we want in the way we want to do it. We then get forced to make or request changes that may or may not ever be made. I've even seen times when teams have a question and can't find anyone to answer it.

On the flip side, commercial tools cost money. Tools are expensive to build because they require developers who want to be paid big bucks. Therefore, tools tend to be costly. The good news is that a commercially supported tool usually has better documentation. If it doesn't, usually you can call the company and get an answer for what problem you are facing. If you want a new feature, usually the company will add it because you are a paying customer, and they want you to be happy so they can keep your business.

I've also found that the quality of commercial products can be higher than open source. There is often an argument that more developers review open source software and therefore is of higher quality. I've not seen that fact come to fruition. There's a lot of harmful open source code out there. Some of it is downright scary!

Now, hear me out before you come after me with your pitchfork. There is a place for both open source and commercial tools in our development environments. I often use a mix of both. When the open source software has the right features and level of quality, I use it. When not, I use a commercial tool. I've occasionally had open source tools I love suddenly disappear or the support stop because the primary backer moved on or got bored. I'm just suggesting that you be careful, analyze your options, and choose the one that best fits your needs.

# Final Thoughts

Tools are a critical component to helping embedded software teams succeed. Unfortunately, there has been a big push in the software industry that everything should be free. If you must pay for a tool, a software stack, etc., then some universal law is being broken. This mentality couldn't be further from the truth and contributes to and increases the risk of failure.

I don't sell tools, but I have found that they dramatically enrich my ability to help my clients. A good tool will provide a lot of value and a great return on investment to the developer, team, and business that use them correctly. In this chapter, we have explored how to evaluate if a tool will provide you with a return on investment, in addition to exploring several tools that I use. Unfortunately, we've only scratched the surface, and there are many more exciting and valuable tools that we have not covered.

Don't join the herd mentality that paying for a tool is terrible. Professional tools are what make you a professional and help you to deliver the quality that is expected from a professional. So don't wait; upgrade your tools if necessary and get back to coding!

| ACTION ITEMS |
|:---:|

To put this chapter's concepts into action, here are a few activities the reader can perform to start finding and using the right tools for their job:

- Identify a tool that you use every day. Then, calculate the value and the return on investment of that tool.

- Identify a tool that you would like to have in your toolbox. Next, calculate the value and the return on investment for that tool. Then, put together a tool request for your management and present your request for the tool as discussed in this chapter.

- Look through each tool category that we have discussed. Are there any tools missing in your toolbox? If so, research and identify the right tool for you and pursue it!

# PART IV

# Next Steps and Appendixes

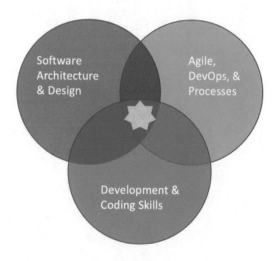

# AFTERWORD

# Next Steps

## Where Are We Now?

Throughout this book, we've explored the three areas that must be balanced for an embedded software team to deliver their software successfully:

- Software Architecture and Design (Part 1)

- Agile, DevOps, and Processes (Part 2)

- Development and Coding Skills (Part 3)

I call these three areas the embedded software triad because if you don't successfully balance them, the development cycle will be a boondoggle. Success might come, but it will be through painful and overexerted effort.

In each part of the book, we looked at the fundamentals necessary for you to start understanding what each area entails and some of the techniques required to help you be successful. Obviously, given the time constraints and the size that a typical book can be, we couldn't cover everything. However, we have covered enough for you to successfully figure out where you are today and what you need to do to start to balance your development cycle.

You may be a single developer or perhaps the leader of a team of developers. No matter what the case may be, completing this book is just the start of your journey. I've shared some of my insights, skills, and experiences working with teams worldwide and in many industries successfully developing embedded software. While I hope this has been helpful to you, we're now at a point where you need to decide how you will take what we've discussed and apply it.

© Jacob Beningo 2022

J. Beningo, *Embedded Software Design*, https://doi.org/10.1007/978-1-4842-8279-3_16

# Defining Your Next Steps

You can take several approaches to decide where you go from here. But, no matter which direction you take, there are several questions that we need to answer before you race off into the sunset.

First, you need to understand where you are today. The embedded software triad is meant to help you understand how to break up everything that we must know and deal with and visualize it in a simple chart. As we have seen, balanced teams are more likely to succeed, so a good first step is to evaluate where you currently are. Is there an area that you are ignoring? Are there skillsets that you are missing? If you haven't already done so, you might consider taking my survey, `www.surveymonkey.com/r/7GP8ZJ8`, which will give you a high-level overview of where you are. You may find that taking a quick 50-question survey doesn't give you the level of detail you need. If that is the case, and you'd like professional assistance, you can always reach out to me, and I can help point you in the direction that may make the most sense for you and your team.

Next, once you know where you are, it's time to decide where you want to go. You can look at your personalized survey results and determine where you may need to add more focus. You can also survey your team or just look back over the chapters in this book and see where your passions lay. Perhaps improving testing is exciting, and the idea of improving your software quality is at the top of your list.

My recommendation is not to try to bite off too much at once. Instead, start by defining no more than three goals that you would like to accomplish in the next three months. Once you know what those improvements are, put the time and budget necessary to make incremental improvements. Once you achieve those initial goals, review where you are, define where you want to be next, and then repeat.

The one truth that I have seen so far in my career is that the developers, managers, and teams that focus on consistent and incremental improvements are the ones that succeed the most. I've often seen them grow beyond many of their own wildest expectations. Of course, discipline is required, but the potential results can be staggering.

# Choosing How You Grow

When I was a kid, I loved hockey. I watched every Red Wing game I could and read everything I could about hockey. I read books about hockey skills and how to improve my hockey skills. Consuming all these materials helped me to understand the game and how to play it. But unfortunately, they didn't do anything to help me improve my hockey skills because I wasn't applying what I was reading!

I had to get skates, a hockey stick, and a puck to improve my hockey skills and go out and play. Playing hockey and getting the experience tremendously helped me take my book knowledge and apply it to the real world of hockey. Practicing and playing hockey on my own helped to improve my skills, but I did discover that in some areas, no matter what I did, I couldn't get better! For example, try as I might, I could not figure out how to hockey stop! My version of stopping was to just plow into another player or the boards. Not very skillful!

I spent years on my own trying to learn to stop. (Seriously, no joke. It was pretty embarrassing.) Then, when I got to college, I decided to take a hockey class for fun and additional exercise. Lo and behold, the teacher discovered I couldn't stop. For the first few weeks of the course, he coached me one on one how to properly shift my momentum and better utilize the edges on my skates, and, finally, I was able to stop! I had been attempting to figure out how to stop for years, using advice from friends and family, and it just took a coach a few weeks to get me there.

I know you're wondering what my hockey skill woes do with embedded software or getting a product to market. My story demonstrates the different ways you can learn and improve your team and the rate at which you can expect results. You can engage your team and go it alone. You will improve and gain the processes and skills you need to succeed. However, you may sometimes struggle or find that you are progressing at the rate you want.

The best athletes worldwide have coaches who watch what they do and give them expert feedback. I know companies whose policy is that every manager has an internal or external coach. I'm a solo consultant and business owner. I have mentors and coaches who are colleagues and even paid business coaches that help give me feedback and whom I can bounce ideas off. I've learned from my hockey experience years ago that I sometimes need external assistance and can't do it all alone for rapid, continued progress. (Or if I can, I can do it faster with some guidance.)

I suggest that you figure out the best route for you and your team to meet your company's goals. For example, you might be able to set up a good, disciplined process and skill enhancement program on your own. Or, you may know you have issues but aren't sure how to resolve them. I've often seen that the teams making waves in the industry are the ones that are willing to admit they're drinking their own Kool-Aid and need some external advice and feedback to keep them on the right path. I've sometimes even seen where a team internally says they need A, but management doesn't agree until a third party comes in and says that the company needs A.

I'm not trying to pitch you anything. I've personally benefited and seen my clients benefit from mentoring and advice. I suggest you carefully examine what it will take for you and your team to succeed. There are many ways that you can ensure you and your team grow and are on the right path. For example, you can pair junior and senior engineers together. You can pair managers in cross-function teams. There are many online communities as well that engineers can join to get questions asked and interact with experts all over the world[1] (and yes, even consultants).

We often look at our embedded software through the lens of developers. At the end of the day, we are contributing to the success or failure of the business we work for. Do what needs to be done and is in the company's best interest. When you do that, you'll find that you end up with higher-quality software, controllable software budgets, and predictable times to market. Your end customers will love it, and management will too!

## Final Thoughts

There are a lot of resources, both free and commercial, that you can use to help take your embedded software development cycle and skills to the next level. However, as we close the book, I have a few recommendations.

First, don't overlook the Appendix in this book. There are several tutorials that I believe you'll find exciting and will help you apply some of the topics discussed in this book. So make sure that you take full advantage of it!

Next, there are many great resources that you can find to continue your embedded software journey. For example, you may find the following sites to be helpful to you:

- www.beningo.com

- www.embeddedonlineconference.com

- www.embeddedrelated.com

- www.embedded.com

- www.designnews.com

---

[1] Think LinkedIn groups, embeddedrelated.com forum, etc.

I may be a little biased toward these sites since I regularly contribute content. However, I feel that these resources have some of the best free and paid materials in the embedded software industry. So don't hesitate to take advantage of them.

Finally, if you ever discover that you are unsure where to go from here or get stuck, you can always reach out to me at jacob@beningo.com. I'll do what I can to direct you to any additional resources or discuss how I can be of assistance.

I've appreciated your time and attention throughout this book. I hope you have found some new insights, acquired new ideas, and will drive your embedded software development cycle and skills to the next level. Until next time, happy coding, and good luck!

## ACTION ITEMS

Our journey through this book has ended, but your embedded software journey may just be beginning. After that, it's up to you to choose your next adventure. The following are a few more summary thoughts for you to consider for what may be next:

- Where are you today? Take time to figure out where you and your team are at.

  - Are you lacking skills in any area of the embedded triad?

  - Is your development cycle out of balance?

  - What challenges continue to trip you up, cause you stress, or deliver late?

- Where are you going?

  - What changes do you need to make to be more successful?

  - What does a balanced development cycle look like to you?

  - What improvement can you make to improve the impact of your software on your customers?

- How are you going to get there?

  - Are you going to go it alone?

  - Engage with your team?

  - Leverage an external expert to coach and advise you.

# Correction to: RTOS Application Design

## Correction to:

**Jacob Beningo, Embedded Software Design,**
**https://doi.org/10.1007/978-1-4842-8279-3_4**

The original version of the book was inadvertently published without incorporating the author's proof corrections. The chapter has now been corrected and approved by the author.

---

An updated original version of this chapter can be found at
https://doi.org/10.1007/978-1-4842-8279-3_4

J. Beningo, *Embedded Software Design*, https://doi.org/10.1007/978-1-4842-8279-3_17

# Security Terminology Definitions

Security is not the core focus of this book, but security is essential for many teams that want to develop embedded software. Throughout the book, there were several security-related terms mentioned that I was not able to go into deeper detail on. Appendix A provides a high-level definition for several of these terms.

## Definitions

**Access control** – The device authenticates all actors (human or machine) attempting to access data assets. Access control prevents unauthorized access to data assets. Counters spoofing and malware threats where the attacker modifies firmware or installs an outdated flawed version.

**Authenticity** – Indicates that we can verify where and from whom the data asset came.

**Arm TrustZone** – Arm Cortex-M hardware technology for creating hardware-based isolation in microcontrollers. Cortex-A processors also have a version of TrustZone, but the implementation and details are different.

**Attestation**[1] – Attestation provides a mechanism for software to prove its identity to a remote party such as a server. Attestation data is signed by the device and verified using a public key by the request server.

**Chain-of-trust** – Established by validating each hardware and software component in a system from power-on through external communications. Each component's integrity and authenticity is checked before allowing it to operate.

**Confidentiality** – Indicates that an asset needs to be kept private or secret.

---

[1] https://courses.cs.washington.edu/courses/csep590/06wi/finalprojects/bare.pdf

© Jacob Beningo 2022
J. Beningo, *Embedded Software Design*, https://doi.org/10.1007/978-1-4842-8279-3

**Denial of service** – An interruption in an authorized user's access to a device or computer network caused by malicious intent.

**Escalation of privilege**[2] – A network attack that is used to gain unauthorized access to systems.

**Firmware authenticity** – The device verifies firmware authenticity before boot and upgrade. Counters malware threats.

**Impersonation**[3] – Is when a malicious actor pretends to be a legitimate user or service to gain access to protected information.

**Integrity** – Indicates that the data asset needs to remain whole or unchanged.

**Malware**[4] – A file or code, typically delivered over a network, that infects, explores, steals, or conducts virtually any behavior an attacker wants.

**Man-in-the-middle**[5] – A form of active wiretapping attack in which the attacker intercepts and selectively modifies communicated data to masquerade as one or more of the entities involved in a communication association.

**MPU** – Memory protection unit. Used to separate memory into memory regions and limit permissions for those regions.

**NSPE** – Nonsecure processing environment. An NSPE is the "feature-rich" application execution environment in a secure microcontroller-based system.

**PPU** – Peripheral protection unit.

**PSA**[6] – The Platform Security Architecture (PSA) is a family of hardware and firmware security specifications, as well as open source reference implementations, to help device makers and chip manufacturers build best-practice security into products.

**PSA Certified**[7] – A security certification scheme for Internet of Things (IoT) hardware, software, and devices. It was created by seven stakeholder companies as part of a global partnership. The security scheme was created by Arm Holdings, Brightsight, CAICT, Prove & Run, Riscure, TrustCB, and UL.

**Repudiation** – An attack that can occur against the system or application in which the system does not have adequate controls in place to detect that the attack has occurred.

---

[2] www.cynet.com/network-attacks/privilege-escalation
[3] https://powerdmarc.com/what-is-an-impersonation-attack
[4] www.paloaltonetworks.com/cyberpedia/what-is-malware
[5] https://csrc.nist.gov/glossary/term/man_in_the_middle_attack
[6] www.ietf.org/id/draft-tschofenig-rats-psa-token-09.html
[7] https://en.wikipedia.org/wiki/PSA_Certified

**Root-of-Trust** – This is an immutable process or identity which is used as the first entity in a trust chain. No ancestor entity can provide a trustable attestation (in Digest or other form) for the initial code and data state of the Root-of-Trust.

The Root-of-Trust often includes security functions such as initialization, software isolation, secure storage, firmware updates, secure state, cryptography functions, attestation, audit logs, and debug capabilities.

**Secure communication** – The device authenticates remote servers, provides confidentiality (as required), and maintains the integrity of exchanged data. Counters man-in-the-middle (MitM) threats.

**Secure state** – Ensures that the device maintains a secure state even in case of failure to verify firmware integrity and authenticity. Counters malware and tamper threats.

**Secure storage** – The device maintains confidentiality (as required) and integrity of data assets. Counters tamper threats.

**SMPU** – Shared memory protection unit.

**SPE** – Secure processing environment. A hardware isolated execution environment including memory and peripherals that contains the Root-of-Trust and secure application services.

**SRAM PUFs**[8] – SRAM physical unclonable function is a physical entity embodied in a physical structure. PUFs utilize deep submicron variations that occur naturally during semiconductor production and which give each transistor slightly random electric properties – and therefore a unique identity.

**Tamper**[9] – An intentional but unauthorized act resulting in the modification of a system, components of systems, its intended behavior, or data.

---

[8] www.intrinsic-id.com/sram-puf

[9] https://csrc.nist.gov/glossary/term/tampering

# 12 Agile Software Principles

Over the past several decades, the agile movement has become a major methodology within the software industry. Unfortunately, there have been many offshoots where teams think they are "doing Agile" but are doing nothing of the sort. There are lots of mixed messages, and unfortunately the core message is often lost among the noise. I thought it would be helpful to remind developers of the 12 core agile principles that are outlined on the Agile Manifesto.

## 12 Agile Software Principles[10]

1. Our highest priority is to satisfy the customer through early and continuous delivery of valuable software.

2. Welcome changing requirements, even late in development. Agile processes harness change for the customer's competitive advantage.

3. Deliver working software frequently, from a couple of weeks to a couple of months, with a preference to the shorter timescale.

4. Business people and developers must work together daily throughout the project.

---

[10] https://agilemanifesto.org/principles.html

© Jacob Beningo 2022
J. Beningo, *Embedded Software Design*, https://doi.org/10.1007/978-1-4842-8279-3

5. Build projects around motivated individuals. Give them the environment and support they need and trust them to get the job done.

6. The most efficient and effective method of conveying information to and within a development team is face-to-face conversation.

7. Working software is the primary measure of progress.

8. Agile processes promote sustainable development. The sponsors, developers, and users should be able to maintain a constant pace indefinitely.

9. Continuous attention to technical excellence and good design enhances agility.

10. Simplicity – the art of maximizing the amount of work not done – is essential.

11. The best architectures, requirements, and designs emerge from self-organizing teams.

12. At regular intervals, the team reflects on how to become more effective, then tunes and adjusts its behavior accordingly.

# Hands-On – CI/CD Using GitLab

## An Overview

In Part 2, we discussed several modern processes involving DevOps and testing. Appendix C is designed to give you a little hands-on experience putting together a CI/CD pipeline to compile and test embedded software for an STM32 microcontroller. Appendix C will guide you on how to

- Set up Docker

- Configure an image for CI/CD

- Integrate code with a CI/CD pipeline

- Run unit tests

- Deploy built code to an STM32 development board

The example that we are about to walk through is exactly that. It is not meant for production code. The example will provide you with some hands-on experience and help to solidify the skills we've discussed throughout this book.

## Building STM32 Microcontroller Code in Docker

Docker provides a mechanism to set up our build, analysis, and testing environments in a container, making it easier to integrate into a CI/CD pipeline. Developers can set up a container whether they are running Windows, Linux, or macOS. One environment

© Jacob Beningo 2022
J. Beningo, *Embedded Software Design*, https://doi.org/10.1007/978-1-4842-8279-3

that can be used across multiple platforms. Containers also have the added benefit that new developers to the team don't need to spend a bunch of time setting up software and trying to duplicate the build environment. Instead, as part of the repo, the new developer only needs to run the image build commands, and the latest image with the code will be built and ready to go.

Before we walk through how to set up a Docker container for the STM32, I would recommend that you do the following:

- Install the latest version of Docker[11] and Docker Compose[12] on your development machine.

- Install the latest version of STM32CubeIDE.[13]

- Create an empty Git repo to save our project and files to.

- Find any STM32 development board that you have lying around the office (optional: only needed if you plan to deploy your code to a board).

Once you've prepared these items, you'll be ready to walk through setting up our build process within a Docker container.

# Installing GCC-arm-none-eabi in a Docker Container

The STM32 is an Arm Cortex-M–based 32-bit architecture. Typically, you would install your compiler as part of the IDE toolchain. However, we will install the GCC-arm-none-eabi compiler as part of a Docker image. For this example, I will use the October 2021 release of the GCC-arm-none-eabi compiler. I would recommend reviewing what the latest version is and using that one.

Our Dockerfile will have four sections:

1. The base image that we build on top of

2. Support tool installation

---

[11] https://docs.docker.com/get-docker/

[12] https://docs.docker.com/compose/install/

[13] www.st.com/en/development-tools/stm32cubeide.html

3.  GCC-arm-none-eabi compiler installation (and configuring our path)

4.  Setting up our working directory

The entire Dockerfile contents to create these sections and set up our image can be seen in Listing C-1. If you plan to follow along, I recommend creating the Dockerfile in the root of your Git repo and then pasting it into the code listing. In our setup script, you will notice that we are using Linux within our image. If you are running Windows, you don't need to worry about using Cygwin or any other Linux emulator. The Linux distribution is being run within the container that is running on top of Windows. We only need to worry about installing the tools we want within our image.

*Listing C-1.* The complete Dockerfile source for creating a Docker image for the STM32 build process

```
FROM ubuntu:latest
ENV REFRESHED_AT 2022-06-01

# Download Linux support tools
RUN apt-get update && \
    apt-get clean &&  \
    apt-get install -y \
        build-essential \
        wget \
        curl \
        git

# Set up a development tools directory
WORKDIR /home/dev
ADD . /home/dev

RUN wget -qO- https://developer.arm.com/-/media/Files/downloads/gnu-rm/
10.3-2021.10/gcc-arm-none-eabi-10.3-2021.10-x86_64-linux.tar.bz2 | tar -xj

# Set up the compiler path
ENV PATH $PATH:/home/dev/gcc-arm-none-eabi-10.3-2021.10/bin

WORKDIR /home/app
```

Let's now break the Dockerfile down and examine what each section is doing and why.

The first line in the Dockerfile specifies the base image on which we will be basing our image. We will use the latest version of Ubuntu for our base image using the following Docker command:

```
FROM ubuntu: latest
```

---

**Best Practice**    It is possible that updates to the latest image could break your image. You may want to consider specifying a specific. For example, specify ubuntu 18.04 using FROM  ubuntu:18.04.

---

Next, we need several support tools in Linux to download, unzip, install, and then configure the Arm compiler. We will install these tools in one RUN command as shown in Listing C-2. If additional tools were needed, we could continue to add the commands to install them to the bottom of this RUN statement.

*Listing C-2.*  The Dockerfile command to download and install Linux support tools

```
# Download Linux support tools
RUN apt-get updated && \
    apt-get clean && \
    apt-get install -y \
        build-essential \
        wget \
        curl
```

Before we install GCC-arm-none-eabi, we want to specify where to install it within the Docker image. It is common to install it in a directory such as /home/dev. We can do this in our Dockerfile using the code shown in Listing C-3.

*Listing C-3.*  Specify where the tool directory is located

```
# Set up a development tools directory
WORKDIR /home/dev
ADD . /home/dev
```

Now, we are ready to add the command to install the compiler using Listing C-4.

*Listing C-4.* The Dockerfile command to run to install the Arm compiler within the Docker container

```
RUN wget -qO- https://developer.arm.com/-/media/Files/downloads/gnu-rm/
10.3-2021.10/gcc-arm-none-eabi-10.3-2021.10-x86_64-linux.tar.bz2 | tar -xj
```

---

**Note**   I am requesting a specific compiler version, 10.3-2021.10. You may want to update to the latest version when you read this. The latest version numbers can be found at `https://bit.ly/3LX8zqt`.

---

With the compiler command now included, we want to set up the PATH variable to include the path to the compiler. This will allow us to use make to compile our application. The command to add to the Dockerfile can be seen in Listing C-5.

*Listing C-5.* Set up the compiler path in the Dockerfile

```
# Set up the compiler path
ENV PATH $PATH:/home/dev/gcc-arm-none-eabi-10.3-2021.10/bin
```

Finally, we can set our working directory for our application code to the /home/app directory using

```
WORKDIR /home/app
```

We now have a Dockerfile that can be built to create our image. Our image can then be run to create our container where we have a virtual environment from which to build our code.

# Creating and Running the Arm GCC Docker Image

We now have a Dockerfile that contains all the commands necessary to create a Docker image, but the image does not exist yet. To create the image and then run it, there are a few commands that we need to run first. The first command is the Docker build command. The build command reads in our Dockerfile and then creates a Docker image based on the commands in the file. For example, to build the Dockerfile we just created, we would open a terminal in our Git repo root directory and type the following command:

```
docker build -t beningo/GCC-arm .
```

Notice that in the build command, we are using the -t option to tag our image. In the example, we are tagging the image as beningo/GCC-arm. The resulting output in the terminal from the build command should look something like Figure C-1.

```
docker build -t beningo/gcc-arm .
[+] Building 54.6s (8/8) FINISHED
 => [internal] load build definition from Dockerfile                    0.0s
 => => transferring dockerfile: 734B                                    0.0s
 => [internal] load .dockerignore                                       0.0s
 => => transferring context: 2B                                         0.0s
 => [internal] load metadata for docker.io/library/ubuntu:latest        0.0s
 => CACHED [1/4] FROM docker.io/library/ubuntu:latest                   0.0s
 => [2/4] RUN apt-get update &&     apt-get clean &&     apt-get instal 24.6s
 => [3/4] RUN wget -qO- https://developer.arm.com/-/media/Files/download 26.0s
 => [4/4] WORKDIR /home/app                                             0.0s
 => exporting to image                                                  3.8s
 => => exporting layers                                                 3.8s
 => => writing image sha256:a651eee9482f512619ee78cda7d53c07bfc4dbf655fef 0.0s
 => => naming to docker.io/beningo/gcc-arm                              0.0s

Use 'docker scan' to run Snyk tests against images to find vulnerabilities and l
earn how to fix them
beningo@Jacobs-MacBook-Pro stm32-example % ▯
```

***Figure C-1.*** *The terminal output from running the Docker build command on our Dockerfile*

The first time you build an image, it may take several minutes to create and run all the commands. For example, when I built the Dockerfile, it took 58.2 seconds the first time I built it. However, if you build it a second time, it will run much faster because only changes to the image will need to execute.

Once we have an image built, we are ready to run the image. The image can be started using the following command:

```
docker run --rm -it --privileged -v "$(PWD):/home/app"
beningo/gcc-arm:latest bash
```

The run command looks a bit complicated and overwhelming, but it's relatively straightforward. First, `--rm` is telling Docker that when we exit the container, we want to remove the image. Docker images usually require gigabytes of hard drive space, so we don't want to accumulate a whole bunch of them! Next, `-it` instructs Docker to allocate a pseudo-TTY connected to the container's stdin, creating an interactive bash shell in the container.[14] Next, the `--privileged` flag gives all capabilities to the container. Finally, the `-v` option mounts the current working directory into the container. This allows us to access the host file system, which is useful when we want to compile our code and easily access the generated object and binary files. All the stuff that comes after `-v` specifies where the mount is located and what Docker image file tag we are running.

At this point, we now have a Docker image, a running container, and a command prompt that allows us to use the container we have running. The next thing we want to do is build something! However, to do so, we need a test project. So, let's now create an STM32 test project that we can build using our Docker container.

# Creating an STM32 Test Project

To test our Docker build environment, there are several options available to us. First, we just create a few C modules and write a makefile. To test the environment, we don't necessarily need to build a microcontroller, but what's the fun in that? For this example, we will leverage the STM32 project generator tool STM32CubeMx. The STM32CubeMx tool will allow us to create a makefile-based project that we can put right into the root directory of our Git repo.

As I mentioned earlier, any STM32 board around your office will work. I have an STM32L475 IoT Discovery board that I use in my RTOS courses. So, for this example, that is what I'm going to use as my target. The exact target here isn't important; it's the general process that I want you to focus on. I recommend creating a new STM32CubeMx project named Blinky and initializing your development board to the default settings. I won't discuss how to do this because the toolchain walks you through it.

STM32CubeMx allows developers to configure development boards and STM32 microcontroller projects easily. Once the project is created, developers can use the project settings to choose their development environment. For example, developers can use a makefile, CubeIDE, Keil, and IAR Embedded Workbench. The trick is to set

---

[14] https://docs.docker.com/engine/reference/commandline/run/

up the project to be makefile based. If you set the project up for a different toolchain, you can change it under the Project Manager in the Code Generation section shown in Figure C-2.

***Figure C-2.*** *The STM32CubeMx tool can be configured for several toolchains. For example, under Project Manager ➤ Code Generator, developers can set the Toolchain/IDE to Makefile*

From this point, an example project can be created by clicking the generate code button. Once done, my project directory, the root of my repo, will look something like Figure C-3. Notice that my STM32 project is stored in a project directory; I have a .gitignore file to ignore those pesky object files from being committed, along with the Dockerfile and a top-level Makefile. You'll also notice that I have a .gitlab-ci.yml file. Later in the chapter, I will show you how to create this file; for now, just ignore it.

| Name | Last commit |
|------|-------------|
| 📁 Project/makefile/Blinky | Updated Makefile |
| ◈ .gitignore | Add .gitignore |
| 🐋 .gitlab-ci.yml | Update .gitlab-ci.yml |
| 🗋 CHANGELOG | Add CHANGELOG |
| 🐳 Dockerfile | Added STM32CubeMx generated files |
| 🗋 Makefile | Updated Makefile |
| M↓ README.md | Initial commit |

***Figure C-3.*** *A representation of the root directory after creating the STM32 project*

## Compiling the STM32 Makefile Project

We now have installed a Docker container with GCC-arm-none-eabi compiler and an example baseline STM32 project. Within the Docker image, which you should have opened in a terminal earlier, we want to navigate to the root directory of our STM32 project. For me, this is done using

```
cd Project/makefile/Blinky/
```

Once I'm in the directory, I can use the STM32 generated makefile to compile the project using

```
make all
```

The resultant output should look something like Figure C-4. You should notice that we successfully built our STM32 project from within a Docker image. While this doesn't seem super impressive now, this is precisely what we needed to do to start building our CI/CD pipeline.

```
=hard –specs=nano.specs –TSTM32L475VGTx_FLASH.ld  –lc –lm –lnosys  –Wl,–Map=buil
d/Blinky.map,––cref –Wl,––gc-sections –o build/Blinky.elf
arm–none–eabi–size build/Blinky.elf
   text    data     bss     dec    hex filename
  21232      32    9696   30960   78f0 build/Blinky.elf
arm–none–eabi–objcopy –O ihex build/Blinky.elf build/Blinky.hex
arm–none–eabi–objcopy –O binary –S build/Blinky.elf build/Blinky.bin
root@2ed2feef1d3d:/home/app/Project/makefile/Blinky# 
```

*Figure C-4.* *Successful compilation of the STM32 project within the Docker container*

---

**Note**   To exit the Docker container, type EXIT in the terminal.

---

## Configuring the Build CI/CD Job in GitLab

Now that we have a Docker environment successfully building our project, we are ready to create the build job in our Embedded DevOps pipeline. Obviously, before configuring the pipeline, you'll have to go through and create a GitLab account and follow their setup instructions. If you plan to follow along, you can find the instructions at `https://docs.gitlab.com/runner/install/`. Alternatively, you could use a tool like Jenkins and use the general process, but you'll be responsible for figuring out the details.

# Creating the Pipeline

Once you have your GitLab runner set up, log in to GitLab and navigate to your repo. On the left side of the interface, you'll find a CI/CD menu option that includes a pipeline menu. Click the pipeline. You should now see that you have options for creating your pipeline. GitLab, by default, does not have a C template option, but they do have a C++ option. While the C++ template is a good start, we want to understand the entire CI/CD setup process. Therefore, I recommend starting with the "Hello World with GitLab CI."

By clicking the template, GitLab will present a default .gitlab-ci.yml file for you to review and commit to your repository. Commit this file, and then let's explore its contents. The code can be seen in Listing C-6.

***Listing C-6.*** The "Hello World with GitLab CI" yml template file for a three-stage pipeline

```
stages:            # List of stages for jobs and their order of execution
  - build
  - test
  - deploy

build-job:         # This job runs the build stage first
  stage: build
  script:
    - echo "Compiling the code..."
    - echo "Compile complete."

unit-test-job:     # This job runs in the test stage.
  stage: test      # Runs if build stage completes successfully.
  script:
    - echo "Running unit tests... This will take about 60 seconds."
    - sleep 60
    - echo "Code coverage is 90%"
```

```
lint-test-job:    # This job also runs in the test stage.
  stage: test     # Run at the same time as unit-test-job
  script:
    - echo "Linting code... This will take about 10 seconds."
    - sleep 10
    - echo "No lint issues found."

deploy-job:       # This job runs in the deploy stage.
  stage: deploy   # It only runs when *both* jobs in the test stage complete
                    successfully.
  script:
    - echo "Deploying application..."
    - echo "Application successfully deployed."
```

The first thing to notice with our pipeline YAML file is that it creates three stages: build, test, and deploy. Each job created is assigned a stage in the pipeline to run. Each job also contains a script that specifies what commands should be executed during that stage. If you carefully examine the example YAML, you'll notice the only calls made in the script are to echo and sleep. In time, we will replace these calls with our own.

---

**Note**    Each stage runs sequentially. If a job in the stage fails, the next stage will not run. If a stage has more than one job, those jobs will execute in parallel.

---

After committing the YAML file to your repository, you can navigate the CI/CD pipeline menu again. You'll notice that it now displays the status of the CI/CD runs! For example, after the last commit, the pipeline began executing and, when completed, looked something like Figure C-5. Notice that we get an overall pass on the left, a repo description and hash, and icons for each stage in our pipeline. If one of the stages failed, that stage would be red and have an x in it to notify us where the pipeline failed.

⊘ passed          Update .gitlab-ci.yml file
⏱ 00:01:36        #424797727  ⅌ main  ⦿ 34c44eff ⊞

*Figure C-5.* *The GitLab pipeline status shows that the pipeline passed all three stages successfully*

417

You can click each check mark, or x mark, to review the output for each job. For example, if I click the first check mark that corresponds to the build process, you will see something like the details shown in Figure C-6. Notice that you can see our echo script commands being executed. We also get a confirmation at the end that the job succeeded. If the job had failed, this is where we would go to get the details about what went wrong.

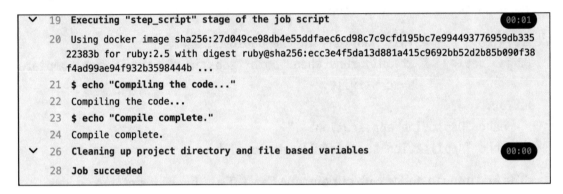

*Figure C-6.* *A partial output from the successful build job*

## Connecting Our Code to the Build Job

Getting the example up and running is a significant first step, but we want to replace the "Hello World" scripts with connections to our code. To get our code connected, we're going to have a little bit of work to do.

The first step to connecting our code to the CI/CD pipeline is to make our Docker image available to GitLab. The easiest way to do this at this stage is to create a public image repository on Docker Hub. First, you'll need to go to https://hub.docker.com/ and create an account if you haven't already done so. Once your account is created, from your terminal, log in to Docker using the command:

```
docker login -u YOUR_USER_NAME
```

Enter your password, and the terminal should tell you that you are logged in successfully.

Next, we want to tag and push our Docker image to the Docker Hub. You'll need to make sure that your image is tagged correctly. For example, if my Docker Hub username is beningo, then I want my image tag to be beningo/gcc-arm. If my username is smith, I want the tag to be smith/gcc-arm. You can tag your image using the following command in the terminal:

```
docker tag IMAGE_NAME YOUR-USER-NAME/IMAGE_NAME
```

From here, we just need to push the image to Docker Hub using the following commands:

```
docker push YOUR_USER_NAME/gcc-arm
```

You'll find that it will take a couple of minutes for the image to be pushed up to Docker Hub.

---

**Beware**   We are posting our Docker image to the public, which anyone can use! You may want to create a private repository and/or use your server for actual development.

---

The next step to connecting our code to the CI/CD pipeline is to update our YAML file. Two changes need to be made. First, we must tell GitLab what Docker image to use for the pipeline. We must add the code found in Listing C-7 to the top of our YAML file before the build job dictionary. Second, we need to update the build stage script to build our code using make! We can use make to build our code in several different ways.

***Listing C-7.***  The YAML file entry to tell GitLab to use our gcc-arm Docker image

```
default:
  image: beningojw/gcc-arm:latest
```

First, we can update the YAML build stage script to match that shown in Listing C-8. In this case, we are using commands to cd into our project directory where the Makefile for the STM32 project is, cleaning it, and then running make all. Running make this way is acceptable, although it is not the most elegant solution. A better solution would be to create a high-level makefile in the root of your project directory. You can then use .phony to create make targets for

- Cleaning
- Building
- Testing
- Making, running, and pushing the Docker image
- Etc.

***Listing C-8.*** Adding commands to move into the Blinky directory and
execute make

```
build-job:        # This job runs the build stage first
  stage: build
  script:
    - echo "Compiling the code..."
    - cd Project/makefile/Blinky/
    - make clean
    - make all
    - echo "Compile complete."
```

Finally, we are ready to run our pipeline. The pipeline is executed when we commit
code to the Git repo. Thankfully, we have just made changes to our YAML file! So,
commit your code, log in to your GitLab project, and navigate the pipeline. You may
need to wait a few minutes for the pipeline to run, but once the build process completes,
you can click it and should see something like Figure C-7.

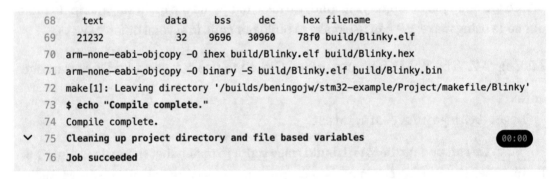

***Figure C-7.*** *A successful build pipeline using our Docker image and custom*
*code base*

A great advantage to CI/CD is that you can be notified when a pipeline passes and
fails. For example, Figure C-8 shows the contents of the email I received when I ran
the build process successfully. If a build were to fail, I would also receive an email, but
instead of a nice green happy email, I'd receive an angry red email! I don't like to receive
failed emails because it tells me I messed up somewhere, and now I must go back and
fix what I broke. The good news is that I know within a few minutes rather than days
or weeks.

***Figure C-8.*** *The email notification received when the CI/CD pipeline runs successfully*

# Build System Final Thoughts

Embedded DevOps has become a critical process for embedded software developers. Unfortunately, teams often put off using these modern processes because they fear them. They present an unknown to the team that requires some up-front research and development time to get it right. There's no reason to fear Embedded DevOps. So far, we quickly got a Docker container up and running to manage our build process and connected it to a CI/CD framework. Let's now look at how we can add CppUTest into the mix so that we can set up testing and regression testing in our CI/CD pipeline.

## Leveraging the Docker Container

The docker-compose run command causes Docker to load the CppUTest container and then make all. Once the command has been executed, it will leave the Docker container. In the previous figure, that is the reason why we get the ERROR: 2. It's returning the error code for the exit status of the Docker container.

It isn't necessary to constantly use the "docker-compose run CppUTest make all" command. A developer can also enter Docker container and stay there by using the following command:

```
docker-compose run --rm --entrypoint /bin/bash cpputest
```

By doing this, a developer can then simply use the command "make" or "make all." The advantage of this is that it streamlines the process a bit and removes the ERROR message returned when exiting the Docker image from the original command. So, for example, if I run the Docker command and make, the output from the test harness now looks like what is shown in Figure C-9.

```
[root@e0384cff4bf3:/home/src# make
compiling MyFirstTest.cpp
Linking rename_me_tests
Running rename_me_tests
..
tests/MyFirstTest.cpp:23: error: Failure in TEST(MyCode, test1)
        Your test is running! Now delete this line and watch your test pass.

..
Errors (1 failures, 4 tests, 4 ran, 10 checks, 0 ignored, 0 filtered out, 0 ms)

make: *** [/home/cpputest/build/MakefileWorker.mk:458: all] Error 1
```

***Figure C-9.*** *The output from mounting the Docker image and running the test harness*

To exit the Docker container, all I need to do is type exit. I prefer to stay in the Docker container, though, to streamline the process.

# Test-Driving CppUTest

Now that we have set up the CppUTest starter project, it's easy to go in and start using the test harness. We should remove the initial failing test case before we add any tests of our own. This test case is in /tests/MyFirstTest.cpp. The file can be opened using your favorite text editor. You'll notice from the previous figure that the test failure occurs at line 23. The line contains the following:

```
FAIL("Your test is running! Now delete this line and watch your test pass.");
```

FAIL is an assertion that is built into CppUTest. So, the first thing to try is commenting out the line and then running the "make" or "make all" command. If you do that, you will see that the test harness now successfully runs without any failed test cases, as shown in Figure C-10.

```
[root@001496277db5:/home/src# make
compiling MyFirstTest.cpp
Linking rename_me_tests
Running rename_me_tests
....
OK (4 tests, 4 ran, 9 checks, 0 ignored, 0 filtered out, 0 ms)
```

***Figure C-10.*** *A successfully installed CppUTest harness that runs with no failing test cases*

Now you can start building out your unit test cases using the assertions found in the CppUTest manual. The developer may decide to remove MyFirstTest.cpp and add their testing modules or start implementing their test cases. It's entirely up to what your end purpose is.

# Integrating CppUTest into a CI/CD Pipeline

Now that you've had the chance to play around a bit a test harness and experiment with Test-Driven Development a bit, it's a good time to discuss how to integrate CppUTest and your test cases into your CI/CD pipeline. Recall that when we set up our GitLab CI/CD pipeline, within our YAML file we had a section for a unit-test-job. We're now going to connect that job with our unit tests, but before we do that, we need to first add CppUTest to our Docker image.

## Adding CppUTest to the Docker Image

In order to run our unit tests from within GitLab, we need our Docker image to have CppUTest installed. The easiest way to install CppUTest is to open our Dockerfile and add the code snippet that can be found in Listing C-9. Take a moment to examine the few lines of code. The first thing you should notice is that we are cloning the CppUTest git repository from the latest tagged version. We are then running three commands to configure and install CppUTest per the CppUTest documentation.

***Listing C-9.*** The Docker commands necessary to install CppUTest v4.0 from source

```
# Install and configure CppUTest
WORKDIR /home/cpputest
RUN git clone --depth 1 --branch v4.0 \
                  https://github.com/cpputest/cpputest.git .
RUN autoreconf . -i
RUN ./configure
RUN make install
ENV CPPUTEST_HOME=/home/cpputest
```

---

**Tip**   Before you rebuild your Docker image, navigate to `https://github.com/cpputest/cpputest` and look under tags. Replace v4.0 with the latest version of CppUTest.

---

If you were to run make image to rebuild the Docker image, you would discover that image fails to build. The problem is that we don't have the autoconf tool installed. You would then also find that you don't have libtool installed either. Within the Dockerfile, under the section where we are downloading and installing Linux support tools, add the command to install autoconf and libtool. When you are done, your Linux support tool section should look something like Listing C-10.

***Listing C-10.*** The Dockerfile Linux support tool installation section

```
# Download Linux support tools
RUN apt-get update && \
    apt-get clean &&  \
    apt-get install -y autoconf && \
    apt-get install -y libtool  && \
    apt-get install -y \
        build-essential \
        wget \
        curl \
        git
```

You can now rebuild the Docker image by running make image.

---

**Beware**    Whenever you update your Docker image, you'll need to make sure that you push it up to Docker Hub so that GitLab can access the latest image.

---

# Creating a Makefile for CppUTest

Within our Docker environment, we are using make to manage our builds. Optimally, we would like to create a rule inside our makefile that we can use to call a secondary makefile that will manage all our tests. Before we create this rule, we need to create our CppUTest.mk file.

The best place to create your custom makefile for your project is to start with James Grenning's cpputest-starter-project[15] on GitHub. If you examine the repo, you'll find that there is a makefile that we can use as a template. I would recommend copying this file to your repo root directory and renaming it cpputest.mk. Once done, open the file and take a quick look through the contents of the makefile. As you can see, this can work as is, but for a custom application we will need to make several changes.

Let's start with specifying our output. Change the value of COMPONENT_NAME from rename_me to Tests/main. On the input side, you'll notice that we have to specify PROJECT_HOME_DIR and CPPUTEST_HOME. In my makefile, I've updated these lines of code to

```
CPPUTEST_HOME := /home/cpputest
PROJECT_HOME_DIR = /home/app
```

These are the locations where CPPUTEST and our project are located within the Docker container.

Next, we need to tell CppUTest where our source files are located. The source code can be found within the Firmware folder. We certainly don't want to have to list all the source files manually. Instead, we can leverage the shell find command to ease the burden. There are several different ways that we could provide CppUTest our source file. One way would be to define our sources in our top-level makefile. The problem

---

[15] https://github.com/jwgrenning/cpputest-starter-project

with this method is that we may pull in files that we are not interested in testing such as middleware or vendor-supplied modules (although perhaps we should test these more than we typically do …). For now, we will be interested in only adding the source files in our application folder. The code to search our Firmware directory for headers, source, and assembly files can be found in Listing C-11.

---

**Beware**   Automatically searching for files is a great automation, but you can run into issues if you plan to use test doubles, mocks, and other test items.

---

***Listing C-11.*** cpputest.mk updates to add our application source code to the test build

```
SRC_FILES := $(shell find Firmware/Application -type f -name '*.c')
```

With our source files now added, we will want to specify our test directory. At this point, we don't have any files to add, but we do want the TEST_SRC_DIRS to include our Tests directory. Changing these couple lines of code to the following will do the trick:

```
TEST_SRC_DIRS += Tests
```

At this point, we are also not using any mocks, so these lines can be left as is. However, I would recommend enabling GCOV after the extension line. GCOV can be used to tell us how much test coverage we have on our source modules. GCOV can be enabled by including the following line:

```
CPPUTEST_USE_GCOV = Y
```

Next, we need to specify our include directories. By default, the cpputest.mk template has a bunch of additional include directories that we won't have in our own code. I would recommend deleting the lines of code and replacing them with the ones shown in Listing C-12. You'll notice that we are referencing a variable named HEADER_FOLDERS that we have not yet defined. In the top-level makefile, you can add the definition for the variable as follows:

```
SRC_DIR := Firmware
HEADERS := $(shell find $(SRC_DIR) -type f -name '*.h')
HEADER_FOLDERS := $(shell dirname $(HEADERS) | sort --unique)
```

***Listing C-12.*** The code necessary to include our include directories and those for CppUTest

```
INCLUDE_DIRS += $(HEADER_FOLDERS)
INCLUDE_DIRS += $(CPPUTEST_HOME)/include
INCLUDE_DIRS += $(CPPUTEST_HOME)/include/Platforms/Gcc
```

A little bit further into cpputest.mk, you'll notice the following lines of code:

```
CPPUTEST_OBJS_DIR = test-obj
```

```
CPPUTEST_LIB_DIR = test-lib
```

These lines specify where the outputs from CppUTest will be placed. For now, I'm leaving them set as default, which means there will be a folder named test-lib and test-obj that will be created in the root folder of our repository when we run CppUTest. If you change the default paths, make sure you read the comments carefully. Some strange behavior can ensue.

The next section in the makefile allows us to customize what warnings and errors we would like to be enabled and disabled within the CppUTest build. I would recommend scheduling some time to examine this section more. For now, we will just leave the defaults as is.

Our makefile is now complete. The only thing missing is that we want to be able to run CppUTest from our top-level makefile. The code in Listing C-13 shows how to create a rule to run CppUTest and also kick off gcov.

***Listing C-13.*** The rule syntax to run CppUTest and gcov using our custom makefile

```
# Run CppUTest unit tests
tests: $(DIST_DIR)
        $(MAKE) -j CC=gcc -f cpputest.mk gcov
```

---

**Tip**    CC=gcc is used to tell CppUTest to use gcc. Without this, it's possible for confusion to occur between our C modules and the C++ test modules, resulting in a build failure.

---

Before we can run our tests, we need to add some application code to the Firmware/ Application folder. If you would like, you can try out TDD yourself and write your own module, or you can look at Appendix D for a few example source modules. For this initial test, I will just be adding the led.h and led.c modules to Firmware/Application and then adding LedTests.cpp to the Tests folder. Once that is done, from a terminal, you can navigate to the root of the repo and execute the following command:

```
make tests
```

If everything was configured correctly, which I'm sure it was, then you should see CppUTest and gcov run! Listing C-14 shows what you should expect to see as an output. Notice that our tests compiled, and we see the results of our tests and code coverage.

***Listing C-14.*** The successful execution of CppUTest

```
root@2267b7da4eee:/home/app# make tests
mkdir Distribution
make -j CC=gcc -f cpputest.mk gcov
make[1]: Entering directory '/home/app'
compiling AllTests.cpp
compiling LedTests.cpp
compiling MyFirstTest.cpp
compiling led.c
Building archive test-lib/libTests/main.a
a - test-obj/Firmware/Application/led/led.o
Linking Tests/main_tests
Running Tests/main_tests
....
OK (4 tests, 4 ran, 4 checks, 0 ignored, 0 filtered out, 0 ms)

/home/cpputest/scripts/filterGcov.sh gcov_output.txt gcov_error.txt gcov_
report.txt Tests/main_tests.txt
  Lines executed:100.00% of 6
100.00%   Firmware/Application/led/led.c
See gcov directory for details
make[1]: Leaving directory '/home/app'
root@2267b7da4eee:/home/app#
```

---

**Tip**   If your output shows that a test has failed, review the output, and resolve the failed test. Hint: Comment out the line of code!

---

# Configuring GitLab to Run Tests

Now that we have CppUTest up and running within our Docker environment, we can update our .gitlab-ci.yml to run our tests as part of the CI/CD pipeline. The adjustments that need to be made are straightforward.

First, open your YAML file and navigate to the area that defines the unit-test-job. You'll notice that we have a few echo statements that are simulating us running real unit tests. Remove those lines of code and replace them with the code found in Listing C-15. As you can see, all we do is add a call to make tests, and everything goes from there!

***Listing C-15.***   The changes to .gitlab-ci.yml that will run our tests as part of the CI/CD pipeline

```
unit-test-job:    # This job runs in the test stage.
  stage: test     # It only starts when the job in the build stage completes
                    successfully.
  script:
    - echo "Running unit tests ..."
    - make tests
```

With the adjustment made, it's now time to commit the code to the repo and let our pipeline run! After committing the code, go into the GitLab dashboard and open the pipelines. For me, it took about three and a half minutes for the full pipeline to run. Once completed, I could see that all the jobs passed successfully. Clicking the unit test job, you would see something like Figure C-11 as the results. The output is essentially the same as when we execute it locally in our Docker image, except that now it is running on a Docker image on GitLab and as part of our CI/CD pipeline.

```
20  Using docker image sha256:894bd6cc0d43ebf1b85a08c5d9ff1d0c08a06e5b1dad0823d5877149bf0f7c31 for b
    eningojw/gcc-arm:latest with digest beningojw/gcc-arm@sha256:1ef035d3c54f2f3befb7c76c16ef6ac22b28
    2ba44be6626a4c3f45c33d0d8409 ...
21  $ make tests
22  mkdir Distribution
23  make -j CC=gcc -f cpputest.mk gcov
24  make[1]: Entering directory '/builds/beningojw/stm32-example'
25  compiling AllTests.cpp
26  compiling LedTests.cpp
27  compiling MyFirstTest.cpp
28  compiling led.c
29  Building archive test-lib/libTests/main.a
30  a - test-obj/Firmware/Application/led/led.o
31  Linking Tests/main_tests
32  Running Tests/main_tests
33  ....
34  OK (4 tests, 4 ran, 4 checks, 0 ignored, 0 filtered out, 1 ms)
35  /home/cpputest/scripts/filterGcov.sh gcov_output.txt gcov_error.txt gcov_report.txt Tests/main_t
    ests.txt
36     Lines executed:100.00% of 6
37  100.00%   Firmware/Application/led/led.c
38  See gcov directory for details
39  make[1]: Leaving directory '/builds/beningojw/stm32-example'
41  Cleaning up project directory and file based variables                              00:01
43  Job succeeded
```

*Figure C-11.*  *The successful execution of the CppUTest unit tests on GitLab*

# Unit Testing Final Thoughts

Testing is a critical process to developing modern embedded software. As we have seen so far, there isn't a single test scheme that developers need to run. There are several different types of tests at various levels of the software stack that developers need to develop tests for. At the lowest levels, unit tests are used to verify individual functions and modules. Unit tests are most effectively written when TDD is leveraged. TDD allows developers to write the test, verify it fails, then write the production code that passes the test. While TDD can appear to be tedious, it is actually a very effective and efficient way to develop embedded software.

Once we've developed our various levels of testing, we can integrate those tests to run automatically as part of a CI/CD pipeline. Connecting the tests to tools like GitLab is nearly trivial. Once integrated, developers have automated regression tests that easily run with each check-in, double-checking and verifying that new code added to the system doesn't break any existing tests. Let's now explore how we can deploy our software to the STM32 development board.

# Adding J-Link to Docker

When we deploy our compiled binary to our target device, we need to decide what mechanism we will use to program the target. There are several options available such as OpenOCD and SEGGER J-Link. As much as I love open source tools, I will use the SEGGER J-Link tool and executable.

To use J-Link, we must first install the J-Link tools as part of our Docker image. We can do this by adding the commands in Listing C-16 to the Dockerfile and then rebuilding the image using "make image." If you run the Docker container using "make environment," you can navigate to the /opt/SEGGER/JLink to see all the installed tools. Figure C-12 shows what you will find.

*Listing C-16.* Dockerfile commands to install and add J-Link to the path

```
# Download and install JLink tools
RUN wget --post-data 'accept_license_agreement=accepted' https://www.
segger.com/downloads/jlink/JLink_Linux_x86_64.deb \
&& DEBIAN_FRONTEND=noninteractive TZ=America/Los_Angeles apt install -y
./JLink_Linux_x86_64.deb \
&& rm JLink_Linux_x86_64.deb

# Add Jlink to the path
ENV PATH $PATH:/opt/SEGGER/JLink
```

The tool that we will be using to flash the STM32 development board is JLinkExe. There are quite a few other exciting tools that come with the J-Link that are beyond the scope of our current discussion. I would highly recommend that you take some time to examine those tools and learn how you can use them to become a more effective embedded developer. For example, JLinkSWOViewerExe can be used to watch incoming SWO data during a debug session. JScopeExe can be used to trace an RTOS application.

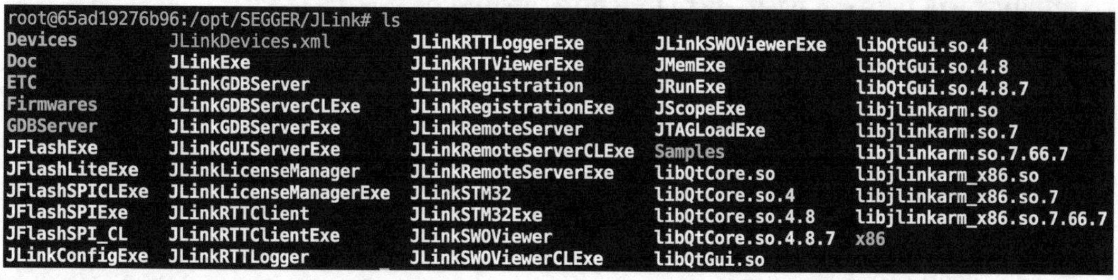

```
root@65ad19276b96:/opt/SEGGER/JLink# ls
Devices              JLinkDevices.xml        JLinkRTTLoggerExe        JLinkSWOViewerExe       libQtGui.so.4
Doc                  JLinkExe                JLinkRTTViewerExe        JMemExe                 libQtGui.so.4.8
ETC                  JLinkGDBServer          JLinkRegistration        JRunExe                 libQtGui.so.4.8.7
Firmwares            JLinkGDBServerCLExe     JLinkRegistrationExe     JScopeExe               libjlinkarm.so
GDBServer            JLinkGDBServerExe       JLinkRemoteServer        JTAGLoadExe             libjlinkarm.so.7
JFlashExe            JLinkGUIServerExe       JLinkRemoteServerCLExe   Samples                 libjlinkarm.so.7.66.7
JFlashLiteExe        JLinkLicenseManager     JLinkRemoteServerExe     libQtCore.so            libjlinkarm_x86.so
JFlashSPICLExe       JLinkLicenseManagerExe  JLinkSTM32               libQtCore.so.4          libjlinkarm_x86.so.7
JFlashSPIExe         JLinkRTTClient          JLinkSTM32Exe            libQtCore.so.4.8        libjlinkarm_x86.so.7.66.7
JFlashSPI_CL         JLinkRTTClientExe       JLinkSWOViewer           libQtCore.so.4.8.7      x86
JLinkConfigExe       JLinkRTTLogger          JLinkSWOViewerCLExe      libQtGui.so
```

*Figure C-12.* The SEGGER J-Link tools installed in the Docker container

# Adding a Makefile Recipe

Now that our Docker image has the necessary tools to program the development board, we need to add a recipe to the makefile that can use JLinkExe to program our binary image to the development board. A simple recipe that can be used is found in Listing C-17. Notice that we have named the recipe "deployToTarget." We have chosen this name over a simpler deploy to allow deployment to multiple environments. For example, we might want to deployToTarget, deployToProduct, deployToDevKit, or deployToSimulator. We aren't going to go into all these cases, but you can see how we want the flexibility to add more recipes as our pipeline becomes more sophisticated.

***Listing C-17.*** A makefile recipe for deploying a binary to a target device

```
## Program the board with JLinkExe
deployToTarget: binaries
        JLinkExe -NoGui 1 -CommandFile $(CONF_DIR)/program.jlink
```

We should also discuss a few additional points in the deployToTarget recipe. First, we make a call to binaries before we invoke a call to JLinkExe. The idea here is to make sure that we compile the code to the latest binary before we deploy it. We also don't have to make binaries before making deployToTarget manually. We just make deployToTarget, and everything is done for us.

---

**Note**    Contrary to popular belief, programmers aren't lazy! We have too much on our plates and must constantly automate and simplify our processes!

---

Next, the recipe executes the JLinkExe application in console mode without launching the graphical user interface. We are going to load the binary using a command file. A command file contains the commands for batch mode/auto execution.[16] The commands will perform a variety of functions, including

- Setting the target device

- Setting the J-Link speed

- Erasing the target

---

[16] https://wiki.segger.com/J-Link_Commander#Using_J-Link_Command_Files

- Loading the binary

- Running the target

Listing C-18 shows an example command file for an STM32L475.

***Listing C-18.*** Example command file for an STM32L475

```
device STM32L475RE
si 1
speed 4000
erase
loadfile Firmware/Project/build/Blinky.hex
r
g
exit
```

As part of the recipe, you'll also need to define CONF_DIR. To simplify things, I created a Config folder at the root of my repo. I placed the program.jlink file there. I then defined CONF_DIR := Config.

# Deploying Through GitLab

Now that we have a makefile that can successfully deploy our binary to our target device, we have just one more step until we can seamlessly deploy the device to our target; we need to update our GitLab pipelines. As you may recall, by looking at our .gitlab-ci.yml file, we have a deploy-job that isn't doing much more than pretending to be deployed to the target. Therefore, we can update our YAML file deploy-job to the code in Listing C-19.

***Listing C-19.*** GitLab YAML file updates to deploy our binary to target as part of the CI/CD pipeline

```
deploy-job:      # This job runs in the deploy stage.
  stage: deploy  # It only runs when *both* jobs in the test stage complete
                   successfully.
  script:
    - echo "Deploying application..."
    - make deployToTarget
    - echo "Application successfully deployed."
```

At this point, you should be able to test that you can successfully program the target. Remember that the target must be connected to the computer on which you have your GitLab runner on. Otherwise, GitLab won't be able to detect the target and will lead you to a series of errors about not being able to connect to the J-Link or the target. The result will be a failed pipeline.

## Deployment Final Thoughts

We've just seen that deploying our firmware to a development board through CI/CD does not need to be overly complicated. The setup we just explored can be used to automate hardware-in-loop testing. Deploying software to a fleet of devices is more complicated; however, from the basics we have just explored, you should be able to build out an automated test suite to run on your development or prototype hardware.

# Hands-On TDD

In Chapter 8, we discussed testing and Test-Driven Development (TDD). Appendix D is designed to show you the solutions that I came up with for designing a heater module using TDD. Your code and tests may turn out completely different than mine due to how you utilize the C programming language. In any event, Appendix D will show you an example solution.

## The Heater Module Requirements

The requirements provided for the heater module include

- The module shall have an interface to set the desired state of the heater: HEATER_ON or HEATER_OFF.

- The module shall have an interface to run the heater state machine that takes the current system time as a parameter.

- If the system is in the HEATER_ON state, it may not be on for more than HeaterOnTime before transitioning to the HEATER_OFF state. HeaterOnTime can be between 100 and 1000. Any number less than 100 results in 100; any number greater than 1000 results in 1000.

- If the heater is on, and requested on while active, then the HeaterOnTime shall reset and start to count over.

## Designing the Heater Module

So far, we have just a few simple requirements. From these requirements, we can begin to map out what our code is going to have to do. I often like to start by drawing out my design. Figure D-1 provides an example what the state machine design might look

© Jacob Beningo 2022
J. Beningo, *Embedded Software Design*, https://doi.org/10.1007/978-1-4842-8279-3

like. As you can see, the design doesn't have to be complicated. We have just a simple state diagram that shows the two major states and the conditions that cause transitions between the states.

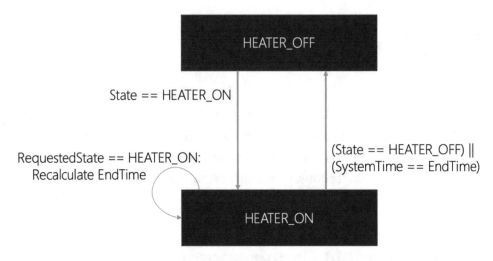

***Figure D-1.***  *The heater state machine design*

Once I have my initial design in place, looking at the interfaces that I am going to need is a good next step. For example, after reviewing the requirements, I can see that there are three functions we are going to need:

- HeaterSm_Init

- HeaterSm_Run

- HeaterSm_Set

---

**Note**    I prefer to be verbose with my naming; however, I shorthand state machine as Sm. The interface could be HeaterStateMachine, but even with autocorrect, it feels too verbose for me. (There is also a 32-character limit in ANSI-C which I get worried about exceeding.)

---

You may be wondering why I have not defined a HeaterSm_Get function. The HeaterSm_Run function will return the current state. In fact, to avoid confusion, I may map out an initial prototype of the interfaces as shown in Listing D-1.

**Listing D-1.** Heater module interface

```
void            HeaterSm_Init(uint32_t const TimeOnMax)
HeaterState_t HeaterSm_Run(uint32_t const SystemTime)
void.           HeaterSm_Set(HeaterState_t const State)
```

One of the tenants of TDD is that we don't write any code until a test forces us to. Technically, I have not written any code yet. I'm just mapping out what I think the interface for the module will be. The tests themselves will help me to flush these design elements out.

# Defining Our Tests

Before I start to write my tests or my module, I also like to create a simple list of tests that will need to be performed on the module to make sure it meets requirements. For example, my initial test list for the heater module will look something like the following:

- The heater state initializes to HEATER_OFF.

- The heater can be set to the HEATER_ON state.

- The heater can be set to the HEATER_ON state and time out to the HEATER_OFF state.

- The heater can be set to the HEATER_ON state and commanded to the HEATER_OFF state.

- The current heater state can be retrieved from the run state.

You may wonder how I came up with this test list. The initial tests are listed based on reviewing the design. As I learned from James Grenning, my initial test list is not meant to be extensive. Typically, I will spend one to two minutes listing out all the tests that come to my mind immediately. I'm just trying to list out my starting point. As I develop my tests and code, additional tests may come to mind.

# Writing Our First Test

For our first test, I will assume that you set up CppUTest in Chapter 8. You saw upon executing the test harness that you had a failed test caused by the FAIL macro in MyFirstTest.cpp. Copy the MyFirstTest.cpp and rename it to HeaterTests.cpp. Inside HeaterTests.cpp, rename the TestGroup to HeaterGroup. Also, set up the first test to now be in the group HeaterGroup and leave it as test1 for now. Make sure that the FAIL macro is included and run your test harness. You should see the following:

```
tests/HeaterTests.cpp:37: error: Failure in TEST(HeaterGroup, test1)
        Your test is running! Now delete this line and watch your
        test pass.
Errors (1 failures, 2 tests, 2 ran, 1 checks, 0 ignored, 0 filtered
out, 0 ms)
```

We now know that our new test group has been added successfully. We can comment out our failed macro and get to writing our first test.

Let's start with the initial state of the heater. We want to initialize the system with the heater off. I'm going to start by renaming my test1 test case to InitialState. We get the current state of the system by running the HeaterSm_Run function. So, let's implement that now. Remember, for TDD we implement the minimum necessary for the test case. In this instance, I would create the enum for the HeaterState_t as shown in Listing D-2.

***Listing D-2.*** Definitions for the heater states

```
typedef enum
{
    HEATER_OFF,
    HEATER_ON,
    HEATER_END_STATE
}HeaterState_t;
```

I would also implement the prototype and an empty function for HeaterSm_Run. I want the test case to fail first, so I will return HEATER_ON from the function as shown in Listing D-3. Now, I know that looking at Listing D-3 is going to drive you crazy! It drove me crazy too when I first started with TDD. The minimum amount of production code required to make the test fail/pass is to just return the desired state! Many of you would want to create a variable right off the bat, but so far, that would be more code than is necessary for what the test cases are calling for.

***Listing D-3.*** Minimum code required to pass the heater on test

```
HeaterState_t HeaterSm_Run(uint32_t const SystemTimeNow)
{
    return HEATER_ON;
}
```

Our very first test case will look something like Listing D-4. Notice we are just checking that the test returns HEATER_OFF. When we do run our test harness, initially the test should show us that it failed. The reason is that HeaterSm_Run is hard-coded to return HEATER_ON. Once you see that the test fails, update HeaterSm_Run to return HEATER_OFF.

***Listing D-4.*** Test case to verify the initial state of the heater

```
TEST(HeaterGroup, InitialState)
{
  CHECK_EQUAL(HeaterSm_Run(0), HEATER_OFF);
}
```

Congratulations! You've just created your first test! We followed the TDD microcycle in this process. We added a test. We verified the test failed, which tells us that our test case would detect the failure case! If we just wrote a test that passes, we have no way of knowing if our test case will catch the failure case if something breaks. We made a small change to make the test pass. All the tests passed. At this point, there is no need to refactor. In fact, we are now ready to write our next text case!

# Test Case – Setting to HEATER_ON

I know that leaving HeaterSm_Run to a hard-coded HEATER_OFF state is driving you crazy. Don't worry though. We will soon be forced to update this function. Let's tackle writing the test case to change the heater state from HEATER_OFF to HEATER_ON.

We start by adding a new test case to our test harness. I'm going to call mine StateOFF_to_ON. I'm going to start by implementing the test which can be seen in Listing D-5. I'll then run my tests. What you will discover is that there is a compilation error. It should tell you that HeaterSm_StateSet was not declared! The reason for this is that we have not yet added the interface. We can now go ahead and do so. (This is our small change.) We now should see the code compiling, but our test case is failing.

***Listing D-5.*** Example test for checking that the off to on transition works correctly

```
Test(HeaterGroup, State_OFF_to_On)
{
   HeaterSm_StateSet(HEATER_ON);
   CHECK_EQUAL(HeaterSm_Run(0), HEATER_ON);
}
```

Our test case is failing because our HeaterSm_StateSet function is not setting the state of the state machine! We need a shared variable to hold the state between the HeaterSm_StateSet and HeaterSm_Run functions. We can do this by creating a static variable at module scope within heater_sm.c. The variable would be defined like this:

```
static HeaterState_t HeaterState = HEATER_OFF;
```

We can then set the state of the variable within HeaterSm_StateSet as shown in Listing D-6.

***Listing D-6.*** Code to set the heater state

```
void HeaterSm_StateSet(HeaterState_t const State)
{
    HeaterState = State;
}
```

You can now run the test case. Unfortunately, you will find that it fails as shown in the following:

```
tests/HeaterTests.cpp:48: error: Failure in TEST(HeaterGroup, State_
OFF_to_ON)
        expected <0>
        but was  <1>
```

```
difference starts at position 0 at: <            1           >
                                                 ^
```

Errors (1 failures, 3 tests, 3 ran, 2 checks, 0 ignored, 0 filtered out, 0 ms)

The problem here is that our test case is failing because HeaterSm_Run has a hard-coded return value! We now need to make a small change to have HeaterSm_Run return the state of the variable HeaterState. Make that change now. Rerun your tests.

What on Earth is going on? My tests are still failing! The problem now is that our test harness gives no guarantee as to the order in which our tests are executing! You'll find that the test that fails now is InitialState! The reason is that State_OFF_to_ON runs and changes the state to HEATER_ON, and then the InitialState test runs and finds it is not HEATER_OFF. This is actually good! Our tests are discovering potential issues with our modules and our tests. What we need to do is after each test, put the heater module back into a known good state. In the TEST_GROUP section of our HeaterTests.cpp module, there is a function called teardown. If we update this function with the code listed in Listing D-7, the heater state will be restored after each test.

***Listing D-7.*** Code to reset the heater state at the end of each test

```
void teardown()
{
    HeaterSm_StateSet(HEATER_OFF);
}
```

Now if you run your tests, you should see that all three pass as shown in the following:

OK (3 tests, 3 ran, 2 checks, 0 ignored, 0 filtered out, 0 ms)

Continue to work through your tests and develop your production code. Remember to take this slow and follow the TDD microcycle. I'm not going to continue to walk you through all the tests. In the next two sections, you can find my final test cases and module code. After that, we will have a quick closing discussion.

# Heater Module Production Code

The header file for heater_sm.h will look something like Listing D-8.

***Listing D-8.*** The heater_sm.h file

```
/*****************************************************************
* Title                  :   Heater State Machine
* Filename               :   heater_sm.h
* Author                 :   Jacob Beningo
* Origin Date            :   07/30/2022
* Version                :   1.0.0
* Notes                  :   None
*****************************************************************/
/** @file heater_sm.h
 *  @brief This module contains the heater state machine.
 *
 *  This is the header file for application control of the heaters.
 */
#ifndef HEATER_SM_H_
#define HEATER_SM_H_

/*****************************************************************
* Includes
*****************************************************************/
#include <stdbool.h>

/*****************************************************************
* Preprocessor Constants
*****************************************************************/

/*****************************************************************
* Configuration Constants
*****************************************************************/

/*****************************************************************
* Macros
*****************************************************************/
```

```
/**********************************************************************
* Typedefs
**********************************************************************/
/**
 * Defines the potential states for the heaters.
 */
typedef enum
{
    HEATER_OFF,          /**< Represents the heater off state */
    HEATER_ON,           /**< Represents the heater on state */
    HEATER_END_STATE     /**< Maximum state value used for bounds
checking */
}HeaterState_t;

/**********************************************************************
* Variables
**********************************************************************/

/**********************************************************************
* Function Prototypes
**********************************************************************/
#ifdef __cplusplus
extern "C"{
#endif

uint32_t HeaterSm_Init(uint32_t const TimeOnMaxValue);
HeaterState_t HeaterSm_Run(uint32_t const SystemTimeNow);
void HeaterSm_StateSet(HeaterState_t const State);

#ifdef __cplusplus
} // extern "C"
#endif

#endif /*HEATER_SM_H_*/


/*** End of File ****************************************************/
```

The heater_sm.c module code will look something like Listing D-9.

**Listing D-9.** The heater_sm.c module

```
/**********************************************************************
* Title                  :    Heater State Machine
* Filename               :    heater_sm.c
* Author                 :    Jacob Beningo
* Notes                  :    None
**********************************************************************/
/**********************************************************************/
/** @file heater_sm.c
 *  @brief This module contains the heater state machine code.
 */
/**********************************************************************
* Includes
**********************************************************************/
#include <stdint.h>                   // For portable types
#include <stdbool.h>
#include "heater_sm.h"                // For this modules definitions.

/**********************************************************************
* Module Preprocessor Constants
**********************************************************************/

/**********************************************************************
* Module Preprocessor Macros
**********************************************************************/

/**********************************************************************
* Module Typedefs
**********************************************************************/

/**********************************************************************
* Function Prototypes
**********************************************************************/
```

```
/*********************************************************************
* Module Variable Definitions
*********************************************************************/
/**
 * The current state value for the heaters.
 */
static HeaterState_t HeaterState = HEATER_OFF;

/**
 * Holds the state of the requested heater state
 */
static HeaterState_t HeaterStateRequested = HEATER_OFF;

/**
 * Tracks the time intervals since the HEATER_ON state was entered.
 */
static uint16_t TimeOnMax = 0;

/**
 * Stores the time to end heater control.
 */
static uint32_t EndTime = 0;

/**
 * Used to indicate if the heater off time needs to be updated.
 */
static bool UpdateTimer = false;

/*********************************************************************
* Function Definitions
*********************************************************************/
/*********************************************************************
* Function : HeaterSm_Init()
*//**
* @section Description Description:
*
*   This function is used to initialize the heater state machine
    parameters with
```

```
*   system level parameters.
*
* PRE-CONDITION:   None
*
* POST-CONDITION: The heater state machine parameters will be initialized.
*
* @param                 uint32_t TimeOnMax - The maximum time the heater will
                         remain on.
*
* @return               None.
*
* @see Heater_On
* @see Heater_Off
*
****************************************************************/
uint32_t HeaterSm_Init(uint32_t const TimeOnMaxValue)
{
    if(TimeOnMaxValue < 100)
    {
        TimeOnMax = 100;
    }
    else if(TimeOnMaxValue >= 1000)
    {
        TimeOnMax = 1000;
    }
    else
    {
        TimeOnMax = TimeOnMaxValue;
    }

    return TimeOnMax;
}
```

```
/**********************************************************************
* Function : HeaterSm_Run()
*//**
* @section Description Description:
*
*  This function is used to progress and run the heater state machine.
   It should
*  be called periodically.
*
* PRE-CONDITION:  HeaterSm_Init has been executed.
*
* POST-CONDITION: The heater state machine will be executed and either
   the heater
*                 enabled or disabled.
*
* @param           uint32_t - The current system time.
*
* @return          None.
*
**********************************************************************/
HeaterState_t HeaterSm_Run(uint32_t const SystemTimeNow)
{
    static HeaterState_t HeaterStateTemp = HEATER_OFF;

    // Manage state transition behavior
    if ((HeaterState != HeaterStateRequested) || (UpdateTimer == true))
    {
        HeaterState = HeaterStateRequested;

        if (HeaterState == HEATER_ON)
        {
            EndTime = SystemTimeNow + TimeOnMax;
            UpdateTimer = false;
        }
```

```
        else
        {
            EndTime = SystemTimeNow;
        }
    }

    // Turn off the heater if we've reached the turn-off time.
    if(SystemTimeNow >= EndTime)
    {
        HeaterState = HEATER_OFF;
        HeaterStateRequested = HEATER_OFF;
    }

    return HeaterState;
}
/*******************************************************************
* Function : HeaterSm_StateSet()
*//**
* @section Description Description:
*
*  This function is used to request a state change to the heater state
   machine.
*
* PRE-CONDITION:   None.
*
* POST-CONDITION: The HeaterStateRequest variable will be updated with the
   desired state.
*
* @param            HeaterState_t - The desired heater state.
*
* @return           None.
*
* @see Heater_StateSet
*
*******************************************************************/
```

```
void HeaterSm_StateSet(HeaterState_t const State)
{
    HeaterStateRequested = State;

    // If the heater is already enabled, we need to let the state
       machine know
    // that the time has just been updated.
    if (State == HEATER_ON)
    {
        UpdateTimer = true;
    }
}

/************** END OF FUNCTIONS ***********************************/
```

# Heater Module Test Cases

The test cases that were used to drive the creation of the production code for the heater_sm module can be found in Listing D-10. Note, your tests may look slightly different, but should in general contain at least this set of tests.

*Listing D-10.* The heater_sm module test cases

```
#include "CppUTest/TestHarness.h"

extern "C"
{
  #include "heater_sm.h"
  #include "task_heater.h"
}

// Test initial state of state machine
// Test all state transitions of state machine
// Make sure the heater can turn on
// Make sure heater can turn off
// Turn heater on and let it time out
// Turn heater on and reset the count down
```

```
TEST_GROUP(HeaterGroup)
{
    void setup()
    {

    }

    void teardown()
    {
      HeaterSm_StateSet(HEATER_OFF);
      HeaterSm_Run(0);
    }
};

TEST(HeaterGroup, InitialState)
{
  CHECK_EQUAL(HeaterSm_Run(0), HEATER_OFF);
}

TEST(HeaterGroup, TimeoutValue)
{
  CHECK_EQUAL(HeaterSm_Init(0), 100);
  CHECK_EQUAL(HeaterSm_Init(500), 500);
  CHECK_EQUAL(HeaterSm_Init(1001), 1000);
}

TEST(HeaterGroup, State_OFF_to_ON)
{
    HeaterSm_StateSet(HEATER_ON);
    CHECK_EQUAL(HeaterSm_Run(0), HEATER_ON);
}

TEST(HeaterGroup, State_ON_to_OFF)
{
    HeaterSm_StateSet(HEATER_ON);
    CHECK_EQUAL(HeaterSm_Run(10), HEATER_ON);

    HeaterSm_StateSet(HEATER_OFF);
    CHECK_EQUAL(HeaterSm_Run(20), HEATER_OFF);
}
```

```
TEST(HeaterGroup, Timeout)
{
    // Turn the heater on for 100 counts
    HeaterSm_Init(100);
    HeaterSm_StateSet(HEATER_ON);
    CHECK_EQUAL(HeaterSm_Run(10), HEATER_ON);

    // Check at the 99 count that the heater is still on
    CHECK_EQUAL(HeaterSm_Run(109), HEATER_ON);

    // Check at the 101 count that the heater is now off
    CHECK_EQUAL(HeaterSm_Run(110), HEATER_OFF);
}

TEST(HeaterGroup, TimeoutReset)
{
    // Turn the heater on for 100 counts
    HeaterSm_Init(100);
    HeaterSm_StateSet(HEATER_ON);
    CHECK_EQUAL(HeaterSm_Run(10), HEATER_ON);

    // Reset the counter by setting the state to HEATER_ON
    HeaterSm_StateSet(HEATER_ON);
    CHECK_EQUAL(HeaterSm_Run(50), HEATER_ON);

    // Check at the 99 count that the heater is still on
    CHECK_EQUAL(HeaterSm_Run(149), HEATER_ON);

    // Check at the 101 count that the heater is now off
    CHECK_EQUAL(HeaterSm_Run(150), HEATER_OFF);
}
```

# Do We Have Enough Tests?

At this point, you should have created a module that roughly has around seven tests with around fourteen different checks. When I run the test harness, I get the following output:

```
OK (7 tests, 7 ran, 14 checks, 0 ignored, 0 filtered out, 0 ms)
```

The module seems to be doing everything that I need it to do. I used my tests to drive the production code I wrote. I used the TDD microcycle at each point to make small changes. I verified that every test failed and then made incremental adjustments until the tests passed.

The question that arises now is "Do I have enough tests?". I used TDD to drive my code, so I should have all the coverage I need. I really should prove this though. The first check I can perform is to make a Cyclomatic Complexity measurement. This will tell me the minimum number of tests I need. Running pmccabe on the heater_sm.c module results in the following output:

```
Modified McCabe Cyclomatic Complexity
|   Traditional McCabe Cyclomatic Complexity
|         |      # Statements in function
|         |          |    First line of function
|         |          |        |    # lines in function
|         |          |        |        |    filename(definition line number):function
|         |          |        |        |         |
3         3          6        85       17       heater_sm.c(85): HeaterSm_Init
5         5          11       121      29       heater_sm.c(121): HeaterSm_Run
2         2          3        169      11       heater_sm.c(169): HeaterSm_StateSet
```

I should have at least three tests for HeaterSm_Init, five tests for HeaterSm_Run, and two tests for HeaterSm_StateSet.

This is where things can get a bit tricky. First, I only have seven tests in my test harness. Seven is less than ten, so I might be tempted to think that I have not reached my minimum number of tests yet. However, within those seven tests, there were fourteen checks. The fourteen checks more closely correlate to what Cyclomatic Complexity refers to as a test. In fact, Cyclomatic Complexity really is just telling us the minimum tests to cover all the branches.

A useful tool to help us close the gap on this is to enable gcov and see if we have 100% branch coverage. To enable gcov, in your cpputest.mk file, find the line with CPPUTEST_USE_GCOV and set it to

```
CPPUTEST_USE_GCOV = Y
```

Once you've done this, you can run gcov with the cpputest.mk file by running gcov afterward. For example, I have a makefile that has a "make tests" command that looks like the following:

```
test: $(DIST_DIR) ## Run CPPUTest unit tests
     $(MAKE) -j CC=gcc -f cpputest.mk gcov
```

When I run my test harness with gcov, the output states the following:

```
100.00%   firmware/app/heaters/heater_sm.c
```

As you can see, my 14 checks are covering 100% of the branches in the heater_sm module. In fact, I have four additional tests that are checking for boundary conditions as well.

# TDD Final Thoughts

If you've followed along with this brief tutorial, congratulations! You've just taken your first steps toward using Test-Driven Development to improve your embedded software! As you've seen, the process is not terribly difficult. In fact, even if it added just a few extra minutes, we now have a bunch of tests that can be run with each commit to ensure that we have not broken any of our code.

What's even more interesting, you've also seen in this example how we can create application code without being coupled to the hardware! We wrote an application code state machine, but we didn't care what I/O would control the heater. We simply wrote hardware-independent code to determine the state and then returned it. If we wanted to run this on-target hardware, we just need some code that takes the output from the state machine and knows which I/O line to control.

I hope that in this quick tutorial, you've seen the power of what TDD can offer you. I hope it has shown you how you can decouple yourself from the hardware and rapidly develop application code in an environment that is much easier to work in. Compiling code on a host and debugging is far easier than having to deploy to a target and step through the code. Yes, it's a bit different, but with some practice, you'll be using TDD efficiently and effectively to improve the quality of your embedded software.

# Index

## A

Access control, 65, 401

Activity synchronization
  bilateral rendezvous, 136
  coordinating task execution, 134
  multiple tasks, 137, 138
  unilateral rendezvous (interrupt to
      task), 135, 136
  unilateral rendezvous (task to
      task), 135

Agile software principles, 405, 406

Analysis tools, 165, 250, 385, 387–390

Application business architecture, 33

Application domain decomposition
  cloud domain, 52
  execution domain, 50
  privilege domain, 48, 49
  security domain, 49, 50

Application programming interface
      (API), 277

Architect secure application, 72, 74, 75

Architectural design patterns
  Event-driven, 42–44
  layered monolithic, 39, 41, 42
  microservice, 45–47
  unstructured monolithic
      architecture, 38, 39

Architectural tools
  requirements solicitation and
      management, 369
  storyboarding, 369, 370

UML diagramming and modeling,
      370, 371

Architecture
  bottom-up approach, 33, 34, 36
  characteristics, 36–38
  top-down approach, 33, 34, 36

Arm TrustZone, 71, 401

Armv8-M architecture, 77

Artificial intelligence and real-time
      control, 145

Assertions, 254
  definition, 282–284
  Dio_Init function, 283
  INCORRECT use, 287
  instances
      initialization, 292
      microcontroller drivers, 292, 293
      real-time hardware
          components, 293
  setting up
      assert_failed, 289–291
      assert macro definition, 288, 289
  uses, 285, 287

Assert macro, 283, 285, 288–289

Attestation, 72, 81, 401

Authenticity, 64, 80, 401

Autogenerating task configuration
  Python code, 322–331
  source file templates, 317–322
  YAML configuration files, 315, 317

Automated unit testing, 24

© Jacob Beningo 2022
J. Beningo, *Embedded Software Design*, https://doi.org/10.1007/978-1-4842-8279-3

# M

# N

# O

Printed in the United States
by Baker & Taylor Publisher Services